Lecture Notes in Mathematics 1919

Editors:
J.-M. Morel, Cachan
F. Takens, Groningen
B. Teissier, Paris

René A. Carmona · Ivar Ekeland
Arturo Kohatsu-Higa · Jean-Michel Lasry
Pierre-Louis Lions · Huyên Pham
Erik Taflin

Paris-Princeton Lectures on Mathematical Finance 2004

Editorial Committee:

R.A. Carmona, E. Çinlar,
I. Ekeland, E. Jouini,
J.A. Scheinkman, N. Touzi

Authors

René A. Carmona
Bendheim Center for Finance
Department of Operations Research
  and Financial Engineering
Princeton University, Princeton
NJ 08544, USA
e-mail: rcarmona@princeton.edu

Arturo Kohatsu-Higa
Graduate School of Engineering Sciences
Osaka University
Machikaneyama Cho 1–3
Osaka 560-8531, Japan
e-mail: kohatsu@sigmath.es.osaka-u.ac.jp

Pierre-Louis Lions
Collège de France
11, place Marcelin Berthelot
75005 Paris, CX 05, France
e-mail: pierre-louis.lions@college-de-france.fr

Erik Taflin
Chair in Mathematical Finance
EISTI, Ecole Internationale des Sciences
  du Traitement de l'Information
Avenue du Parc
95011 Cergy, France
e-mail: taflin@eisti.fr

Ivar Ekeland
Canada Research Chair
  in Mathematical Economics
University of British Columbia
Department of Mathematics
1984 Mathematics Road
V6T 1Z2 Canada
e-mail: ekeland@math.ubc.ca

Jean-Michel Lasry
CALYON
9, Quai du Président Paul Doumer
92920 Paris-La Défense Cedex, France
e-mail: jeanmichel.lasry@ca-indosuez.com

Huyên Pham
Laboratoire de Probabilités
  et Modèles Aléatoires
CNRS, UMR 7599
Université Paris 7 and Institut Universitaire
  de France
Paris CX 05, France
e-mail: pham@math.jussieu.fr

[The addresses of the volume editors
appear on page IX]

Library of Congress Control Number: 2007930225

Mathematics Subject Classification (2000): 91B28, 91B70, 60G49, 49J55, 60H07, 90C46

ISSN print edition: 0075-8434
ISSN electronic edition: 1617-9692
ISBN 978-3-540-73326-3 Springer Berlin Heidelberg New York
DOI 10.1007/978-3-540-73327-0

This work is subject to copyright. All rights are reserved, whether the whole or part of the material is concerned, specifically the rights of translation, reprinting, reuse of illustrations, recitation, broadcasting, reproduction on microfilm or in any other way, and storage in data banks. Duplication of this publication or parts thereof is permitted only under the provisions of the German Copyright Law of September 9, 1965, in its current version, and permission for use must always be obtained from Springer. Violations are liable for prosecution under the German Copyright Law.

Springer is a part of Springer Science+Business Media
springer.com
© Springer-Verlag Berlin Heidelberg 2007

The use of general descriptive names, registered names, trademarks, etc. in this publication does not imply, even in the absence of a specific statement, that such names are exempt from the relevant protective laws and regulations and therefore free for general use.

Typesetting by the authors and SPi using a Springer LATEX macro package

Cover design: *design & production* GmbH, Heidelberg

Printed on acid-free paper   SPIN: 12083206   41/SPi   5 4 3 2 1 0

Preface

This is the third volume of the Paris-Princeton Lectures in Mathematical Finance. The goal of this series is to publish cutting edge research in self-contained articles prepared by well known leaders in the field or promising young researchers invited by the editors. Particular attention is paid to the quality of the exposition, and the aim is at articles that can serve as an introductory reference for research in the field.

The series is a result of frequent exchanges between researchers in finance and financial mathematics in Paris and Princeton. Many of us felt that the field would benefit from timely exposés of topics in which there is important progress. René Carmona, Erhan Cinlar, Ivar Ekeland, Elyes Jouini, José Scheinkman and Nizar Touzi serve in the first editorial board of the Paris-Princeton Lectures in Financial Mathematics. Although many of the chapters involve lectures given in Paris or Princeton, we also invite other contributions. Given the current nature of the collaboration between the two poles, we expect to produce a volume per year. Springer Verlag kindly offered to host this enterprise under the umbrella of the Lecture Notes in Mathematics series, and we are thankful to Catriona Byrne for her encouragement and her help in the initial stage of the initiative.

This third volume contains five chapters. In the first chapter, René Carmona demonstrates how the HJM approach to the construction of dynamic models can be used for different financial markets. The original proposal of Heath, Jarrow and Morton was framed for the world of Treasury bonds, but its applicability was extended soon after its publication. However implementation of the same modeling philosophy in the case of credit and equity markets had to wait. Purely for pedagogical reasons, this chapter starts with a review of the original HJM approach to fixed income markets. Then, the recent works of Sidenius, Pitterbarg and Andersen and Schoenbucher on credit portfolio modeling are presented. Finally, the last part of the chapter explains how Carmona and Nadtochiy developed the program outlined a few years ago by Derman and Kani for equity markets.

The second chapter, by Ivar Ekeland and Erik Taflin, also develops the HJM framework. The emphasis here is on the optimal management of bond portfolios, that is, on Merton's problem of expected utility maximization. The authors introduce a special class of assets, the rollovers, which have a constant time to maturity, and describe the bond portfolios as a combination of rollovers rather than a combination of zero-coupon bonds, as is usual in the literature. The advantage of this approach is that a rollover does not mature, in contrast to a bond. By considering bond portfolios as a combination of rollovers, one brings them close to stock portfolios, which do not mature either, and one paves the way for a unified theory of money markets and equity markets. In addition, the authors derive explicit formulas for the optimal portfolios, at least in the case when the drift and volatility are deterministic processes. These advantages come at a price: a mathematical setting must be found, which will accommodate the curves describing the term structure of interest rates, and allow them to vary randomly, subject to possibly infinitely many sources of noise, while remaining sufficiently smooth. The Musiela parametrization (that is, taking time to maturity instead of date of maturity as the relevant variable), although it is natural in that context, considerably complicates matters, for it introduces an additional (and discontinuous) term in the equations of motion. Roughly speaking, that chapter complements the preceding one: taken together, they highlight the flexibility and generality of the HJM model, as well as its many unexplored consequences.

The third one is written by Arturo Kohatsu-Higa and is based on a short course given in Paris in November and December 2004. It reviews recent results on models for insider trading based on the theory of enlargement of filtrations. After reviewing the case of an insider having extra information given by the knowledge of the distribution of certain random variables, the author concentrates on the case when the insider benefits from almost sure additional information, leading naturally to the use of anticipative calculus. This review can be viewed as a natural companion to the chapter delivered by Fabrice Baudouin in the first volume of the series. The style of this chapter is purposely pedagogical, emphasizing discrete time approximations as a way to illustrate the differences between anticipative and non-anticipative calculus, and exercises with solutions as a way to isolate proofs of technical results.

The fourth chapter is contributed to by Pierre Louis Lions and Jean Michel Lasry. It is concerned with the influence of hedging on the dynamics of the underlying asset price. The problem is completely solved in the case of a *large* investor. The results are based on a detailed analysis of liquidation and indifference prices studied via the solutions of non-standards stochastic control problems. This paper is more of a technical nature, and reads more like a research article than an introductory review. However, the importance of the issue and the originality of the mathematical approach fully justify its inclusion in the series.

The last chapter is concerned with applications of large deviations to problems in finance and insurance. It is contributed by Huyén Pham. Large deviation approximations and importance sampling methods are discussed in the context of option pricing. Credit risk losses and portfolio benchmarking are also analyzed with the tools of large deviations.

The Editors
Paris / Princeton
January 31, 2007

Contents

HJM: A Unified Approach to Dynamic Models for Fixed Income, Credit and Equity Markets
René A. Carmona 1

Optimal Bond Portfolios
Ivar Ekeland and Erik Taflin 51

Models for Insider Trading with Finite Utility
Arturo Kohatsu-Higa 103

Large Investor Trading Impacts on Volatility
Pierre-Louis Lions and Jean-Michel Lasry 173

Some Applications and Methods of Large Deviations in Finance and Insurance
Huyên Pham 191

Editors

René A. Carmona
Paul M. Wythes '55 Professor of Engineering and Finance
ORFE and Bendheim Center for Finance
Princeton University
Princeton NJ 08540, USA
email: rcarmona@princeton.edu

Erhan Çinlar
Norman J. Sollenberger Professor of Engineering
ORFE and Bendheim Center for Finance
Princeton University
Princeton NJ 08540, USA
email: cinlar@princeton.edu

Ivar Ekeland
Canada Research Chair in Mathematical Economics
Department of Mathematics, Annex 1210
University of British Columbia
1984 Mathematics Road
Vancouver, B.C., Canada V6T 1Z2
email: ekeland@math.ubc.ca

Elyes Jouini
CEREMADE, UFR Mathématiques de la Décision
Université Paris-Dauphine
Place du Maréchal de Lattre de Tassigny
75775 Paris Cedex 16, France
email: jouini@ceremade.dauphine.fr

José A. Scheinkman
Theodore Wells '29 Professor of Economics
Department of Economics and Bendheim Center for Finance
Princeton University
Princeton NJ 08540, USA
email: joses@princeton.edu

Nizar Touzi
Centre de Mathématiques Appliquées
Ecole Polytechnique Paris
F-91128 Palaiseau Cedex, FRANCE
email: nizar.touzi@polytechnique.edu

HJM: A Unified Approach to Dynamic Models for Fixed Income, Credit and Equity Markets*

René A. Carmona

Bendheim Center for Finance
Department of Operations Research & Financial Engineering,
Princeton University, Princeton, NJ 08544, USA
email: rcarmona@princeton.edu

Summary. The purpose of this paper is to highlight some of the key elements of the HJM approach as originally introduced in the framework of fixed income market models, to explain how the very same philosophy was implemented in the case of credit portfolio derivatives and to show how it can be extended to and used in the case of equity market models. In each case we show how the HJM approach naturally yields a consistency condition and a no-arbitrage condition in the spirit of the original work of Heath, Jarrow and Morton. Even though the actual computations and the derivation of the drift condition in the case of equity models seems to be new, the paper is intended as a survey of existing results, and as such, it is mostly pedagogical in nature.

Keywords: Implied volatilty surface, Local Volatility surface, Market models, Arbitrage-free term structure dynamics, Health–Jarrow–Morton theory
Mathematics Subject Classification (2000) 91B24
JEL Classification (2000) G13

1 Introduction

The motivation for this paper can be found in the desire to understand recent attempts to implement the HJM philosophy in the valuation of options on credit portfolios. Several proposals appeared almost simultaneously in the literature on credit portfolio valuation. They were written independently by N. Bennani [3], J. Sidenius, V. Piterbarg and L. Andersen [26] and P. Shönbucher [41], the latter being most influential in the preparation of the present survey. After a sharp increase in volume and liquidity due to the coming of age of the single tranche synthetic CDOs, markets for these credit portfolios came to a stand still due to the lack of dynamic models needed to price forward starting contracts, options on options, ... So the need

---
* This research was partially supported by NSF DMS-0456195.

for dynamic models prompted these authors to build analogies between the original HJM approach to interest rate derivatives and derivatives on credit portfolio losses. The common starting point of these three papers is the lithany of well documented shortcomings of the market standard for the valuation of Collaterized Debt Obligations (CDOs). The standard Gaussian copula model is intrinsically a *one period* static model which cannot be used to price forward starting contracts. The valuation by expectation of these forward starting contracts require the analysis of a term structure of forward loss probabilities. The HJM modeling of the dynamics of the forward instantaneous interest rates, suggests how to choose dynamic models for these forward loss probabilities. The three papers mentioned above try to take advantage of this analogy with various degrees of generality and success.

The goal of this paper is to review the salient features of the HJM modeling philosophy as they can be applied to three different markets: the fixed income markets originally considered by Heath, Jarrow and Morton, the credit markets and the equity markets. In each of the three cases considered in this paper, the financial market model is based on a set of financial securities which are assumed to be liquidly traded. A basic assumption is that the price of each such security is observable, and any quantity of the security can be sold or bought at this observed price. These prices are used to encapsulate what the market is telling the modeler, and the thrust of the HJM modeling approach is to postulate dynamical equations for the prices of all these liquid instruments and to check that the multitude of all these equations do not introduce inconsistencies and arbitrage opportunities in the market model.

The classical HJM approach is reviewed in Section 3. Our informal presentation does not do justice to the depth of the original contribution [14] of Heath, Jarrow and Morton. It is meant as a light introduction to the modeling philosophy, our main goal being to introduce notation which are used throughout the paper, and to emphasize the crucial steps which will recur in the discussion of the other market models. Section 5 is devoted to the discussion of the recent works [26] of Sidenius, Pitterbarg and Andersen and [41] Schoenbucher on the construction of dynamic models for credit portfolios in the spirit of the HJM approach. These two papers are at the root of our renewed interest in the HJM modeling philosophy. It is while reading them that we realized the impact they could have on the classical equity models. The latter are usually calibrated to market prices by constructing an implied volatility surface, or equivalently a local volatility surface as advocated by Dupire and Derman and Kani in a series of influential works [19][16]. As we explain in Section 6, the construction of these surfaces is only the first step in the construction of a dynamic model. A dynamic version of local volatility modeling was touted by Derman and Kani in a paper [17] mostly known for its discussion of implied tree models. Motivated by the fact that the technical parts of [17] dealing with continuous models are rather informal and lacking mathematical proofs, Carmona and Nadtochiy developed in [7] the program outlined in [17]. On the top of providing a rigorous mathematical derivation of the so-called drift condition, they also provide calibration and Monte Carlo implementation recipes, and they analyze the classical Markovian spot models as well as stochastic volatility models in a generalized HJM framework. We present their results in the last section of this paper.

Acknowledgements. I would like to thank Dario Villani and Kharen Musaelian for introducing me to the intricacies of the credit markets. Their insights were invaluable: what they taught me cannot be found in textbooks !!!

2 General Mathematical Framework

This section is very abstract in nature. Its goal is to set the notation and the stage for the discussion of a common approach to three different markets.

2.1 Mathematical Notation

Throughout this paper we assume that $(\Omega, \mathcal{F}, \mathbb{P})$ is a probability space and $\{\mathcal{F}_t\}_{t \geq 0}$ is a right continuous filtration of sub-σ-fields of $\mathcal{F}$, $\mathcal{F}_0$ containing all the null sets of $\mathbb{P}$. Most often, we assume that this filtration is a Brownian filtration in the sense that it is generated by a Wiener process $\{W_t\}_{t \geq 0}$. We allow this Wiener process to be multi-dimensional, and in fact, it can even be infinite dimensional. The facts from infinite dimensional stochastic analysis which are actually needed to prove the results discussed in this paper in the infinite dimensional setting can be found in many books and published articles. Most of them can be derived without using too much functional analysis. For the sake of my personal convenience, I chose to refer the interested reader to the book [9] for definitions and details about those infinite dimensional stochastic analysis results which we rely upon.

In order to compute cash flow *present values*, we use a discount factor which we denote by $\{\beta_t\}_{t \geq 0}$. The latter is a non-negative adapted stochastic process. Typically we use for β_t the inverse of the bank account B_t which is defined as the solution of the ordinary (possibly random) differential equation:

$$dB_t = r_t B_t \, dt, \qquad B_0 = 1, \tag{1}$$

where the stochastic process $\{r_t\}_{t \geq 0}$ has the interpretation of a short interest rate. In this case we have

$$\beta_t = e^{-\int_0^t r_s \, ds}. \tag{2}$$

Notice that $\{\beta_t\}_{t \geq 0}$ is multiplicative in the sense that

$$\beta_{s+t}(\omega) = \beta_s(\omega)\beta_t(\theta_s \omega), \qquad \omega \in \Omega,$$

where $\{\theta_t\}_{t \geq 0}$ is a semigroup of shift operators on Ω. For the sake of illustration, we should think of the ω's in Ω as functions of time, in which case $[\theta_s \omega](t) = w(s+t)$.

We shall assume that $\mathbb{P}$ is a pricing measure. This means that the market price at time $t = 0$ of any liquidly traded contingent claim which pays a random amount ξ at time T, say p_0, is given by (notice that the pay-off ξ is implicitly assumed to be a $\mathcal{F}_T$ integrable random variable):

$$p_0 = \mathbb{E}\{\beta_T \xi\}$$

where $\mathbb{E}\{\cdot\} = \mathbb{E}^\mathbb{P}\{\cdot\}$ denotes the expectation with respect to the probability measure $\mathbb{P}$. In other words, $\mathbb{P}$ is a pricing measure if prices of contingent claims are given by $\mathbb{P}$-expectations of present values of their future cashflows.

If we also assume that the market is free of arbitrage, then the price p_t at time $t < T$ of the same contingent claim is necessarily given by the conditional expectation

$$p_t = \frac{1}{\beta_t}\mathbb{E}\{\beta_T \xi | \mathcal{F}_t\}$$

which shows that $\{\beta_t p_t\}_{t\geq 0}$ is a $\mathbb{P}$-martingale in the filtration $\{\mathcal{F}_t\}_{t\geq 0}$. In other words, if $\mathbb{P}$ is a pricing measure, the discounted prices are $\mathbb{P}$-martingales.

Notice that we do not assume that such a pricing measure is unique. In other words, we allow for incomplete market models in our discussion.

2.2 Liquidly Traded Instruments

We next assume that our economy is driven by a set of liquidly traded instruments whose prices at time t, we denote by P_t^α. We can think of the vector $\mathbf{P}_t = (P_t^\alpha)_{\alpha \in \mathcal{A}}$ of these observable prices as a state vector for our economy. We will not make the completeness assumption that

$$\mathcal{F}_t = \sigma\{\mathbf{P}_s;\ 0 \leq s \leq t\}, \qquad t \geq 0.$$

These instruments are fundamental for the analysis of the market, and a minimal requirement on a dynamical model of the economy will be that such a model provides prices for forward starting contracts and European call and put options on these basic instruments. In particular, at each time t, we should be able to compute the quantity

$$\mathbb{E}\{\beta_T (P_T^\alpha - K)^+ | \mathcal{F}_t\} \tag{3}$$

for every maturity $T > t$ and strike $K > 0$. Since a measure μ on the half line $\mathbb{R}_+$ is entirely determined by the knowledge of its call transform, i.e. the values of the integrals

$$\int_{\mathbb{R}_+} (x - K)^+ \mu(dx),$$

for $K > 0$, the knowledge at time t, of the prices of all the call options completely determines the distributions under the conditional measure $\mathbb{P}_t$, of all the random variables P_T^α for all $T > t$ and all $\alpha \in \mathcal{A}$.

Here, for each $t > 0$, we define the random measure $\mathbb{P}_t$ as the (regular version of the) conditional distribution given $\mathcal{F}_t$ of the discounted version of $\mathbb{P}$. In other words, $\mathbb{P}_t$ is characterized by the requirement that the equality

$$\mathbb{E}\{\beta_{t+T} \Phi \Psi \circ \theta_t\} = \mathbb{E}\{\beta_t \Phi \mathbb{E}^{\mathbb{P}_t}\{\beta_T \Psi\}\}$$

holds for all bounded random variables Φ and Ψ which are $\mathcal{F}_t$ and $\mathcal{F}_T$ measurable respectively.

Remark. Notice that if instead of simply requiring the knowledge of the prices of all the European call options we were to also require the knowledge of the prices of all the path dependent options, then for each $\alpha \in \mathcal{A}$, the entire (joint) distribution under $\mathbb{P}_t$ of $(P_T^\alpha)_{T\geq t}$ would be determined. In the situation of interest to us, only the one-dimensional marginal distributions of $\mathbb{P}_t$ are determined by the prices we can observe.

2.3 Dynamic Market Model

All the information about the market model should be contained in the specification of a pricing measure $\mathbb{P}$. However, as we explained earlier, it seems that a reasonable market model should

- *be consistent with the prices of the liquidly traded instruments quoted on the market*, in other words, the numerical values P_t^α observed on the market should be recovered as conditional expectations under the pricing measure $\mathbb{P}$ of the discounted cashflows of the corresponding instruments;
- *allow for the pricing of forward starting contracts (e.g. European call options on call options* using the identified liquidly traded instruments as underlyers. In other words, it should provide a way to compute the time evolution of the conditional (random) measures $\mathbb{P}_t$, or at least its marginal distributions.

The first bullet point involves simply reproducing the prices of the basic liquid instruments at time $t = 0$. It usually goes under the name of calibration. The restriction of the measure $\mathbb{P}$ to $\mathcal{F}_0$ is typically trivial and the computation of these prices involves only regular expectations with respect to $\mathbb{P}$ which can be computed at time $t = 0$. So this first bullet point does not seem to involve the dynamics of the stochastic evolution of the characteristics of the market model: it looks like a static requirement for a one period model.

On the other hand, the second bullet point involves information about the model (and hence the pricing measure $\mathbb{P}$) of a more dynamic nature. For this reason, if will appear to be preferable to specify this dynamic information about $\mathbb{P}$ by specifying $\{\mathbb{P}_t\}_{t\geq 0}$ as a stochastic process in the space of probability measures on the possible future time evolutions of the vectors $\{\mathbf{P}_{t+s}\}_{s\geq 0}$ of basic instruments. This is the main thrust of the HJM approach to fixed income market models as it was originally introduced by Heath, Jarrow and Morton, and this is the point of view we take to review in the remaining part of this paper, recent developments in modeling credit and equity markets.

3 The Classical HJM Approach

The goal of this section is purely of a pedagogical nature. It is not intended as a rigorous *exposé* of the original work of Heath, Jarrow and Morton: it is merely an informal discussion aimed at a very general audience. In the case of fixed income markets (also called interest rate derivatives markets), the simplest form of interest rate is the spot rate whose value at time t we denote by r_t. As we will emphasize in several instances, any market model needs to provide with the distribution of the stochastic process $\{r_t\}_{t\geq 0}$, even if its role is limited to the introduction of the bank account and the discount factor as in the previous section. Many market models have been based on the specification of the dynamics of this process. For this reason they are called *short rate models*. Despite the limitations which we are about to document, they remain very popular, mostly because of their versatility and the existence of closed form formulae for the prices of many liquidly traded instruments.

There are several sets of liquid interest rate derivatives actively traded and quoted daily. Coupon bearing bonds, caps, floors, swaptions, are some of them. But because most of them can be viewed as portfolios of zero coupon bonds, or European options on zero coupon bonds, and because this section aims at recasting classical material (which can be found in most financial mathematics textbooks) into the framework adopted in the paper, we find convenient to choose, for the set of liquidly traded securities, the ensemble of all the zero coupon non-defaultable bonds.

For the sake of definiteness, we denote by $B(t,T)$ the price at time t of such a zero coupon bond with maturity T. We shall often use the term "Treasury" (which essentially means that the bond will not default) interchangeably with "non-defaultable". The entire face value will be paid at time T by the issuer of the bond to the buyer as long as $T > t$. So at time $t = 0$, all the prices $B(0,T)$ can be observed and the entire curve

$$T \hookrightarrow B_0(T) = B(0,T) \tag{4}$$

is known. So as stated in the first bullet point of Subsection 2.3 above, a first requirement for a model given by a pricing measure $\mathbb{P}$ is to reproduce these prices exactly.

As we are about to see, this innocent looking condition cannot always be satisfied by the short interest rate models which need to be re-calibrated frequently to satisfy, at least approximatively this requirement. Indeed, short interest rate models are endogenous term structure models as the initial term structure of zero coupon bond prices (4) is an output of the model instead of being an input observed in the market place. This last point is one of the main components of the HJM approach.

Since the cash flows of a zero coupon bond reduce to paying its nominal amount (which we conveniently normalize to 1) at time T, the price has to be given by

$$B_0(T) = \mathbb{E}\{\beta_T\} = \mathbb{E}\{e^{-\int_0^T r_s ds}\}, \tag{5}$$

recall that $\beta_0 = 1$. So if the parameters of the pricing measure $\mathbb{P}$ allow for the computation of the expectation in the above right hand side, the value of this expectation will have to coincide with the observed price $B_0(T)$ if we want to satisfy the first bullet point above.

Using Instantaneous Forward Rates Instead. For reasons that will become clear later, if the zero coupon prices $B(t,T)$ are (or assumed to be) smooth in the maturity variable T, it is more convenient to work with the forward rates defined by

$$f(t,T) = -\frac{\partial}{\partial T} \log B(t,T) \tag{6}$$

rather than the bond prices directly. Since the bond prices can be recovered from the forward rates

$$B(t,T) = e^{-\int_t^T f(t,u)\,du} \tag{7}$$

the term structure of interest rates can be given equivalently by the forward curves. In particular, observing all the bond prices $B_0(T)$ at time $t = 0$ is equivalent to observing all the forward rates $f_0(T)$, and the initial forward rate curve

$$T \hookrightarrow f_0(T)$$

can be the object of the calibration efforts (in the case of short rate models) or it can serve as initial condition (in the case of HJM models).

3.1 Short Rate Models

Since the prices of the basic instruments of the market can be computed as expectations over the short interest rate, recall formula (5), the simplest prescription for a pricing measure $\mathbb{P}$ is to describe the dynamics of the short rate process. Typically, a short rate model assumes that under the pricing measure $\mathbb{P}$, the short interest rate r_t is the solution of a stochastic differential equation of the diffusion form (i.e. Markovian):

$$dr_t = \mu^{(r)}(t, r_t)\, dt + \sigma^{(r)}(t, r_t)\, dW_t \tag{8}$$

where the drift and volatility terms are given by real-valued (deterministic) functions

$$(t, r) \hookrightarrow \mu^{(r)}(t, r) \quad \text{and} \quad (t, r) \hookrightarrow \sigma^{(r)}(t, r)$$

such that existence and uniqueness of a strong solution hold. For the sake of illustration, we consider only one specific example. Indeed, the goal of this section is not to present the theory of short rate models. They are mentioned only as motivation for the introduction of the HJM modeling approach.

We choose the **Vasicek** model because of its simplicity, but for the purpose of the present discussion, a **CIR** model of the square root diffusion could have done as well. In the case of the Vasicek model, the dynamics of the short rate are given by the stochastic differential equation:

$$dr_t = (\alpha - \beta r_t)\, dt + \sigma\, dW_t. \tag{9}$$

This equation is simple enough (linear) to be solved explicitly. The solution is given by

$$r_t = e^{-\beta t} r_0 + (1 - e^{-\beta t})\frac{\alpha}{\beta} + \int_0^t e^{-\beta(t-s)} \sigma\, dW_s. \tag{10}$$

$\{r_t\}_{t \geq 0}$ is a Gaussian process whenever r_0 is, and at each time $t > 0$ there is a positive probability that r_t is negative. Despite this troubling feature (not only can an interest rate be negative in this model, but it is almost surely unbounded below!), this model is very popular because of its tractability and because a judicious choice of the parameters can make this probability of negative interest rate quite small. The tractability of the model is due to the fact that the random variable $\int_0^t r_s\, ds$ is Gaussian with mean and variance which can be explicitly computed from the parameters α, β and σ of the model, and from this fact, one gets an explicit formula for the expectation (5) giving the price of the zero coupon bonds. We get:

$$B_0(T) = e^{a(T)+b(T)r_0} \tag{11}$$

where r_0 is the current value of the short rate, and where the functions $a(T)$ and $b(T)$ are given by:

$$b(T) = -\frac{1}{\beta}\left(1 - e^{-\beta T}\right) \tag{12}$$

and

$$a(T) = \frac{4\alpha\beta - 3\sigma^2}{4\beta^3} + \frac{\sigma^2 - 2\alpha\beta}{2\beta^2}T + \frac{\sigma^2 - \alpha\beta}{\beta^3}e^{-\beta T} - \frac{\sigma^2}{4\beta^3}e^{-2\beta T}. \tag{13}$$

Alternatively, if we use the forward curve instead of the zero coupon bond curve we get:

$$f(t,T) = re^{-\beta(T-t)} + \frac{\alpha}{\beta}\left(1 - e^{-\beta(T-t)}\right) - \frac{\sigma^2}{2\beta^2}\left(1 - e^{-\beta(T-t)}\right)^2, \tag{14}$$

from which we get an expression for the initial forward curve $T \hookrightarrow f_0(T)$ by setting $t = 0$. Notice that such a forward curve converges to the constant $(2\alpha\beta - \sigma^2)/2\beta^2$ when $T \to \infty$. This limit can be given the interpretation of a *long rate* (as opposed to the short rate) when $\sigma^2 < 2\alpha\beta$. In any case, a Vasicek forward curve flattens and becomes horizontal for large maturity T. The graph of a typical example of a forward curve given by the Vasicek model is given in the left pane of Figure 1. We used the parameters $\alpha = 13.06$, $\beta = 2.5$ and $\sigma = 2$ to produce this plot. We clearly see the flattening of the curve on the right part of the plot.

Rigid Term Structures for Calibration

As we explained earlier, choosing values for the parameters of the model (α, β and σ in the Vasicek model discussed in this section) in order for the model to reproduce the observed forward curve is what is usually called calibration of the model. Since the Vasicek model depends upon three parameters, three quoted prices, say $B_0(T_1)$, $B_0(T_2)$ and $B_0(T_3)$ for three different maturities T_1, T_2 and T_3 should in principle be enough to determine these parameters. But unfortunately, the curve $T \hookrightarrow B_0(T)$ constructed from formulae (11), (12), and (13) and three parameter values derived from three bond prices does not always look like the curve produced by the market quotes, and most importantly, it changes with the choices of the three maturities T_1, T_2 and T_3. For the sake of illustration, we give in the right pane of Figure 1 the plot of the market zero-coupon forward curve on 3/28/1996, and we super-impose on the same graph the plot of the best least squares fit among the possible forward curves produced by the Vasicek model. This optimal Vasicek forward curve was obtained for the values $\alpha = 13.06$, $\beta = 2.401$ and $\sigma = 1.724$ of the parameters. The fact that a Vasicek forward curves flattens for large maturity makes it impossible to match the typical increase in T found in most practical instances.

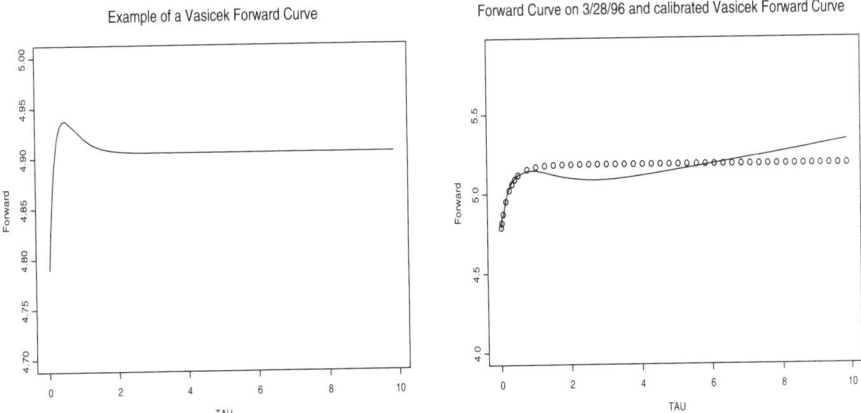

Fig. 1. Typical forward curve produced by the Vasicek model (left) and calibrated Vasicek forward curve (dotted line) to the zero-coupon forward curve on 3/28/1996

A Possible Fix

Several solutions have been proposed to the undesirable rigidity of the initial term structure curves produced by short rate models. The most popular one is to force some of the coefficients to be time dependent in order for the model to match any market forward curve $T \hookrightarrow f_0(T)$. This is especially simple and useful in the case of the Vasicek model for if the time dependent coefficients are deterministic, the solution process remains Gaussian, and closed form solutions for the values of the forward rates and zero coupon prices can still be derived. To be more specific, formula (10) becomes

$$r_t = e^{-\int_0^t \beta_s ds} r_0 + \int_0^t e^{-\int_s^t \beta_u du} \alpha_s ds + \int_0^t e^{-\int_s^t \beta_u du} \sigma_s dW_s. \quad (15)$$

and since the conditional distribution of the integral $\int_s^t f_u du$ is Gaussian, bond prices

$$B(t,T) = \mathbb{E}\{e^{-\int_t^T r_s ds} | \mathcal{F}_t\}$$

can still be derived from the expression of the Laplace transform of the Gaussian distribution.

This strategy was successfully implemented in the case of the Vasicek model (9) by Hull and White. These two authors proposed to leave the volatility σ and the mean reversion rate β constant, and to replace the parameter α by a deterministic function $t \hookrightarrow \alpha(t)$. In this case, the solution r_t is given by the formula

$$r_t = e^{-\beta t} r_0 + \int_0^t e^{-\beta(t-s)} \alpha_s ds + \sigma \int_0^t e^{-\beta(t-s)} dW_s, \quad (16)$$

and the forward rate is given by the formula

$$f(t,T) = e^{-\beta(T-t)} r_t + \int_t^T e^{-\beta(T-s)} \alpha_s ds - \frac{\sigma^2}{2\beta^2}[1 - e^{-\beta(T-t)}]^2. \quad (17)$$

If we replace in this formula t by 0 and $T - t$ by t, we get simple formulae for the initial forward curve and its derivative. From there one easily sees that it is possible to choose the function $t \hookrightarrow \alpha(t)$ to obtain any given (smooth) forward curve. To be specific, if we denote by $\overline{f}_0(T)$ the forward rate observed on the market at time $t = 0$ for maturity T, then choosing

$$\alpha_t = \overline{f'_0}(t) + \beta \overline{f_0}(t) - \frac{\sigma^2}{2\beta}(1 - e^{-\beta t})(3e^{-\beta t} - 1)$$

will force the initial forward curve $T \hookrightarrow f_0(T)$ produced by the Vasicek model with this time dependent coefficient $\alpha(t)$ to coincide with the market (observed) forward curve $T \hookrightarrow \overline{f}_0(T)$. The model is now compatible with the current *observed* forward curve, it is *calibrated* to the market.

Model calibration is everyday practice in quantitative finance, and procedures similar to the Hull-White modification of the Vasicek model are regarded as useful. But despite their popularity with practitioners, these calibration techniques remain problematic for several reasons.

Firstly, this fix is short lived for in general the adequacy of the modified model is limited to a short period. Indeed, the next time we check the forward curve given by the market, it will most likely not agree with the forward curves implied by the model, hence the need to recalibrate and changing the stochastic differential equation as we need to change its coefficients. The relevant question is then, when to recalibrate, and there is no theoretical answer to that question in general. The calibration procedure described above limits the usefulness of the model to a short time period, and de facto, turns a dynamic model into a *one period model*.

But there are other reasons to go beyond the short rate models. Indeed, specifying a short rate model amounts to specifying the (stochastic) dynamics of the whole forward curve by specifying the (stochastic) dynamics of the left-hand point of the curve. Indeed, $r_t = f(t,t)$, and this rigidity is confirmed by the fact that given any two maturities T_1 and T_2 the correlation coefficient between the *random variables* $df(t, T_1)$ and $df(t, T_2)$ is necessarily equal to 1!

Factor Models, Consistency and No-Arbitrage

Short rate models are particular cases of factor models of the terms structure of interest rates. They correspond to the case when the number of factors is one, and the sole factor is the short interest rate itself. More general factor models have been considered, and no-arbitrage conditions in the spirit of the discussion of this section have been derived at various levels of generality. See for example [9] or Proposition 2.2. of [24] for a sample condition.

Notation. We introduce a special notation $\tau = T - t$ for the time-to-maturity of a bond, or yield, or forward, etc. The forward rates (as well as the bond prices) are defined accordingly in terms of this new variable.

$$\tilde{f}_t(\tau) = f(t, t+\tau), \qquad \tau \geq 0. \tag{18}$$

Expressing the forward rates at time t in terms of time-to-maturity τ instead of time-of-maturity T has the advantage of forcing all the forward curves $\tilde{f}_t$ to be defined on the same domain $[0, \infty)$. This convenient notation is often called the Musiela notation.

We concentrate later on the no-arbitrage condition for general HJM models. For the time being, we discuss it in the context of factor models built from parametric families of forward or yield curves. These families are usually introduced in the following way. We start from a function G from $\Theta \times [0, \infty)$ into $[0, \infty)$ where Θ is an open set in $\mathbb{R}^d$ which we interpret as the set of possible values of a vector θ of parameters $\theta_1, \cdots, \theta_d$. In this way, for each $\theta \in \Theta$ the *curve* $G(\theta, \cdot) : \tau \hookrightarrow G(\theta, \tau)$ can be viewed as a possible candidate for the forward curve. For the sake of illustration we give the classical example of the Nelson-Siegel family defined by

$$G(\theta, \tau) = \theta_1 + (\theta_2 + \theta_3 \tau) e^{-\theta_4 \tau}, \qquad \tau \geq 0. \tag{19}$$

The parameters θ_1 and θ_4 are assumed to be positive. θ_1 represents the asymptotic (long) forward rate, $\theta_1 + \theta_2$ gives the left end point of the curve, namely the short rate, while θ_4 gives an asymptotic rate of decay. The set Θ of parameters is the subset of $\mathbb{R}^4$ determined by $\theta_1 > 0$, $\theta_4 > 0$ and $\theta_1 + \theta_2 \geq 0$ since the short rate should not be negative. The parameter θ_3 is responsible for a hump when $\theta_3 > 0$, or a dip when $\theta_3 < 0$. Other parametric families have been used, the most popular one being the Svensson's family. See [9] and the references therein.

We now introduce factor models from the notion of parametric family formalized above. We assume that we are given a parametric family G as before and we suppose that $\Theta = \{\Theta_t\}_{t \geq 0}$ is a d-dimensional semi-martingale with values in the parameter space Θ. We then set

$$\tilde{f}_t(\tau) = G(\theta_t, \tau), \qquad t \geq 0, \tau > 0.$$

Recall that τ represents the time to maturity. The d components θ_t^j of θ_t are interpreted as economic factors driving the dynamics of the term structure of interest rates. Assuming further that G is twice continuously differentiable in the variables θ^j, we can use Itô's formula and derive the dynamics of $f_t(\tau)$.

As we assume that the measure $\mathbb{P}$ is the measure used by the market to compute prices, in this context, the absence of arbitrage is equivalent to the the fact that all the discounted bond prices $\{B^*(t, T)\}_{t \in [0,T]}$ are local martingales. Recall the discussion of Section 2. Here, the discounted bond price at time t for maturity T is given by

$$B^*(t, T) = \beta_t B(t, T) = e^{-\int_0^t r_s ds} B(t, T) = e^{-\int_0^t f(s,s) ds} e^{-\int_t^T f(t,u) du} \tag{20}$$

since we are using the inverse of the bank account as discount factor. For each fixed $T > 0$, the process $\{B^*(t, T)\}_{0 \leq t \leq T}$ is a local martingale if the drift in its Itô's stochastic differential is 0. Such a condition takes a particularly simple form in the case of a factor model defined as above and when the factors θ_t^j form a d-dimensional Markov diffusion. We refrain from giving the details as we are about to discuss the same condition in a more general setting. The interested reader is referred to [9] p.70 or [25] for details. In the literature on the classical HJM approach to fixed income markets, a pair (G, Θ) satisfying the no-arbitrage condition is said to be *consistent*. Again, see for example [25] and [9]. For the sake of consistency (!!), we use the same terminology in the present situation. The context will make clear whether we mean consistency with a spot model or absence of arbitrage for a factor model.

3.2 The Heath–Jarrow–Morton Approach

In the far reaching paper [14], Heath, Jarrow and Morton proposed to solve the above dilemma by modeling directly the dynamics of the entire term structure of interest rates, in other words, by modeling the dynamics of the forward curve. This seemingly minor change has dramatic consequences: *it kills two birds with one stone* in the sense that both bullet points of Subsection 2.3 are taken care of by this change. Indeed, calibration merely reduces to feeding the initial condition to the dynamical equation (this takes care of the first bullet point), and the time evolution of the conditional probabilities $\mathbb{P}_t$ follows from the same dynamical equation.

In order to be more specific, we consider a pricing measure $\mathbb{P}$, we choose the basic instruments to be at each instant t the discounted bond prices $\{B^*(t,T)\}_{T\geq t}$ defined by equation (20) above, and we assume that for each fixed maturity T, these discounted bond prices form a continuous local martingale for $\mathbb{P}$. The martingale property is our way to guarantee that such a market model is free of arbitrage opportunities. We explain below that enforcing this martingale property in a model leads to a constraint which is known under the name of drift condition. In their original proposal, Heath, Jarrow and Morton suggested to work with the forward rates $\{f(t,T)\}_{t\in[0,T]}$ instead of the actual bond prices. So instead of starting from a dynamical equation of the form

$$dB^*(t,T) = \sum_{i=1}^d \tau^{(i)}(t,T)dW_t^{(i)} \tag{21}$$

for some predictable processes $\{\tau^{(i)}(t,T)\}_{t\in[0,T]}$, they assume that the dynamics of the forward rates $\{f(t,T)\}_{t\in[0,T]}$ are given by stochastic differential equations of the form

$$df(t,T) = \alpha(t,T)dt + \beta(t,T)\cdot dW_t, \tag{22}$$

where the processes $\{\alpha(t,T)\}_{t\in[0,T]}$ and $\{\beta(t,T)\}_{t\in[0,T]}$ are assumed to be predictable with respect to the filtration generated by the Wiener process. Notice that both β and W can be multivariate (i.e. vector valued) in which case the above equation can be understood in developed form as

$$df(t,T) = \alpha(t,T)dt + \sum_{j=1}^d \beta^{(j)}(t,T)dW_t^{(j)}. \tag{23}$$

Notice that, as we mentioned in the introduction, the theory allows for $d = \infty$. See for example [8] or the contribution of Ekeland and Taflin to this volume. Note also that elementary stochastic calculus manipulations can be used to derive an equation of the form (23) from a starting point like (21), and conversely, it is easy to go from (21) to (23).

Finally, we note that the dynamics (23) are given by a large number of stochastic equations, one for each maturity T. Equivalently, this can be rewritten as a single equation for a function of T, in other words a semi-martingale given by the solution of a stochastic differential equation with values in a space of functions of T. Still another possibility is to view the forward rate $f(t, T)$ as a random field parameterized by t and T. The reader interested in the interactions between these three points of view is referred to [8].

A Spot Consistency Condition

In most cases of interest, the limit $\lim_{T \searrow t} f(t, T)$ exists almost surely for each fixed t, and as we already mentioned, this limit can be naturally identified with the short interest rate r_t. Such a process $\{r_t\}_{t \geq 0}$ defined as the left hand point of the forward curve is a semi-martingale and its stochastic differential can resemble a stochastic differential equation of the form we used to define short interest rate models, though it turns out that this is generally not the case. This definition of the short rate can also be viewed as a consistency restriction between the specification of the dynamics of the forward curve and the possible prescription of stochastic dynamics for the short rate. It is expressed as:

$$r_t = f(t, t). \tag{24}$$

The Original HJM Drift Condition

The discounted bond prices $B^*(t, T)$ can be written in terms of the instantaneous forward rates $f(t, T)$ as

$$B^*(t, T) = e^{-\int_0^t r_s \, ds} e^{-\int_t^T f(t, u) \, du}$$

and computing their stochastic differentials using the dynamic equation (23), and setting the resulting drift to zero gives another restriction on the coefficients α and β of (23). As explained in most financial mathematics textbooks, this constraint can be written as:

$$\alpha(t, T) = \beta(t, T) \cdot \int_t^T \beta(t, s) ds = \sum_{j=1}^d \beta^{(j)}(t, T) \int_t^T \beta^{(j)}(t, s) ds. \tag{25}$$

The above formula shows that the drift is completely determined once the volatilities have been chosen. It was discovered by Heath, Jarrow and Morton [14] and is widely known as the HJM drift condition.

Summary of the Approach

In order to highlight the main components of the HJM modeling philosophy we summarize the preceding discussion in a short list of a few bullet points.

- At any time t, we coded the prices of the liquidly traded instruments (i.e. the zero coupon bonds) by a forward curve.
- We prescribed stochastic dynamics for the elements of the code-book under the pricing measure $\mathbb{P}$.
- We derived a *consistency* condition which holds if the model has to coexist with a short rate model.
- We derived a condition guaranteeing the absence of arbitrage (the discounted prices of all the liquidly traded instruments are local martingales) which took the form of a *drift condition*.

These results are quite satisfactory from the theoretical point of view. However, the business of choosing the number d of factors and the actual volatility processes $\beta^{(j)}(t,T)$ still remains. This issue is especially thorny as the prices of the liquidly traded instruments are supposed to go in the initial condition and not in the choice of the volatility factors. There is no generally accepted solution to this difficult problem. The most popular approach relies on prices of more exotic instruments and the analysis in principal components for the determination of d and the $\beta^{(j)}(t,T)$'s. See for example [9].

4 First Extensions to Equity Markets

Before switching gear and extending the HJM approach to more complex code-books as in the case of credit and equity markets discussed in the following sections, we review two extensions of the HJM approach to the equity markets when the complexity of the code-book is the same as in the classical case described in the previous section where the liquidly traded instruments were coded with a mere one dimensional curve.

4.1 Realized Variance and Variance Swaps

The goal of this first subsection is to illustrate the HJM framework based on the stochastic dynamics of a family of curves with the example of a class of instruments traded on equity desks. It appears that, when dealing with equity models, both in this section and in Section 6, discounting does not play any significant role except for complicating the nature of the formulae. so without any loss of generality, we assume in both sections that the short interest rate is zero and hence that the bank account B_t and the discount factor β_t are identically equal to 1.

Variance swaps on a stock (or an index) promise the payment of the realized variance of the log-returns of the underlier to the holder of the swap. They are popular ways for investors to gain pure exposure to variance, or to hedge volatility products.

Their prices are given by the expectation of this realized variance up to maturity. Assuming that they can be observed, instead of working from a model of the *spot variance* itself, we follow the approach proposed in [5] by Buehler who chooses to work directly with the dynamics of the entire implied variance swap curve, very much in the spirit of the HJM approach to the term structure of interest rates reviewed in the previous section.

To be specific, we define the annualized variance of a stock or index $S = \{S_t\}_{t \geq 0}$ over a period of n consecutive trading days $0 = t_0 < t_1 < \cdots < t_n = T$ by

$$\hat{V}_n = \frac{252}{n} \sum_{i=1}^{n} \left(\log \frac{S_{t_i}}{S_{t_{i-1}}} \right)^2$$

where 252 represents the number of trading days in one year. We follow the market practice of not subtracting the mean of the daily log-returns. For that reason, $\hat{V}_n$ is not exactly a variance. In any case, a (mean zero) variance swap with maturity T and strike K is a contract which pays $\hat{V}_n - K$ at time T. Since the strike K appears merely as an additive factor, the following analysis will be done by assuming, without any loss of generality, that $K = 0$.

If we assume that the dates $t_0, t_1, \cdots, t_n$ form a partition of the fixed time interval $[0, T]$, and that the mesh of this partition (i.e. the number $\sup_{i=1,\cdots,n} |t_i - t_{i-1}|$) goes to 0, the realized variance $\hat{V}_n$ converges towards the quadratic variation $\langle \log S \rangle_T$ of the logarithm of the underlier. So for the purpose of the mathematical analysis of these instruments, we assume that a variance swap with maturity T pays the realized quadratic variation $\langle \log S \rangle_T$, and we denote by $V_t(T)$ the price at time t of such an instrument.

In this subsection, we assume that there exists a liquid market of variance swaps on the underlier S. This assumption may be far-fetched for most stocks, but it is quite realistic for the major stock indexes. In particular, at time $t = 0$, the prices $V_0(T)$ of variance swaps for all maturities $T \geq 0$ can be observed. At each time t, we use $\{V_t(T); T \geq t\}$ for the set of prices of the liquid instruments on which we base our financial market model, and we define a dynamic market model by specifying the stochastic time evolution of this set of prices.

If as before we assume that the market chose a pricing measure $\mathbb{P}$, and if the underlier spot price S satisfies

$$dS_t = S_t \sigma_t dB_t, \qquad t \geq 0,$$

for some Wiener process $\{B_t\}_{t \geq 0}$ and an adapted process $\{\sigma_t\}_{t \geq 0}$, then since we assume that $V_t(T)$ is the price of a liquidly traded instrument, requiring absence of arbitrage implies that:

$$V_t(T) = \mathbb{E}\{\langle \log S \rangle_T | \mathcal{F}_t\} = \mathbb{E}\left\{ \int_0^T \sigma_s^2 ds \Big| \mathcal{F}_t \right\}.$$

Throughout this paper, we assume that interest rate is 0 (and $\beta_t \equiv 1$) whenever we discuss equity markets.

As in the case of the HJM approach to the term structure of interest rates, we assume that for each fixed t, $V_t(T)$ is a smooth function of the maturity T, and we define the forward variance $v_t(T)$ as its derivative with respect to maturity. In analogy with the term structure of interest rates, was coded by the instantaneous forward curve, we capture the term structure of realized variance by the forward variance curve $\tilde{v}_t$ defined by:

$$\tau \hookrightarrow \tilde{v}_t(\tau) = v_t(t+\tau), \qquad \tau \geq 0$$

where by definition $v_t(T) = \partial_T V_t(T)$. Notice that

$$\tilde{v}_t(0) = v_t(t) = \sigma_t, \qquad \mathbb{P}-a.s.$$

gives a simple form of the HJM spot consistency condition (24).

Notice also that, with the above notation, for each fixed T we must have:

$$v_t(T) = \partial_T V_t(T) = \mathbb{E}\{\sigma_T^2 | \mathcal{F}_t\}, \qquad 0 \leq t \leq T,$$

which shows that for each fixed T, the process $\{v_t(T)\}_{0 \leq t \leq T}$ is a martingale. Consequently, modeling its dynamics can be done by specifying that it has a semi-martingale decomposition of the form

$$dv_t(T) = \alpha_t(T)dt + \beta_t(T)\,dW_t$$

with $\alpha_t(T) \equiv 0$. So in this particular case, the HJM drift condition takes a trivial form.

The reader interested in factor models for the forward variance $v_t(T)$ and their consistency with no-arbitrage, as well as pricing and hedging of variance swaps in this setting is referred to [5].

Remark. The HJM framework also has bee applied to the commodity markets where most of the trading is done via forward contracts. So like in the case of the fixed income models reviewed in the previous section, the commodity forward markets can be characterized by the set of liquidly traded instruments formed by the forward contracts with a specific set of maturities. So the term structure of forward contracts is captured by a code-book of curves (functions of the date of maturity of the contracts), but since these forward contracts are traded, they must be martingales under the pricing measure chosen by the market and as in the case of the variance swaps markets, the HJM drift condition guaranteeing no-arbitrage is trivial. The drift condition is non-trivial only when the code-book is formed of non-traded instruments.

4.2 European Call Maturity Term Structure

The discussion of this subsection is motivated by the work [39] of Schönbucher on the term structure of implied volatility for a fixed strike K. Schönbucher's results were recently generalized in [42] by Schweizer and Wissel to the case of a fixed convex pay-off function h, when the hockey-stick function $h(x) = (x-K)^+$ is replaced by a general non-negative convex function. We review the main results of this more general version which includes for example power options whose pay-offs are given by the function $h(x) = x^\gamma$ for some $\gamma \geq 1$.

This analysis is made on the stochastic basis of a d-dimensional Wiener process $\mathbf{W} = \{W_t\}_{t \geq}$ on which the dynamics of the underlying stock price are given by an equation of the form

$$dS_t = S_t[\mu_t dt + \sigma_t dW_t^1] \tag{26}$$

where W_t^1 denotes the first component of $W_t \in \mathbb{R}^d$ and where $\{\mu_t\}_{t \geq}$ and $\{\sigma_t\}_{t \geq 0}$ are adapted stochastic processes to be specified.

In this application, as explained earlier, we assume that the market of liquidly traded instruments is formed by the contingent claims with maturity $T > 0$ and payoff $h(S_T)$ where h is a single non-negative convex function fixed once for all. We denote by $C_t(T)$ the price of such a claim and by $\Sigma_t(T)$ the corresponding implied volatility. Under the assumption of zero interest rate,

$$C_t(T) = \mathbb{E}\{h(S_T)|\mathcal{F}_t\}$$

and $\Sigma_t(T)$ is the unique number σ which recovers the price $C_t(T)$ from the Black-Scholes formula, i.e. the solution of

$$B_h(t, S_t, T, \sigma) = C_t(T)$$

where $B_h(t, S_t, T, \sigma) = \mathbb{E}\{h(S_T)|\mathcal{F}_t\}$ and the expectation is over a geometric Brownian motion with drift $\mu_t \equiv r$ and volatility $\sigma_t \equiv \sigma$. The existence and the uniqueness of such a $\Sigma_t(T)$ are well known in the classical case of the hockey-stick pay-off function $h(x) = (x - K)^+$. We review these facts later in Section 6. In the present situation of a general convex pay-off function h, we need to use a simple no-arbitrage argument which shows that the prices of the call options satisfy almost surely:

$$C_t(T_1) \leq C_t(T_2) \qquad \text{whenever } t \leq T_1 < T_2. \tag{27}$$

Indeed, if this inequality is violated with positive probability, it is possible to set up a costless portfolio at time t which can be re-balanced at time T_1 to provide riskless profit at time T_2, with positive probability. See [42] Proposition 2.1 for details. Moreover,

$$h(S_t) \leq C_t(T) \leq h(0+) + S_t h'(\infty) \qquad \text{for all } t \leq T,$$

the inequalities being strict if h is not affine and the spot process $\{S_t\}_{t \geq 0}$ satisfies a mild non-monotonicity condition. These properties guarantee the existence and the uniqueness of the implied volatility in the general case.

The purpose of this subsection is to analyze dynamic models for which the prices of the liquidly traded instruments are coded by their respective implied volatility. In other words, for each time $t > 0$, we want to use the one-to-one correspondence

$$\{C_t(T); T \geq t\} \leftrightarrow \{\Sigma_t(T); T \geq t\}$$

as a code-book for these prices. We recast the current set-up in the HJM framework described in the previous section by having the implied variance $\Sigma_t(T)^2$ play the same role as the yield to maturity of a discount bond. So in full analogy with the original HJM approach, we replace the implied volatility code book by the codebook of *forward implied variances* $X(t,T)$ defined by

$$X(t,T) = \frac{\partial}{\partial T}\left((T-t)\Sigma_t(T)^2\right) \tag{28}$$

so that we have the familiar expression

$$\Sigma_t(T)^2 = \frac{1}{T-t}\int_t^T X(t,u)du. \tag{29}$$

A dynamic model for our equity market is then determined by prescribing for each maturity T, the dynamics of $X(t,T)$ in the form

$$dX(t,T) = \alpha(t,T)dt + \beta(t,T)dW_t, \qquad 0 \leq t \leq T. \tag{30}$$

In the previous section, we emphasized the simplifications provided by a switch to a notation system based on the time to maturity $\tau = T - t$. With this in mind we set $\tilde{C}_t(\tau) = C_t(t+\tau)$ and $\tilde{X}_t(\tau) = X(t, t+\tau)$ and the dynamic model is defined for each fixed $\tau > 0$ by:

$$d\tilde{X}_t(\tau) = \tilde{\alpha}_t(\tau)dt + \tilde{\beta}_t(\tau)dW_t, \qquad t \geq 0. \tag{31}$$

The fact that $\mathbb{P}$ is a pricing measure (by which we mean that the underlying spot process $\{S_t\}_{t\geq 0}$ and price processes $\{C_t(T)\}_{0\leq t\leq T}$ of the the liquid instruments are both local martingales is essentially equivalent to a *spot consistency* condition

$$\sigma_t = \tilde{X}(t,0), \qquad \mathbb{P}\,a.s. \tag{32}$$

for all $t > 0$ (in full analogy with the classical HJM case), together with a *drift condition*

$$\tilde{\alpha}_t(\tau) = -\frac{\partial^2_{\tau\tau}\tilde{C}_t(\tau)}{\partial_\tau\tilde{C}_t(\tau)}\tilde{\beta}_t(\tau)\int_0^\tau \tilde{\beta}_t(u)du - \frac{1}{2}\partial_\tau\frac{\partial^2_{\tau\tau}\tilde{C}_t(\tau)}{\partial_\tau\tilde{C}_t(\tau)}\tilde{X}_t(\tau)\Big|\int_0^\tau \tilde{\beta}_t(u)du\Big|^2$$
$$-S_t\frac{\partial^2_{S\tau}\tilde{C}_t(\tau)}{\partial_\tau\tilde{C}_t(\tau)}\sigma_t\tilde{\beta}^1_t(\tau) - S_t\partial_\tau\Big(\frac{\partial^2_{S\tau}\tilde{C}_t(\tau)}{\partial_\tau\tilde{C}_t(\tau)}\Big)\tilde{X}_t(\tau)\sigma_t\int_0^\tau \tilde{\beta}^1_t(u)du.$$

Remarks. 1. We stated above that the spot consistency and the drift conditions are *essentially* equivalent to the absence of arbitrage because on the top of some natural technical assumptions, the proof also requires the smoothness of the pay-off function h. See [42] for details.

2. Making explicit the deep and profound relationship between the spot volatility σ_t and the implied volatility Σ_t was done in the more general setting of the full implied volatility surface $(T, K) \hookrightarrow \Sigma_t(T, K)$ by Durrleman in [20]. Despite the fact that

his goals were different, most of the computations involved in the derivations of the results of [42] stated above can be found in one form or another in Durrleman's proofs.

3. The results reviewed in this subsection should also be linked to a recent work of Jacod and Protter who study in [27] the problem of the completion of a market by adding derivative instruments. Indeed, in so doing. they derive conditions very similar to the spot consistency and the drift conditions reviewed aove. As an added bonus, and if the equations were not technical enough, Jacod and Protter work in the more general set-up of semi-martingale dynamics with jumps.

4. The complexity of the drift condition (33) and the technicalities involved in its derivation are the main reason why dynamic models for the entire implied volatility surface have not been pursed. This is in fact the reason why Schönbucher in [39] and Schweizer and Wissel in [42] limit themselves to dynamic models for a cross section of the implied volatility surface. This complexity is also at the root of the point of view taken by Carmona and Nadtochiy in [7] where they give up on the implied volatility code-book and work with dynamic models based on the local volatility code-book instead. We review the main elements of this approach in Section 6.

5 The HJM Approach for Credit Markets

We now explain how the above modeling philosophy can be used in the case of credit markets. For the sake of simplicity, we assume that the time evolution of the discounting factor is independent of all the default processes underlying the credit derivatives we consider in this section. So for all practical purposes, we can assume that $\{r_t\}_{t\geq 0}$ and $\{B(t,T)\}_{0\leq t\leq T}$ are deterministic. The market of Collaterized Debt Obligations (CDOs for short), and especially the market for single tranch synthetic CDOs saw a tremendous growth in the last five years, and because of their increased liquidity, they became a favorite testbed for quantitative research for the credit markets. As they were the main motivation for the works [26] and [41] which we draw from in this section, for the sake of completeness, we review some of their basic characteristics. In this section, we concentrate on the analysis of these instruments and when we say *credit markets* we mean the markets they span. They provide an appropriate set-up in which we test the HJM approach advocated in this paper. The reader interested in a broader perspective on the credit markets is referred to the textbooks of Schönbucher [40], Duffie and Singleton [18] or Lando [29].

5.1 Single Tranche Synthetic CDO Market Data

We review rapidly the major properties of Single Tranche Synthetic CDOs, often abbreviated as STSCDOs. Not only this will serve as motivation for the following developments, but it will also help us set the notation. Even though these instruments are best understood as derivatives on a portfolio of Credit Default Swaps (known as CDSs), for the sake of time and space, we introduce them independently.

Two parties are involved in any single STSCDO transaction: a counterparty seeking protection against defaults of all or part of a set of firms, and a counterparty selling this protection. To be more specific, a CDO swap with maturity T and notional N, on a tranche with attachment point ℓ_1 and detachment point ℓ_2, is a contract over the period $[0, T]$ whereby the protection seller will compensate the protection buyer for all the default losses in the interval $[\ell_1 N, \ell_2 N]$, in exchange for regular coupon payment computed at a fixed rate (the so-called spread) on a loss depreciating notional. We shall give a formal definition below.

But first, for the sake of illustration, we reproduce the following tables giving bid and ask quotes on the 4th and 5th series of the CDX-IG tranches on December 19, 2005. A hand-picked board of professionals selected a pool of firms as a representative snapshot of an homogeneous slice of the market (here IG stands for Investment Grade, but there exist indexes based on pools of high volatility firms, etc.), and portfolios of credit losses are used to construct an index and tranches which are traded on the market. These indexes are maintained and updated from one series to the next. Each series typically comprise $I = 125$ firms

IG4	0 - 3%	3 - 7%	7 - 10%	10 - 15%	15 - 30%
5-year	38 1/4 - 39 1/4	106 - 112	26 - 32	11 - 16	6 - 7 1/2
7-year	51 3/8 - 52 1/8	244 - 254	47 - 54	26 - 32	8 1/2 - 11
10-year	57 1/2 - 59 1/8	598 - 617	118 - 126	58 - 66	16 - 22

IG5	0 - 3%	3 - 7%	7 - 10%	10 - 15%	15 - 30%
5-year	41 1/4 - 42 1/4	107 1/4 - 112	26 - 29	11 - 14	6 1/2 - 9 1/2
7-year	54 3/4 - 55 5/8	290 - 300	45 - 51	27 - 31	7 - 10
10-year	61 3/4 - 62 3/4	685 - 705	118 - 124	61 - 66	17 - 21

The interpretation of these figures is the following. The quote for the equity tranche (0 - 3%) is the upfront payment (as a percentage of the notional) that is paid *in addition* to the minimal of 500 basis points per year. Quotes for all other tranches are in basis points per year.

We explain the meaning of these quotes by explaining in detail the cash flows associated with one of these tranches. For the sake of definiteness we choose the super-senior tranche with attachment points 15 and 30% on the 5yr CDX-IG index series 4. Let us assume that this tranche traded for 7 basis points. In this case, the protection buyer is to pay 0.07% of the notional per year (in quarterly coupon payments made in arrear). In return, she will be compensated for any losses on the portfolio during the five years that are between 15% and 30% of the principal. The losses are computed from the portfolio underlying the index at the original time of the trade.

The quotes for all the other tranches are defined similarly except for the equity tranche for which the buyer of protection pays an upfront fee and a spread of 500 basis points per year. The published quotes give the bid and ask for the upfront fee expressed as a percent of the notional. The percent of the notional that the protection buyer of the equity-tranche has to pay on December 19, 2005 was between 38.25% and 39.25% for five-year protection.

The index is also quoted to indicate the cost of buying full protection against all $I = 125$ names.

5.2 First Mathematical Model

As the largest volume of transactions involve derivative contracts written on synthetic portfolios identified and maintained by the Dow Jones (CDX in the US and iTraxx in Europe), we restrict ourselves to a fixed credit portfolio of I firms, and we denote by τ_i the time of default of firm i. In practice, one is limited to a finite horizon T^* and one only observes $\tau_i \wedge T^*$. Motivated by single tranche synthetic CDOs, we mostly consider instruments with maturities 3, 5, 7 or 10 years, so T^* can safely be assumed to be 10.

We denote by $\{L(t)\}_{t\geq 0}$ the cumulative portfolio loss (appropriately normalized) up to and including time t. We denote by $N(t)$ the nominal of the portfolio at time t so that $N(0)$ denotes the initial nominal. Note that $N(t)$ is a non-increasing function of time and that

$$L(t) = 1 - \frac{N(t)}{N(0)}$$

is a non-decreasing function of time which satisfies

$$L(0) = 0, \quad \text{and} \quad L(t) \leq 1.$$

Since the purpose of the present paper is mostly pedagogical, we make several assumptions with the mere intent to avoid unnecessary technicalities and simplify the notation.

Motivated by the example of the Dow Jones indexes, and especially by the actively traded Investment Grade (IG for short) North America index, we assume that the portfolio is symmetric in the sense that the credit exposure due the possible default of any single firm does not change with the firm in question. So typically, we restrict ourselves to firms included in the CDX and ITraxx indexes published by Dow Jones, and when we discuss CDOs, we consider only single tranche synthetic CDOs on these indexes. So not only do we assume that the individual firm nominal amounts are the same, but we also assume that the recovery rates in case of default are also deterministic, and the same for all the firms. So ignoring an irrelevant scaling factor, for the sake of definiteness we assume that

$$L(t) = \frac{1}{I} \sum_{i=1}^{I} D_i(t)$$

where $D_i(t)$ is the default indicator of firm i defined as:

$$D_i(t) = \mathbf{1}_{\{\tau_i \leq t\}}$$

τ_i being the stopping time giving the time of default of firm i. See for example the discussions in [18] and [29] for what kind of event can trigger or constitute default.

Defined in this way, $L(t)$ represents the relative number of defaults prior to and including t, given the fact that there was no default at time $t = 0$. $\{L(t); t \geq 0\}$ is a stochastic process with non-negative piecewise constant and non-decreasing sample paths with values in the finite set $\mathcal{I} = \{0, 1/I, 2/I, \cdots, (I-1)/I, 1\}$.

CDO Mechanics and Liquidly Traded Instruments

Even though this is not exactly the case (as the membership in these portfolios is reviewed on a regular basis), we shall assume that the set of firms included in the portfolio is fixed and does not change over the life of the derivatives we consider. Moreover, for the sake of simplicity, we shall assume that the discounting factor β_T is deterministic (or independent of the default times) and hence, can be taken out of the expectations.

The prices of the basic instruments playing the role of the prices of the zero coupon bonds, are the tranche and index spreads. To be more specific, we shall assume that for each $i = 0, 1, 2, 3, 4$ the spread $s_i(T)$ is observable for each maturity $T = 1, 3, 5, 7, 10$. By convention, we assume that s_0 is the spread on the index, s_1 is the spread on the equity tranche, s_2 the spread on the lower mezzanine trance, etc. In order to explain how each spread is computed, we introduce the tenor structure

$$T_1 < T_2 < \cdots < T_n$$

of the days on which the coupon payments are to take place, and we continue the analysis of the tranche with attachment point ℓ_1 and detachment point ℓ_2 introduced earlier. Recall that we now assume that the portfolio nominal has been scaled down to 1.

Let us denote by s the rate of the coupon payments, and let us first evaluate the protection payments received by the protection buyer. Recall that, each time a loss L occurs, we assume that the part RL of the loss is recovered independently of the existence of the protection contract.

For notational convenience, for each time t, we define the quantity $L(t, \ell_1, \ell_2)$ by

$$L(t, \ell_1, \ell_2) = (L(t) - \ell_1)^+ - (L(t) - \ell_2)^+.$$

It gives at time t, the cumulative losses in the tranche. Indeed, it is equal to 0 if there were not enough losses to affect the tranche (i.e. if $L(t) < \ell_1$), it is equal to the tranche nominal $\ell_2 - \ell_1$ if the tranche was wiped out by losses (i.e. if $L(t) > \ell_2$), and it gives the lost part of the tranche nominal (i.e. $L(t) - \ell_1$) in the remaining cases (i.e. when $\ell_1 \leq L(t) \leq \ell_2$). So the expected present value (at time $t = 0$) of all the default losses recovered under the protection contract is

$$PL = (1-R) \sum_{i=1}^{n} \beta_{T_i} [\mathbb{E}\{L(T_i, \ell_1, \ell_2)\} - \mathbb{E}\{L(T_{i-1}, \ell_1, \ell_2)\}] \tag{33}$$

We now consider the cashflows to the protection seller. At each coupon payment date T_i, she should receive the interest accumulated over the period $[T_{i-1}, T_i]$ computed on the remaining tranche nominal $(\ell_2 - \ell_1) - L(T_i, \ell_1, \ell_2)$. So the expected present value (at time $t = 0$) of all these coupon payments is

$$IL = s \sum_{i=1}^{n} \beta_{T_i}(T_i - T_{i-1})\mathbb{E}\{(\ell_2 - \ell_1) - L(T_i, \ell_1, \ell_2)\} \qquad (34)$$

The (fair) spread of the tranche at time $t = 0$ is the break even value of s making the expected present values of the two legs (34) and (33) equal to each other. Hence, the spread is given by the formula:

$$s = (1 - R)\frac{\sum_{i=1}^{n} \beta_{T_i} \mathbb{E}\{L(T_i, \ell_1, \ell_2) - L(T_{i-1}, \ell_1, \ell_2)\}}{\sum_{i=1}^{n} \beta_{T_i}(T_i - T_{i-1})\mathbb{E}\{(\ell_2 - \ell_1) - L(T_i, \ell_1, \ell_2)\}} \qquad (35)$$

Since we want to work with as many tranches as possible at once, we give up our notation $\ell_1 < \ell_2$ for the attachment/detachment points limiting the tranche, and for the sake of convenience, we shall from now on use the notation $0 = K_0 < K_1 < \cdots < K_k = 1$ for the end points of the tranche intervals.

Our goal is to extract the values of all the expectations

$$C_{i,j} = \mathbb{E}\{(L(T_i) - K_j)^+\}, \qquad i = 1, 2, \cdots, n, \quad j = 1, \cdots, k, \qquad (36)$$

from the values at time $t = 0$ of the spreads for all the available maturities (typically $\tau_1 = 1$, $\tau_2 = 3$, $\tau_3 = 5$, $\tau_4 = 7$, and $\tau_5 = 10$ years) on the index and all the liquidly traded tranches. This problem is not well posed as there are many more expectations than spread quotes. We use a simple form of regularization method to extract a set of expectations from the set of observable spreads. This is in stark contrast with the situation encountered next section when we discuss dynamic models for the equity markets. There, the expectations are directly observable.

The simplest regularization method leads to the least squares estimation We estimate them by solving the least squares minimization problem

$$\mathbf{C} = [C_{i,j}]_{i,j} = \arg\inf_{\mathbf{C}} \qquad (37)$$

$$\sum_{j,k} w_{j,k} \Big| s_j(\tau_k) - R' \frac{\sum_{T_i \leq \tau_k} \beta_{T_i}[C_{i,j} - C_{i,j-1} - C_{i-1,j} + C_{i-1,j-1}]}{\sum_{T_i \leq \tau_k} \beta_{T_i}(T_i - T_{i-1})(K_j - K_{j-1} - C_{i,j} + C_{i,j-1})} \Big|^2,$$

where $R' = 1 - R$, and where for each maturity τ_k and tranche label j, the weights $w_{j,k}$ are chosen to be increasing in liquidity and decreasing in the size of the bid-ask spread. Unfortunately, this naive idea is unrealistic because of the large discrepancy between the number of reliable observations $s_j(\tau_k)$ and the number of desired C_{ij}. Even Levenberg-Marquard algorithms cannot provide a stable solution. The only known fixes are based on hand-waiving arguments and their reliability is questionable. See nevertheless [35] or [36]. Despite all that, it is common to assume that the numbers

$$C_{i,j} = \int (x - K_j) d\mu_{T_i}(x), \qquad i = 1, 2, \cdots, n, \quad j = 1, \cdots, k, \qquad (38)$$

are known! Here we use the notation μ_T for the distribution of the cumulative loss $L(T)$. As we already mentioned, for any measure μ, the *call transform* C_μ defined by

$$K \hookrightarrow C_\mu(K) = \int (x-K)^+ \, d\mu(x), \qquad (39)$$

completely determines the measure μ. In general, a measure μ cannot be completely recovered from the mere knowledge of $C_\mu(K)$ for a small number of values of K, unless extra information on μ is available, e.g. μ is a finite sum of point masses.

As we will see in the next section, there are many ways to extrapolate these values of $C_{i,j}$ in between the attachment points to obtain for each T_i a convex function of the continuous variable K which coincides with the values derived above for all the $K = K_j$. We postpone the discussion of this point to the review of what is known in the case of equity options in Section 6 below.

So it is commonly assumed that at time $t=0$, one knows the values of all the expectations $\mathbb{E}\{(L(T_i)-K)^+\}$ for all $K > 0$ which is equivalent to the full knowledge of the marginal distributions of the cumulative loss $L(T_i)$ at all the coupon payment times T_i under the distribution $\mathbb{P}$. This is the common starting point of the two papers on dynamic credit portfolio models which we review in this section.

Loss Distribution Dynamics

Having a hold of the marginal distributions of $L(T_i)$ is enough to price many instruments consistently with the spreads quoted on the market on day $t=0$. However, this may not be enough for *forward starting contracts*. Let us consider for example the case of a tranche swaption, i.e. an option to enter a tranche swap contract (with maturity T and attachment/detachment points $\ell_1 < \ell_2$) at a later time $0 < T_0 < T$ at a spread level s fixed today at time $t = 0$. The value today of such an option is given by

$$\beta_{T_0} \mathbb{E}\{PL(T_0) - IL(T_0)\}$$

Here, the protection leg $PL(T_0)$ is the random variable equal to the value of PL computed from formula (33) provided we replace in formula (33) the expectations $\mathbb{E}\{\cdot\}$ by conditional expectations $\mathbb{E}_{T_0} = \mathbb{E}\{\cdot | \mathcal{F}_{T_0}\}$ with respect to the sigma-field $\mathcal{F}_{T_0}$ of the information which will be available at time T_0. Similarly, the investment leg $IL(T_0)$ is the random variable equal to the value of IL computed from formula (34) for the spread $s = $ s and the conditional expectation $\mathbb{E}_{T_0} = \mathbb{E}\{\cdot | \mathcal{F}_{T_0}\}$ instead of the plain expectation with respect to $\mathbb{P}$.

So attempting to price forward starting contracts requires for each future time $t > 0$, to go through the calibration procedure described earlier at time $t = 0$ for the probability structure given by the (unconditional) pricing measure $\mathbb{P}$, using all the information available at time t by replacing $\mathbb{P}$ by its conditional version $\mathbb{P}_t = \mathbb{P}\{\cdot | \mathcal{F}_t\}$.

The above discussion justifies the introduction of the following notation which will be needed to describe dynamical models. For each $t \leq T$, we denote by

$P_t(T, \cdot)$ the distribution of the cumulative loss $L(T)$ conditioned by $\mathcal{F}_t$. In other words,
$$P_t(T, x) = \mathbb{P}\{L(T) \leq x | \mathcal{F}_t\}, \qquad x \in [0, 1].$$
Since $L(T)$ takes only finitely many values, the $I+1$ values $x = i/I$ for $i = 0, 1, \cdots, I$ to be specific, we can talk about its density. We shall use a lower case to denote this density
$$p_t(T, x) = \mathbb{P}\{L(T) = x | \mathcal{F}_t\}, \qquad x \in [0, 1].$$
These distributions will be called the forward loss distributions.

5.3 Two Different Approaches

It is important at this stage to emphasize the main difference between the approach of [26] and the one of [41], as this main difference lies in the choice of the filtration $\{\mathcal{F}_t\}_t$. In [41], the filtration $\{\mathcal{F}_t\}_t$ is the full filtration containing all the information available at time t, including both the economic factors and the default information. In these conditions, even after conditioning by $\mathcal{F}_t$, the above marginal probabilities of the loss distribution are discrete and can take only finitely many values between $L(t)$ and 1, typically the numbers $L(t), L(t) + 1/I, \cdots, 1$. However, in [26] the filtration used for conditioning in the definition of the forward loss distributions is a smaller filtration, say $\{\mathcal{M}_t\}_t$, which at each time t contains only information on economic factors and not necessarily on the actual default times. Intuitively, if one thinks of an intensity model for the time of default τ_i, the knowledge of $\mathcal{M}_t$ will determine the intensity $\lambda_i(t)$ at time t, but no information on the exponential random variable needed to compute the probability of arrival of the i-th default. This lack of information on the default arrival forces an integration with respect to the exponential random variable in order to compute the forward loss probabilities as defined by conditioning with respect to $\mathcal{M}_t$, and this integration justifies the assumption that the forward probabilities as defined above are smooth functions of the variable x. The densities $p_t(T, x)$ then appear to play the same role as the instantaneous forward rates in the classical HJM theory as they are derivatives of the forward rates given by the loss cumulative distribution functions. We come back to this approach at the end of this subsection.

In any case, for each fixed t, and for each $T \geq t$, we denote by $\mu_{t,T}$ the distribution of $L(T)$ under the conditional probability $\mathbb{P}_t$, and as usual, we denote by $\tau = T - t$ the time to maturity. Since the sample paths of the process $\{L(t+\tau)\}_{\tau \geq 0}$ are non-decreasing $\mathbb{P}_t$ – almost surely, for each fixed t, the measures $\{\tilde{\mu}_{t,t+\tau}\}_{\tau \geq 0}$ are non-decreasing in the balayage order in the sense that for every convex function ϕ, it holds
$$\int \phi(x) d\tilde{\mu}_{t,t+\tau_1}(dx) \leq \int \phi(x) d\tilde{\mu}_{t,t+\tau_2}(dx)$$
whenever $\tau_1 \leq \tau_2$. A classical result of Kellerer [28] implies the existence of a Markov process $\{Y_\tau\}_{\tau \geq 0}$ with the marginal distributions $\{\tilde{\mu}_{t,t+\tau}\}_{\tau \geq 0}$. Notice that this process depends upon t, but for the sake of notation we shall not emphasize this fact.

Schönbucher's Approach

In the case of the full filtration $\{\mathcal{F}_t\}_t$, the Markov process $\{Y_\tau\}_\tau$ has finite state space. Hence its distribution is entirely captured by its infinitesimal generator. The latter is a family of $(I+1) \times (I+1)$ Q-matrices indexed by $\tau \geq 0$ as the Markov process is not necessarily time homogeneous. Notice that we use the finite set $\{0, 1/I, 2/I, \cdots, (I-1)/I, 1\}$ as common state space for all these Markov processes instead of limiting the state space to the smaller set $\{L(t), L(t) + 1/I, \cdots, 1\}$ which depends upon the realization of the random loss $L(t)$. Our choice is justified by the need to define dynamic equations which are more easily stated if all the Markov processes have the same state space.

We denote by $\{A_t(\tau); \tau \geq 0\}$ the infinitesimal generator of the Markov process $\{Y_\tau\}_{\tau \geq 0}$, and we denote by $\{a_t(\tau, x, y)\}_{x,y \in \{0, 1/I, 2/I, \cdots, (I-1)/I, 1\}}$ the entries of the Q-matrix $A_t(\tau)$. We shall use this family of Q-matrices as a code-book for the information contained in the forward stochastic model as given by $\mathbb{P}_t$ once calibrated to the observable quotes at time t.

A classical fact from the theory of finite state Markov processes says that for each $\tau > 0$, the off-diagonal entries $a_t(\tau, x, y)$ are non-negative for $x \neq y$ as they have the interpretation of rate of jump from state x to state y. Because of this interpretation, as the sample paths of $L(t+\tau)$ are non-decreasing, the rates $a_t(\tau, x, y)$ should be zero whenever $y < x$, which implies that the matrices $A_t(\tau)$ are upper diagonal. Notice that the last row is identically zero since the state 1 (corresponding to the default of all the firms in the portfolio) is absorbing. Finally, the fact that the matrices $A(\tau)$ form the infinitesimal generator of a Markov process also imply that

$$a_t(\tau, x, x) = -\sum_{y \neq x} a_t(\tau, x, y), \qquad x \in \{0, 1/I, 2/I, \cdots, (I-1)/I, 1\},$$

which shows that the only entries that matter in the characterization of the code-book are the entries in each row, to the right of the diagonal.

Notice that the transition probabilities

$$p_t(\tau_1, \tau_2, x, y) = \mathbb{P}_t\{L(t+\tau_2) = y \,|\, L(t+\tau_1) = x\}$$

contain the same information as the infinitesimal generator matrices $\{A_t(\tau); \tau \geq 0\}$ as the two sets of matrices are related by the forward Kolmogorov equations which read:

$$\frac{\partial}{\partial \tau_2} p_t(\tau_1, \tau_2, x, y) = \sum_{k=0}^{I} p_t(\tau_1, \tau_2, x, k/I) a_t(\tau_1, k/I, y) \tag{40}$$

with initial condition

$$p_t(\tau_1, \tau_2, x, y)|_{\tau_2 = \tau_1} = \mathbf{1}_{\{x=y\}}.$$

Using now the fact that we restrict ourselves to upper triangular Q-matrices, and the fact that the diagonal element of each row is the negative of the sum of the other elements of the row, we see that:

$$\frac{\partial}{\partial \tau_2} p_t(\tau_1, \tau_2, x, y) = \sum_{k=xI}^{yI-1} p_t(\tau_1, \tau_2, x, k/I) a_t(\tau_1, k/I, y)$$

$$- p_t(\tau_1, \tau_2, x, y) \sum_{k=yI+1}^{I} a_t(\tau_1, y, k/I). \qquad (41)$$

Finally it is easy to see that, once in the form (41), these equations can be solved inductively for the transition probabilities. One gets:

$$p_t(\tau_1, \tau_2, x, y) = \qquad (42)$$
$$\begin{cases} 0 & \text{if } y < x \\ \exp[\int_{\tau_1}^{\tau_2} a_t(\tau_1, s, x, x) ds] & \text{if } y = x \\ \sum_{k=xI}^{(y-1)I} \int_{\tau_1}^{\tau_2} p_t(\tau_1, s, x, k/I) \exp[\int_s^{\tau_2} a_t(\tau_1, s, y, y) ds] & \text{if } y > x \end{cases}$$

Notice that for each fixed $t > 0$, the connection between the forward loss distributions $\mathbb{P}_{t,\cdot}\{(t+\tau) = \cdot\} = p_t(\tau, \cdot)$ and the transition probabilities $p_t(\tau_1, \tau_2, \cdot, \cdot)$ of the Markov process $\{Y_\tau\}_{\tau \geq 0}$ is given by the relation

$$p_t(\tau, x) = p_t(0, \tau, L(t), x), \qquad x \in \{0, 1/I, 2/I, \cdots, (I-1)/I\}$$

since the marginal distributions of the Markov process $\{Y_\tau\}_{\tau \geq 0}$ are $\{\tilde{\mu}_{t,t+\tau}\}_{\tau \geq 0}$.

In order to avoid obscuring the main ideas by technicalities, we shall assume that the occurrence of more than one default at a time is impossible. This implies that the cumulative portfolio loss process $\{L(t+\tau)\}_{\tau \geq 0}$ can only jump by the amount $1/I$, and consequently the Q-matrix $A_t(\tau)$ are bi-diagonal in the sense that the only non-zero terms are the diagonal entries $a_t(\tau, x, x)$ and their neighbors $a_t(\tau, x, x+1/I)$ as long as $x < 1$. So under this assumption, the code-book reduces to a set of exactly I functions of time (to maturity) namely

$$\{a_t(\tau, x); \tau \geq 0\}_{x \in \{0, 1/I, 2/I, \cdots, (I-1)/I\}} \qquad (43)$$

where we used the notation $a_t(\tau, x) = -a_t(\tau, x, x) = a_t(\tau, x, x+1/I)$.

It is explained in [41] that this assumption can be restrictive at times, and work-arounds are proposed to develop the same theory without this assumption. However, for the sake of simplicity, we restrict ourselves to models without simultaneous defaults in order to streamline the presentation of this survey.

HJM Dynamics

As explained in the previous section, the crux of the HJM approach to dynamic modeling is the choice of the dynamics of a code-book for the market data in the form of a set of Itô's stochastic differential equations, and the use of observable market data to feed these dynamic equations with an initial condition. Only then should the modeler worry about the consistency of such a model with a stochastic model for the portfolio loss process, and about the existence of possible arbitrages in the model specified in this way. Recall the list in the summary at the end of Subsection 3.2.

These last two issues are considered in the following two subsections. For the time being, we define the dynamics of the code-book by assuming that the forward default rates satisfy the following system of I stochastic differential equations

$$d\, a_t(\tau, x) = \alpha_t(\tau, x) dt + \beta_t(\tau, x) dW_t \qquad (44)$$

where x varies in $x \in \{0, 1/I, 2/I, \cdots, (I-1)/I, 1\}$ and where for each $\tau \geq 0$ and x, $\{\alpha_t(\tau, x)\}_t$ and $\{\beta_t(\tau, x)\}_t$ are adapted processes with values in $\mathbb{R}$ and $\mathbb{R}^d$ respectively.

A Spot Consistency Condition

Consistency holds if the dynamics given by equation (44) can co-exist with a top down model where the time evolution of the system is derived from the dynamics of the cumulative loss process $\{L(t)\}_t$ specified first. The following result gives a necessary condition for this to hold.

Proposition 1. *Let us assume that the process* $\mathbf{L} = \{L(t)\}_{t \geq 0}$ *of cumulative portfolio losses admits transition rates which only jump by* 1. *Then when viewed as a point process,* $\mathbf{L}$ *has an intensity* $\{\lambda_L(t)\}_{t \geq 0}$ *given almost surely by the formula:*

$$\lambda_L(t) = a_t(0, L(t)), \qquad t \geq 0. \qquad (45)$$

The consistency condition (45) is a direct consequence of Aven's theorem [2] and our implicit smoothness assumption on the forward default rates. We reproduce the proof given in [41]. If we fix $t > 0$ and $\epsilon > 0$ we have:

$$\frac{1}{\epsilon} \mathbb{E}_t\{L(t+\epsilon) - L(t)\}$$

$$= \frac{1}{\epsilon} \sum_{n=L(t)}^{I} (\frac{n}{I} - L(t)) p_t(\epsilon, n)$$

$$= \sum_{n=L(t)+1}^{I} (\frac{n}{I} - L(t)) \frac{1}{\epsilon} [p_t(0, \epsilon, L(t), n)$$

$$= \sum_{n=L(t)+1}^{I} (\frac{n}{I} - L(t)) \frac{1}{\epsilon} [p_t(0, 0, L(t), n) + \epsilon \partial_\tau p_t(0, 0, L(t), n) + o(\epsilon)]$$

$$= \sum_{n=L(t)+1}^{I} (\frac{n}{I} - L(t)) [-a_t(0, n) p_t(0, 0, L(t), n)$$

$$+ a_t(0, n-1) p_t(0, 0, L(t), n-1) + O(1)]$$

$$= a_t(0, L(t)) + O(1)$$

where we used Kolmogorov's equation. ∎

Remark. The result of Proposition 1 shows that the jump times of the process **L** (i.e. the default times of the portfolio components) are totally inaccessible. So even though such an assumption was never stated explicitly, we are actually working in the framework of reduced form models (i.e. intensity based models) as opposed to structural models for which the time of default are typically announced by increasing sequences of stopping times. See also the discussion of the SPA approach below.

The HJM Drift Condition

Let us assume that for each $\tau > 0$ and $x \in \{0, 1/I, 2/I, \cdots, (I-1)/I\}$ the stochastic processes $\{a_t(\tau, x)\}_t$ is a non-negative semi-martingal with a decomposition of the form (44).

For each fixed $t > 0$, we can view $a_t(\tau, \cdot)$ as the negative of the diagonal elements of a bidiagonal Q-matrix and by solving the forward Kolmogorov equations as before, we can derive expressions for the transition probabilities $p_t(\tau_1, \tau_2, \cdot, \cdot)$ of a Markov process whose marginal distributions $p_t(\tau, x) = p_t(0, \tau, L(t), x)$ we would like to coincide with the forward loss distributions $\mathbb{P}\{L(t+\tau) = x | \mathcal{F}_t\}$. Notice that in this case, if we fix $T = t + \tau$ and vary t in $[0, T]$, the latter are martingales by construction since they are conditional expectations of a fixed random variable.

If we start from prescription (44) the explicit formulae (42) for the transition probabilities can be used to prove that $\{p_t(0, T, x, y)\}_{0 \leq t \leq T}$ is a semi-martingale for each default levels $x < y$, and we can compute its bounded variation and quadratic variation parts. Substituting $L(t)$ for x, one can show that $p_t(0, \tau, L(t), x)$ also is a semi-martingale, and one can derive its bounded variation and quadratic variation parts in terms of the drift $\alpha_t(\tau, x)$ and volatility $\beta_t(\tau, x)$ of $a_t(\tau, x)$. Now recall that, if the forward default rates $a_t(\tau, x)$ come from an *underlying* loss process $L(t)$, then as we already explained,

$$p_t(0, \tau, L(t), x) = p_t(t + \tau, x) = \mathbb{P}\{L(t+\tau) = x | \mathcal{F}_t\}$$

is necessarily a martingale. Stating that its bounded variation part vanishes leads to the following conclusion.

Proposition 2. *If for each $\tau > 0$ and $x \in \{0, 1/I, 2/I, \cdots, (I-1)/I\}$ the stochastic process $\{a_t(\tau, x)\}_t$ is a non-negative semi-martingale satisfying (44), then the forward loss distributions $\{p_t(T, x)\}_{0 \leq t \leq T}$ are martingales if and only if*

$$p_t(0, T-t, L(t), x)\alpha_t(T, x) = -\beta_t(T, x)v_t(0, T-t, L(t), x), \quad (46)$$

for $x \in \{0, 1/I, 2/I, \cdots, (I-1)/I\}$ where $v_t(\tau_1, \tau_2, x, y)$ is the volatility of the semi-martingale decomposition of the transition probability $p_t(\tau_1, \tau_2, x, y)$ as given by the solution of the forward Kolmogorov's equation.

We refer the interested reader to [41] for the details of the derivation. Condition (46) is called the HJM drift condition because of its striking similarity with the original

HJM drift condition (25). However, a crucial difference needs to be emphasized. While the classical drift condition (25) gives explicitly the drift $\alpha_t(T)$ of the codebook in terms of its volatility $\beta_t(T)$, the above drift condition merely states a relation between drift and volatility of the code $a_t(T, x)$. Indeed, the term $v_t(0, T-t, L(t), x)$ which appears in the right hand side of (46) is a function of the code, and hence of its bounded variation part. In other words, the drift term $\alpha_t(T, x)$ is present in the right hand side of (46) which is only an implicit equation for $\alpha_t(T, x)$. We shall encounter the same problem in our discussion of the HJM approach to equity market models in the next section. However, the situation is easier here. Indeed, because of the finite nature of the state space of the loss process, and because of our assumption of the upper-diagonal nature of the Q-matrices and the fact that their last rows are identically zero, these implicit equations can be solved exactly after finitely many iterations. We refer the interested reader to the details provided in [41].

Volatility Structure Calibration

One of the goals of this review is to emphasize how an HJM modeling approach resolves the calibration issue by encapsulating the market prices of the liquidly traded instruments in the initial condition of the dynamic model, and how the resulting dynamic specifications can be restricted to the volatility term as the drift can be determined from the volatility and the observed market prices. Because everything rides on the particular choice of a volatility structure $\beta_t(\tau, x)$ for the forward default rates, the actual volatility specification is of crucial importance. Unfortunately, it still remains a *touchy business* as there is no clear algorithm providing such a volatility structure, even if it is easy to understand the practical consequences of $\beta_t(\tau, x) \equiv 0$, or $\beta_t(\tau, x)$ having a constant sign, or being very large of $x \approx L(t)$ and small otherwise, ... etc. Unfortunately, this difficulty cannot be resolved without further information about the desired market model, whether this information comes from from prices of exotic derivatives or more qualitative properties that the model should reproduce. We refer to our discussion of the same issue in Section 3 in the case of the fixed income markets.

The SPA Approach

Now we use the smaller market filtration $\{\mathcal{M}_t\}_t$ to condition the time evolutions of the forward loss distributions.

The idea of the SPA approach is to treat the values of the forward loss cumulative distribution functions $P_t(T, x)$ as a family of zero coupon bond prices parameterized by the loss level x, in which case it is natural to introduce the equivalent of instantaneous forward rates by defining

$$f_t(T, x) = -\frac{\partial}{\partial T} \log P_t(T, x) = -\frac{\frac{\partial}{\partial T} P_t(T, x)}{P_t(T, x)} \tag{47}$$

and to construct a dynamic portfolio loss model by specifying a set of stochastic differential equations for these forward loss rates in the form

$$df_t(T,x) = \alpha_t(T,x)dt + \beta_t(T,x)dW_t \tag{48}$$

in full analogy with the HJM prescription (22) used in the fixed income markets. Even though the notation of this approach follow more closely the notation of the classical HJM approach reviewed in Section 3, it is not as natural as the more involved approached based on Markov process codes discussed above. Indeed, the latter will generalize in a straightforward manner to the case of the equity markets discussed in the next section. Moreover, the former cannot be used without introducing an extra layer of technical derivations involving Markov loss processes, obscuring their original claims of simplicity. The interested reader is referred to [26] for details.

6 The HJM Approach to Equity Markets

This section is devoted to the derivation of arbitrage free dynamic stochastic models for the equity markets. We try to incorporate standard features of these markets, and in so doing, we put ourselves in a situation amenable to the HJM philosophy highlighted in the previous sections. This approach to dynamic equity models was originally advocated by Derman and Kani in [17]. The present discussion is based on the recent work of Carmona and Nadotchyi [7] where explicit formulae, rigorous proofs and numerical examples are given.

So as in the cases of fixed income and credit market models reviewed in Section 3, we first identify a set of instruments liquidly traded to which the model needs to be calibrated. The goal of our modeling effort is to characterize important properties (such as for example absence of arbitrage) of the pricing measure $\mathbb{P}$ used by the market by studying the dynamics of these liquidly traded instruments instead of the dynamics of the instruments underlying them. In this way, calibration is taken care of by merely using observed prices as initial conditions for the dynamical equations. As before, the dynamics of the prices of the basic instruments are given by an infinite dimensional stochastic differential equation, or equivalently by a random field.

6.1 Description of the Market

As always, we consider an economy with a perfect frictionless market without bid-ask spreads, with short sales of call and put options allowed in arbitrary sizes, without taxes, etc. In such an idealized market model, it is natural to choose for the set of liquidly traded securities, the ensemble of all the European call options written on underlying instruments spanning the market. For the sake of simplicity, we assume that one single underlyer (e.g. a *stock*) spans the market under consideration. Choosing more underlyers would force the price process to be multivariate and make the notation unnecessarily complicated without changing much to the nature of the results.

Let us denote by $\{S_t\}_{t \geq 0}$ the price process underlying the derivative instruments forming the market. As stated above, for the sake of simplicity we assume that the market comprises only derivatives written on a single underlying instrument, in other words, we assume that S_t is univariate. Again, for the sake of simplicity, we assume that the discount factor is one, i.e. $\beta_t \equiv 1$, or equivalently that the short interest rate is zero, i.e. $r_t \equiv 0$, and that the underlying risky asset does not pay dividends. These assumptions greatly simplify the notation without affecting the generality of our derivations.

We assume that in our idealized market, European call options of all strikes and maturities are liquidly traded, and that their prices are observable. We denote by $C_t(T, K)$ the market price at time t of a European call option of strike K and maturity $T > t$. We assume that today, i.e. on day $t = 0$, all the prices $C_0(T, K)$ are observable. According to the philosophy adopted in this paper, at any given time t, instead of working directly with the price S_t of the underlying asset, we concentrate on the set of call prices $\{C_t(T, K)\}_{T,K}$ as our fundamental market data. This is partly justified by the well documented fact that many observed option price movements cannot be attributed to changes in S_t, and partly by the fact that many exotic (path dependent) options are hedged (replicated) with portfolios of plain (vanilla) call options.

Remarks. 1. It is well known that in order to avoid arbitrage (at least against static strategies) the observed call prices $C_0(T, K)$ should be increasing in T, non-increasing and convex in K, that they should converge to 0 as $K \to \infty$ and that they should recover the underlying price S_0 for zero strike when $K \to 0$. We shall implicitly assume that the observed surface of initial call surface satisfies these properties.

2. **A More Realistic Set-Up.** In the description of our idealized market, we assumed that European call options of all strikes and all maturities were liquidly traded. This assumption is very convenient, though highly unrealistic. Indeed, the knowledge of all the prices $C_t(T, K)$ determine all the marginal distributions of the underlying instruments under the pricing measure $\mathbb{P}$. This information is not available in real life. In practice, the best one can hope for is, for a finite set of discrete maturities $T_1 < T_2 < \cdots < T_n$, one has quotes for the prices of a finite set of call options. In other words, for each $i = 1, 2, \cdots, n$ one has the prices of calls $C_t(T_i, K_{ij})$ for a finite set $K_{i1} < K_{i2} < \cdots < K_{in_i}$.

This more realistic form of the set-up has seen a recent renewal of interest starting with the work of Laurent and Leisen [30]. Our interest in this problematics was triggered by the recent technical reports by Cousot [13] and Buehler [6] who use Kellerer [28] theorem in the same spirit as the present discussion, and by the recent work of Davis and Hobson [15] which relies instead on the Sherman-Stein-Blackwell theorem [43, 44, 4].

We refer the interested reader to [15] and to the references therein.

From now on, we denote by $\tau = T - t$ the time to maturity of the option and we denote by $\tilde{C}_t(\tau, K)$ the price $C_t(T, K)$. In other words

$$\tilde{C}_t(\tau, K) = C_t(t + \tau, K), \qquad \tau > 0. \ K > 0.$$

We assume that the market *prices by expectation* in the sense that the prices of the liquid instruments are given by expectations of the present values of their cashflows with respect to a probability measure. So saying that $\mathbb{P}$ is a pricing measure used by the market implies that for each time $t \geq 0$ we have

$$\tilde{C}_t(\tau, K) = \mathbb{E}\{(S_{t+\tau} - K)^+ | \mathcal{F}_t\} = \mathbb{E}^{\mathbb{P}_t}\{(S_{t+\tau} - K)^+\}.$$

where we denote by $\mathbb{P}_t$ a regular version of the conditional probability of $\mathbb{P}$ with respect to $\mathcal{F}_t$. We denote by $\tilde{\mu}_{t,t+\tau}$ the distribution of $S_{t+\tau}$ for the conditional distribution $\mathbb{P}_t$. It is an $\mathcal{F}_t$- measurable random measure. With this notation

$$\tilde{C}_t(\tau, K) = \int_0^\infty (x - K)^+ \, d\tilde{\mu}_{t,t+\tau}(dx)$$

and for each fixed $\tau > 0$, the knowledge of all the prices $\tilde{C}_t(\tau, K)$ completely determines the distribution $\tilde{\mu}_{t,t+\tau}$ on $[0, \infty)$.

Remarks. 1. Notice that we do not assume uniqueness of the pricing measure $\mathbb{P}$. In other words, our analysis holds in the case of incomplete models as well as complete models.

2. **Notation Convention.** In order to help with the readability of the paper, we use a notation without a tilde or a hat for all the quantities expressed in terms of the variables T and K. But we shall add a tilde for all the quantities expressed in terms of the variables τ and K, and a hat when the strike is given in terms of the variable $x = \log K$.

6.2 Implied Volatility Code-Book

In the classical Black-Scholes theory, the dynamics of the underlying asset are given by the stochastic differential equation

$$dS_t = S_t \sigma dW_t, \qquad S_0 = s_0$$

for some univariate Wiener process $\{W_t\}_t$ and some positive constant σ. In this case, the price $\tilde{C}_t(\tau, K)$ of a call option is given by the Black-Scholes formula

$$BS(S, \tau, \sigma, K) = S_t \Phi(d_1) - K \Phi(d_2) \qquad (49)$$

with

$$d_1 = \frac{-\log M_t + \tau \sigma^2/2}{\sigma \sqrt{\tau}}, \qquad d_2 = \frac{-\log M_t - \tau \sigma^2/2}{\sigma \sqrt{\tau}}$$

where $M_t = K/S_t$ is the moneyness of the option. We use the notation Φ for the cumulative distribution of the standard normal distribution, i.e.

$$\Phi(x) = \frac{1}{\sqrt{2\pi}} \int_{-\infty}^x e^{-y^2/2} \, dy, \qquad x \in \mathbb{R}.$$

The Black-Scholes price is an increasing function of the parameter σ when all the other parameters are held fixed. As a consequence, for every real number C (think of such a number as a quoted price for a call option with time to maturity τ and strike K) in the interval between $(S_t - K)^+$ and S_t there exists a unique number σ for which $\tilde{C}_t(\tau, K) = C$. This unique value of σ given by inverting Black-Scholes formula (49) is known as the implied volatility and we shall denote it by $\tilde{\Sigma}_t(\tau, K)$. This quantity is extremely important as it is used by most if not all market participants as the *currency* in which the option prices are quoted. This practice should not be construed as an endorsement of Black-Scholes model. In order to emphasize this fact, I cannot resist the temptation to characterize the implied volatility by the following statement borrowed from Rebonato's book [37]:

the wrong number to put in the wrong formula to get the right price.

For each time $t > 0$, the one-to-one correspondence

$$\{\tilde{C}_t(\tau, K); \tau > 0, K > 0\} \leftrightarrow \{\tilde{\Sigma}_t(\tau, K); \tau > 0, K > 0\}$$

offers a code-book translating without any loss all the information given by the call prices in terms of implied volatilities, and which we call the Black-Scholes or implied volatility code-book. While Black-Scholes theory predicts a flat profile for the implied volatility surface, one has plenty empirical evidence of the contrary. We refer the reader interested in the empirical properties of the implied volatility surface to the thorough discussion in Rebonato's book [37] and to the references therein. The mathematical analysis of this surface is based on a subtle mixture of empirical facts and arbitrage theories, and it is rather technical in nature. The literature on the subject is vast and it cannot be done justice in a few references. Choosing a few samples for their relevance to the present discussion, we invite the interested reader to consult [10],[20],[31],[33],[21] and the references therein to get a better sense of these technicalities.

Valuation and risk management of complex option positions require models for the time evolution of implied volatility surfaces. [34] and [12] are examples of attempts to go beyond static models, but despite the fact that they consider only a cross section of the surface (say for K fixed), the works of Schönbucher [39] and Schweizer and Wissel [42] are more in the spirit of the HJM approach which we advocate in this section.

At any given time t, absence of (static) arbitrage imposes conditions on the surface of call option prices. As we already mentioned, the surface $\{\tilde{C}_t(\tau, K)\}_{\tau, K}$ should be increasing in τ, non-increasing and convex in K, it should converge to 0 as $K \to \infty$ and recover the underlying price S_0 for zero strike when $K \to 0$. Because of the one-to-one correspondence between call prices and implied volatilities, these conditions can be expressed in terms of properties of the implied volatility surface $\{\tilde{\Sigma}_t(\tau, K)\}_{\tau, K}$ at time t. However inverting Black-Scholes formula (49) is not simple and these conditions become unnecessarily technical. This is one of the reasons why we search for another way to capture the information in the surface of call option prices.

6.3 Choosing Another Option Code-Book

As in the standard framework of the Black-Scholes theory, we start from the dynamics of the underlying asset and we try to identify a code-book for the traded instruments in such a way that the dynamics of the codes could be easily manipulated and most importantly, could be used as a starting point to define the dynamics of the market. Since we assume that the filtration is Brownian, without any loss of generality we can assume

$$dS_t = S_t \sigma_t \, dW_t^1, \qquad S_0 = s_0$$

for some adapted non-negative process $\{\sigma_t\}_{t \geq 0}$. If $t > 0$ is fixed, for any τ_1 and τ_2 such that $0 < \tau_1 < \tau_2$, and for any convex function ϕ on $[0, \infty)$ we have

$$\begin{aligned}
\int_0^\infty \phi(x) \tilde{\mu}_{t,t+\tau_1}(dx) &= \mathbb{E}^{\mathbb{P}_t}\{\phi(S_{t+\tau_1})\} \\
&= \mathbb{E}^{\mathbb{P}_t}\{\phi(\mathbb{E}^{\mathbb{P}_t}\{S_{t+\tau_2}|\mathcal{F}_{t+\tau_1}\})\} \\
&\leq \mathbb{E}^{\mathbb{P}_t}\{\mathbb{E}^{\mathbb{P}_t}\{\phi(S_{t+\tau_2})|\mathcal{F}_{t+\tau_1}\})\} \\
&= \mathbb{E}^{\mathbb{P}_t}\{\phi(S_{t+\tau_2})\} \\
&= \int_0^\infty \phi(x) \tilde{\mu}_{t,t+\tau_2}(dx)
\end{aligned}$$

from which we see that for any given $t > 0$, the probability measures $\{\tilde{\mu}_{t,t+\tau}\}_{\tau > 0}$ are non-decreasing in the balayage order. This implies the existence of a Markov martingale $\{Y_\tau\}_{\tau \geq 0}$ with marginal distributions $\{\tilde{\mu}_{t,t+\tau}\}_{\tau > 0}$. Since the knowledge of all the call prices $\{\tilde{C}(\tau, K)\}_{\tau > 0, K > 0}$ is equivalent to the knowledge of all the distributions $\{\tilde{\mu}_{t,t+\tau}\}_{\tau > 0}$, the Markov martingale $\{Y_\tau\}_{\tau \geq 0}$ is a way to encapsulate the information given by the market at time t by providing the call prices. Obviously, the process $\{Y_\tau\}_{\tau \geq 0}$ contains more information than the mere marginal distributions $\{\tilde{\mu}_{t,t+\tau}\}_{\tau > 0}$ determined by the call option prices. This process can be used to price contracts with path dependent *exotic* pay-offs whose values are not uniquely determined by the *state price densities* of the marginal distributions. The procedure which we just outlined captures perfectly the philosophy and the practice of the market participants: include all the information about the liquidly traded instruments in a model that reproduces all of these prices, and use such a model to price exotic derivatives which cannot be synthesized from the liquid instruments available for trade. As such a model is not uniquely determined by the market prices, there is a lot of freedom in choosing it, and many factors enter the final decision: parsimony, common sense, versatility, basic principles (e.g. maximum entropy, minimum least squares, ...) but in any case, once the choice is made, the only thing left is *hope for the best*.

Notice that, if the process $\{Y_\tau\}_{\tau \geq 0}$ is realized on a Wiener space, then the martingale representation theorem in Brownian filtrations gives that Y_τ can be written as

$$Y_\tau = Y_0 + \int_0^\tau Y_s \tilde{a}(s) \, dB_s$$

and that, because of the Markov property, the predictable process $\{\tilde{a}(s)\}_{s\geq 0}$ can be chosen to be of the form $\tilde{a}(s,\omega) = \tilde{a}_t(s, Y_s(\omega))$ for some function $(s,y) \hookrightarrow \tilde{a}_t(s,y)$ of $(s,y) \in [0,\infty) \times [0,\infty)$ and whose graph can be viewed as a surface over the quadrant $[0,\infty) \times [0,\infty)$. Notice that this surface changes with t in an $\mathcal{F}_t$-measurable way. At each time t, we can choose this surface $\{\tilde{a}_t(\tau,K)\}_{\tau>0,K>0}$ as an alternative code-book for the information contained in the options prices $\{\tilde{C}(\tau,K)\}_{\tau>0,K>0}$. This code-book is different from the Black-Scholes implied volatility code-book $\{\tilde{\Sigma}_t(\tau,K)\}_{\tau>0,K>0}$ given by the implied volatilities of the European call options in question. The deterministic version of the surface $\{\tilde{a}_t(\tau,K)\}_{\tau>0,K>0}$ was introduced in a static framework (i.e. for $t = 0$) simultaneously by Dupire [19] and Derman and Kani [16] though with a different definition, as an alternative to the implied volatility surface. The surface $\{\tilde{a}_t(\tau,K)\}_{\tau>0,K>0}$ has been called the local volatility surface for reasons which will become clear later in the paper. From our point of view, the main reason to work with the local volatility surface instead of the implied volatility surface is the ease with which one can check the presence or absence of static arbitrage. Indeed, as we shall see below, the four conditions (increasing in τ, increasing and convex in K, plus the two boundary conditions) guaranteeing the absence of static arbitrage become merely positivity of the numbers $\tilde{a}_t(\tau,K)$. Replacing difficult conditions to check by such a simple one becomes extremely convenient when we deal with dynamic models. The interested reader is invited to consult [32] for a thorough discussion of the connections between local and implied volatility in the static framework (i.e. at time $t = 0$) of stochastic volatility models.

A dynamic version of local volatility modeling was later touted by Derman and Kani in a paper [17] mostly known for its discussion of implied tree models. Motivated by the fact that the technical parts of [17] dealing with continuous models are rather informal and lacking mathematical proofs, Carmona and Natodchy actually develop the program outlined in [17]. While providing a rigorous mathematical derivation of the so-called drift condition, they also discuss concrete examples and provide calibration and Monte Carlo implementation recipes.

We now derive the property of the local volatility surface which got us interested in its dynamics. Notice that the following derivation is done when the time $t > 0$ and the past up to and including time t are fixed. We give details in the case where for example, we assume that the above marginal distributions $\tilde{\mu}_{t,t+\tau}$ have for each $\tau > 0$ a positive density $\tilde{g}_t(\tau, x)$ (continuous as a function of $x > 0$), which once x is held fixed, are continuously differentiable in the variable τ. Then we can conclude that for each t there exists a function $(\tau, K) \hookrightarrow \tilde{a}_t(\tau, K)$ such that the process

$$dY_\tau = Y_\tau \tilde{a}_t(\tau, Y_\tau) d\tilde{B}_\tau, \quad \tau > 0 \qquad (50)$$

with initial condition

$$Y_0 = S_t$$

is well-defined and has marginal distributions $\tilde{\mu}_{t,t+\tau}$.

We first recall the Breeden-Litzenberger argument which is specific to the *hockey-stick* pay-off function of the European call options. Since the option price with strike K and time to maturity τ is given by

$$\tilde{C}_t(\tau, K) = \int_0^\infty (x - K)^+ \tilde{g}_t(\tau, x) dx$$

we can differentiate both sides twice with respect to K and get:

$$\partial_{KK}^2 \tilde{C}_t(\tau, K) = \tilde{g}_t(\tau, K). \tag{51}$$

Next we apply Itô-Tanaka's formula to (50) and the function $f(y) = (y - K)^+$ (see for example [38]). Note that this function f is convex. It is infinitely differentiable everywhere except at $y = K$ where it has a left and a right derivatives. Obviously $f'(y) = 0$ if $y < K$ and $f'(y) = 1$ if $y > K$. Moreover, the second derivative $f''(y)$ in the sense of distributions is the Dirac point mass at K (also called the *delta function* at K). We get:

$$(Y_\tau - K)^+ = (Y_0 - K)^+ + \int_0^\tau \mathbf{1}_{[K, \infty)}(Y_s) dY_s + \frac{1}{2} L_\tau^K$$

where for each $a \in \mathbb{R}$, $\{L_t^a\}_{t \geq 0}$ is the local time of the semi-martingale $\{Y_t\}_{t \geq 0}$ at a. Using the fact that Y is a martingale satisfying $d\langle Y, Y \rangle_s = Y_s^2 \tilde{a}_t(s, Y_s)^2 ds$, by definition of the local time it holds:

$$L_\tau^K = \lim_{\epsilon \searrow 0} \frac{1}{2\epsilon} \int_0^\tau \mathbf{1}_{(K-\epsilon, K+\epsilon)}(Y_s) d\langle Y, Y \rangle_s$$
$$= \lim_{\epsilon \searrow 0} \frac{1}{2\epsilon} \int_0^\tau \mathbf{1}_{(K-\epsilon, K+\epsilon)}(Y_s) Y_s^2 \tilde{a}_t(s, Y_s)^2 ds,$$

and taking $\mathbb{E}_t$ - expectations on both sides we get:

$$\tilde{C}_t(\tau, K) = (S_t - K)^+ + \frac{1}{2} \lim_{\epsilon \searrow 0} \frac{1}{2\epsilon} \int_0^\tau \int^{\mathbb{R}} \mathbf{1}_{(K-\epsilon, K+\epsilon)}(y) y^2 \tilde{a}_t(s, y)^2 g_t(s, y) dy ds$$
$$= (S_t - K)^+ + \frac{1}{2} \int_0^\tau K^2 \tilde{a}_t(s, K)^2 g_t(s, K) ds.$$

where $g_t(s, y)$ is the density of Y_s for $\mathbb{P}_t$ which is assumed to be continuous in y which justifies taking the limit as $\epsilon \searrow 0$. Finally, taking derivatives with respect to τ on both sides we get:

$$\partial_\tau \tilde{C}_t(\tau, K) = \frac{1}{2} K^2 \tilde{a}_t(\tau, K)^2 \tilde{g}_t(\tau, K). \tag{52}$$

Equating the expressions of the density $\tilde{g}_t(\tau, K)$ obtained in (51) and (52) we get the following expression for the local volatility:

$$\tilde{a}_t(\tau, K)^2 = \frac{2 \partial_\tau \tilde{C}_t(\tau, K)}{K^2 \partial_{KK}^2 \tilde{C}_t(\tau, K)}. \tag{53}$$

Equation (53) determines the local volatility surface $\{\tilde{a}_t(\tau, K)\}_{\tau,K}$ from the values of the call prices $\{\tilde{C}_t(\tau, K)\}_{\tau,K}$. Conversely, if we were to start from a prescription giving the local volatility surface $\{\tilde{a}_t(\tau, K)\}_{\tau,K}$, we would derive the set of call option prices $\{\tilde{C}_t(\tau, K)\}_{\tau,K}$ by solving the partial differential equation (PDE for short)

$$\partial_\tau \tilde{C}(\tau, K) = \frac{1}{2} K^2 \tilde{a}^2(\tau, K) \partial^2_{KK} \tilde{C}(\tau, K), \qquad \tau > 0, \ K > 0 \qquad (54)$$
$$\tilde{C}(0, K) = (S_t - K)^+$$

which is sometimes called the Dupire's PDE because it was first advocated by Bruno Dupire in his groundbreaking work [19] on the volatility smile. We call the one-to-one correspondence given by (53) and (54) the local volatility code-book. The correspondence

$$\{\tilde{C}_t(\tau, K); \tau > 0, K > 0\} \quad \rightleftharpoons \quad \{\tilde{a}_t(\tau, K); \tau > 0, K > 0\}$$

defining our code book is analog, though different from the correspondence given by the Black-Scholes code-book. Indeed, to compute the code from the option prices, we need to compute the right hand side of (53) instead of evaluating the Black-Scholes formula, while in order to recover the option prices from the code we solve the partial differential equation (54) instead of inverting the Black-Scholes formula.

Remark: Statistical Estimation. Recalling the discussion of the remark on a "More Realistic Model", on most every day t, the available data are in the form of a finite set of prices $C_t(T_i, K_{i,j})$ (or possibly of implied volatilities $C_t(T_i, K_{i,j})$) on an irregular grid in the (T, K)-plane. The challenge is to construct a smooth surface $\{C_t(T, K)\}_{T \geq t, K > 0}$ or $\{\Sigma_t(T, K)\}_{T \geq t, K > 0}$ through the observations over the finite grid. This problem is discussed with great care in the small book [23] by Flenger, while the book [1] addresses the same problems in a less statistical and more computational spirit. The interested reader is referred to the review [11] written by Carter and Fouque of Fengler's book for an independent perspective on its content. The world of practitioners and academics is divided into two camps neatly delineated by irreconcilable differences. The first camp argues that, in order to rule out any static arbitrage, the price surface $\{C_t(T, K)\}_{T \geq t, K > 0}$ needs to go through all the observed market prices $C_t(T_i, K_{i,j})$. The second camp does make this strict requirement, claiming that because these prices are not quoted at the same time of the day (i.e. for different values of t), portfolios leading to arbitrage can in principle be constructed mathematically, but they cannot be implemented in practice because of the lack of simultaneity of the quotes, preventing the *wannabe* arbitrager to set up the arbitrage portfolio identified by the mathematical theory. Both arguments are reasonable and quite convincing, and we will not try to take side on this difficult issue.

We can now hint at our implementation of the HJM philosophy in the case of equity markets: as usual, instead of choosing the dynamics of the underlyer S_t and then deriving a set of equations for the prices of the liquidly traded instruments (the

European call option prices in our case), we model directly the dynamics of the prices of the liquidly traded instruments by choosing the dynamics of a specific code-book, and in the present situation, we choose the local volatility code-book.

Another reason for choosing the local volatility code-book over the implied volatility code-book is the fact that the four conditions needed to rule out static arbitrage take a very simple form in the case of the local volatility. Indeed, it is enough to make sure that $\tilde{a}_t(\tau, K)$ is positive to guarantee that $\tilde{C}_t(\tau, K)$ is increasing in T, increasing and convex in K and satisfies the two boundary conditions already discussed. This advantage is priceless when it comes to defining stochastic dynamics.

Remark. As a last remark, we show that, whenever the underlying is known to satisfy an equation of the form

$$dS_t = S_t \sigma_t \, dW_t \tag{55}$$

for some Wiener process $\{W_t\}_t$ and some adapted non-negative process $\{\sigma_t\}_t$, then at each time t, the local volatility $\tilde{a}_t(\tau, K)$ can be viewed as the current expected variance for time to maturity τ and strike K. More precisely, this means that:

$$\tilde{a}_t(\tau, K)^2 = \mathbb{E}_t\{\sigma_{t+\tau}^2 | S_{t+\tau} = K\}. \tag{56}$$

In order to prove this result, it is enough to retrace the steps of the above derivations of (52) and (53) using $S_{t+\tau}$ and its dynamics (55) instead of Y_τ and its own dynamics. This formula is often called Dupire's formula. It is at the origin of the terminology local volatility surface.

6.4 Code-Book Dynamics

We postulate the dynamics of the local volatility surface point by point. For each fixed $T > 0$ and $K > 0$, we assume that the process $\{a_t(T, K)\}_{0 \le t \le T}$ is a semimartingale with decomposition:

$$da_t(T, K) = \alpha_t(T, K)dt + \beta_t(T, K)dW_t, \qquad 0 \le t \le T. \tag{57}$$

for some d-dimensional Wiener process $\{W_t\}_{t \ge 0}$, and some real valued adapted process $\{\alpha_t(T, K)\}_{0 \le t \le T}$ and d-dimensional adapted process $\{\beta_t(T, K)\}_{0 \le t \le T}$ satisfying some mild hypotheses to be specified later. Equivalently, one could specify the dynamics of the local volatility surface parameterized by the time to maturity τ instead of the time of maturity T. In this case, we would assume that

$$d\tilde{a}_t(\tau, K) = \tilde{\alpha}_t(\tau, K)dt + \tilde{\beta}_t(\tau, K)dW_t \tag{58}$$

Using the generalized Îto formula, we see that these two prescriptions are equivalent if and only if

$$\tilde{\alpha}_t(\tau, K) = \alpha_t(t+\tau, K) + \partial_T a(t+\tau, K), \quad \text{and} \quad \tilde{\beta}_t(\tau, K) = \beta_t(t+\tau, K). \tag{59}$$

The results of [7] which we review in this section are proven under the following assumption:

Assumption A

For any fixed $\tau \geq 0$ and $K \geq 0$, $\{\tilde{\alpha}_t(\tau, K)\}_{t \geq 0} \in \mathcal{H}^1_{loc}(\mathbb{R})$ and $\{\tilde{\beta}_t(\tau, K)\}_{t \geq 0} \in \mathcal{H}^2_{loc}(\mathbb{R}^d)$. Moreover, we assume that $\mathbb{P}$ - almost surely, for every $t \geq 0$, the functions $(\tau, K) \hookrightarrow \tilde{\alpha}_t(\tau, K)$ and $(\tau, K) \hookrightarrow \tilde{\beta}_t(\tau, K)$ (and hence $(\tau, K) \hookrightarrow \tilde{a}_t(\tau, K)$) are once continuously differentiable in τ and twice continuously differentiable in K.

Also, $\mathbb{P}$ - almost surely
1) for every $t \geq 0$ and all non-negative numbers τ and K

$$|\alpha_t(\tau, K)| + \|\beta_t(\tau, K)\| \leq \lambda_1(\omega, t)$$

$$0 < \lambda_2(\omega, t) \leq \int_0^t \alpha_u(\tau, K)du + \int_0^t \beta_u(\tau, K)dW_u \leq \lambda_3(\omega, t)$$

for some positive adapted processes λ_1, λ_2 and λ_3.

First Technical Results

The first technical result we need to prove before going any further is the fact that for each $\tau > 0$ and $K > 0$, the process $\{\tilde{C}_t(\tau, K)\}_{t \geq 0}$ is a semi-martingale. This result is quite natural. However, its proof is more technical than we would like, and for the purpose of this presentation, we merely outline the major steps of the proof. Complete details can be found in [7].

Under Assumption A, for each fixed $t > 0$, the stochastic differential equation (50) has a unique solution which we denote $\{Y_{t,\tau}\}_{\tau \geq 0}$. Moreover, since $\tilde{a}_t(\tau, K)$ is bounded above and below away from 0, from Feynman-Kac formula and *Theorem 1.3* of [22], the transition density of $\{\log Y_{t,\tau}\}_\tau$ is the fundamental solution of the backward Kolmogorov's equation

$$\partial_\tau u(\tau, x) = \frac{1}{2}\tilde{a}_t^2(\tau, e^x)\partial_{xx}^2 u(\tau, x) - \frac{1}{2}\tilde{a}_t^2(\tau, e^x)\partial_x u(\tau, x), \qquad \tau > 0,\ x \in \mathbb{R}$$

From this we conclude that the density $g_t(\tau, x)$ of $Y_{t,t+\tau}$ is the fundamental solution of

$$\partial_\tau u(\tau, x) = \frac{1}{2} \cdot K^2 \tilde{a}_t^2(\tau, K)\partial_{KK}^2 u(\tau, K), \qquad \tau > 0,\ K > 0$$

and the price of the vanilla call option is the solution of barrier problem (54). It is well defined, since after the change of variables

$$\widehat{C}(\tau, x) := \tilde{C}(\tau, \exp x), \qquad \tau \geq 0, x \in \mathbb{R} \tag{60}$$

the option partial differential equation becomes

$$\frac{\partial \widehat{C}(\tau, x)}{\partial \tau} = \frac{1}{2}\tilde{a}_t^2(\tau, e^x)\Delta \widehat{C}(\tau, x) - \frac{1}{2}\tilde{a}_t^2(\tau, e^x)\frac{\partial \widehat{C}(\tau, x)}{\partial x}, \qquad T > 0,\ x \in \mathbb{R} \tag{61}$$

with initial condition
$$\widehat{C}(0, x) = (S_t - e^x)^+.$$
The proof that option prices are semi-martingales is done in two steps.

1. We first prove the result by replacing the hockey-stick function appearing in the initial condition by a smooth function. In this case, the solution of equation (61) appears as the uniform limit of the results of a finite difference scheme. It is plain to show that any such explicit scheme provides us at each step with a semi-martingale. The convergence being strong enough, one can pass to the limit and prove that the solution of (61) is also a semi-martingale.
2. The general result is obtained by controlling the limit of the solution of (61) when we approximate the hockey-stick initial condition by a smooth regularization.

The details of these arguments are given in [7].

The conclusion of this subsection, and the starting point of the next one are captured by the fact that for each $\tau > 0$ and $K > 0$ there exist continuous adapted processes $\{\tilde{\mu}_t(\tau, K)\}_{t \geq 0}$ and $\{\tilde{\nu}_t(\tau, K)\}_{t \geq 0}$ such that the following decomposition holds:
$$d\tilde{C}_t(\tau, K) = \tilde{\mu}_t(\tau, K)dt + \tilde{\nu}_t(\tau, K)dB_t. \tag{62}$$
Moreover, the random fields $\{\tilde{\mu}_t(\tau, K)\}_{t,\tau,K}$ and $\{\tilde{\nu}_t(\tau, K)\}_{t,\tau,K}$ satisfy the same assumptions as the random fields $\{\tilde{\alpha}_t(\tau, K)\}_{t,\tau,K}$ and $\{\tilde{\beta}_t(\tau, K)\}_{t,\tau,K}$ appearing in the decomposition of the local volatility $\{\tilde{a}_t(\tau, K)\}_{t,\tau,K}$.

6.5 The HJM Drift Condition

The main goal of this subsection is to derive the following analog of the HJM no-arbitrage analysis.

Theorem 1 (Drift and Consistency Conditions). *The dynamic model of the local volatility surface given by the system of equations*
$$d\tilde{a}_t(\tau, K) = \tilde{\alpha}_t(\tau, K)dt + \tilde{\beta}_t(\tau, K)dW_t, \qquad t \geq 0, \tag{63}$$
with coefficients satisfying assumption A is consistent with a spot price model of the form
$$dS_t = S_t \sigma_t dB_t$$
for some Wiener process $\{B_t\}_t$, and does not allow for arbitrage if and only if the following conditions are satisfied a.s. for all $t > 0$:

- $\tilde{a}_t(0, S_t) = \sigma_t$ \hfill (64)
- $\partial_\tau \tilde{a}_t(\tau, K) \partial_{KK}^2 \tilde{C}_t(\tau, K) =$ \hfill (65)

$$\left(\tilde{a}_t(\tau, K) \tilde{\alpha}_t(\tau, K) + \frac{\|\beta_t(\tau, K)\|^2}{2} \right) \partial_{KK}^2 \tilde{C}_t(\tau, K) + \frac{d}{dt} \langle \tilde{a}_\cdot(\tau, K)^2, \partial_{KK}^2 \tilde{C}_\cdot(\tau, K) \rangle_t$$

where we use the notation $\langle \cdot \quad \cdot \rangle_t$ for the quadratic covariation of two semi-martingales.

Proof:

By construction of the local volatility surface $\{\tilde{a}_t(\tau,K)\}_{\tau>0,K>0}$ we have the equality
$$K^2\tilde{a}_t^2(\tau,K)\partial_{KK}^2\tilde{C}_t(\tau,K) = 2\partial_\tau\tilde{C}_t(\tau,K)$$
which we can rewrite as
$$K^2\tilde{a}_t^2(T-t,K)\partial_{KK}^2\tilde{C}_t(T-t,K) = 2\partial_\tau\tilde{C}_t(T-t,K) \tag{66}$$
for $0 \leq t \leq T$. Both sides are semi-martingales. We use Îto's rule to compute the differential of $\tilde{a}_t^2(\tau,K)$.

$$d\tilde{a}_t^2(\tau,K) = 2\tilde{a}_t(\tau,K)d\tilde{a}_t(\tau,K) + \|\tilde{\beta}_t(\tau,K)\|^2 dt$$
$$= (2\tilde{a}_t(\tau,K)\tilde{\alpha}_t(\tau,K) + \|\tilde{\beta}_t(\tau,K)\|^2)dt + 2\tilde{a}_t(\tau,K)\tilde{\beta}_t(\tau,K)dW_t,$$

However, we also have

$$d\tilde{a}_t^2(T-t,K) = (-2\tilde{a}_t(T-t,K)\partial_\tau\tilde{a}_t(T-t,K) + 2\tilde{a}_t(T-t,K)\tilde{\alpha}_t(T-t,K)$$
$$+ \|\tilde{\beta}_t(T-t,K)\|^2)dt + 2\tilde{a}_t(T-t,K)\tilde{\beta}_t(T-t,K)dW_t, \tag{67}$$

because the effect of replacing τ by $T-t$ in a stochastic differential is merely an argument substitution (τ by $T-t$) in the local martingale part, while a new term, typically a partial derivative with respect to τ, is also added to the drift or bounded variation part of the differential. Consequently

$$d\left(\tilde{a}_t(T-t,K)^2\partial_{KK}^2\tilde{C}_t(T-t,K)\right)$$
$$= \partial_{KK}^2\tilde{C}_t(T-t,K)d\tilde{a}_t^2(T-t,K) + \tilde{a}_t^2(T-t,K)d\left(\partial_{KK}^2\tilde{C}_t(T-t,K)\right)$$
$$+ d\langle\tilde{a}^2(T-\cdot,K),\partial_{KK}^2\tilde{C}.(T-\cdot,K)\rangle_t \tag{68}$$

Since $\tilde{C}_t(\tau,K)$ is a semi-martingale for every fixed $\tau > 0$ and $K > 0$, if we write its decomposition as (recall formula (62))
$$d\tilde{C}_t(\tau,K) = \tilde{\mu}_t(\tau,K)\,dt + \tilde{\nu}_t(\tau,K)dW_t$$
then for each fixed $K > 0$, $\{C_t(T,K) = \tilde{C}_t(T-t,K)\}_{0\leq t\leq T}$ is also a semi-martingale and its decomposition is given by

$$dC_t(T,K) = d\tilde{C}_t(T-t,K) = [\tilde{\mu}_t(T-t,K) - \partial_\tau\tilde{C}_t(T-t,K)]dt + \tilde{\nu}_t(\tau,K)dW_t. \tag{69}$$

1) Let us first assume absence of arbitrage. As we explained earlier, what we mean by that is the fact that the prices of all the liquidly traded assets are local martingales. In particular, for every fixed $T > 0$ and $K > 0$, the process $\{C_t(T,K) = \tilde{C}_t(T-t,K)\}_{0\leq t\leq T}$ is a local martingale. On one hand, this implies that $\partial_\tau\tilde{C}_t(T-t,K)$ is a local martingale, and on the other hand that the bounded variation part of the left hand side of equation (66) is equal to 0. Since developing (68) using (67) gives:

$$d\left(\tilde{a}_t(T-t,K)^2 \partial^2_{KK}\tilde{C}_t(T-t,K)\right)$$
$$= \partial^2_{KK}\tilde{C}_t(T-t,K)d\tilde{a}_t^2(T-t,K) + \tilde{a}_t^2(T-t,K)d\left(\partial^2_{KK}\tilde{C}_t(T-t,K)\right)$$
$$+ d\langle \tilde{a}_\cdot^2(T-\cdot,K), \partial^2_{KK}\tilde{C}_\cdot(T-\cdot,K)\rangle_t$$
$$= \partial^2_{KK}\tilde{C}_t(T-t,K)\left(2\tilde{a}_t(T-t,K)[\tilde{\alpha}_t(T-t,K) - \partial_\tau \tilde{a}_t(T-t,K)]\right.$$
$$\left.+ \|\tilde{\beta}_t(T-t,K)\|^2\right)dt + d\langle \tilde{a}_\cdot^2(T-\cdot,K), \partial^2_{KK}\tilde{C}_\cdot(T-\cdot,K)\rangle_t$$
$$+ d(\text{local martingale})_t$$

and setting the drift component to 0 gives (65).

2) Let us now prove the consistency condition (64). Using the fact that

$$\lim_{\tau \searrow 0} \tilde{C}_t(\tau, K) = (S_t - K)^+ \quad a.s.$$

one can prove that

$$\lim_{\tau \searrow 0} \int_0^t \tilde{\nu}_u(\tau, K) \cdot dW_u = \int_0^t S_u \sigma_u \mathbf{1}_{\{S_u - K \geq 0\}} dB_u$$

for the uniform convergence in probability. This implies that there a.s. exists a sequence $\{\tau_n\}_{n=1}^\infty$ decreasing to 0 for which

$$\lim_{n \to \infty} \int_0^t \tilde{\nu}_u(\tau_n, K) \cdot dW_u = \int_0^t S_u \sigma_u \mathbf{1}_{\{S_u - K \geq 0\}} dB_u$$

which shows that (again because of Tanaka's formula) that

$$\lim_{n \to \infty} \int_0^t \tilde{\mu}_u(\tau_n, K) du = \Lambda_t(K), \quad \text{for any } K > 0 \text{ and } t \in [0, \bar{t}]$$

where $\Lambda_t(K)$ denotes the local time of S_t at K. Since $\tilde{C}_t(T-t,K)$ is a local-martingale in t, we have

$$\tilde{\mu}_u(\tau_n, K) = \partial_\tau \tilde{C}_u(\tau_n, K) = \frac{1}{2} K^2 \tilde{a}_u^2(\tau_n, K) \partial^2_{KK}\tilde{C}_u(\tau_n, K)$$

which in turn implies that for any continuous function h with compact support, we have:

$$\int_0^t h(S_u) S_u^2 [\sigma_u^2 - a_u^2(0, S_u)] du = 0$$

from which we can deduce the consistency condition since h is arbitrary.

3) We now consider the converse. As for the proof of the direct part, the details are technical, so we limit our discussion to the main steps, referring the interested reader to [7] for details. If we denote the drift of $\tilde{C}_t(T-t,K)$ by $\tilde{v}_t(T,K)$, smoothness of $\tilde{C}_t(.,.)$, $\tilde{\mu}_t(.,.)$ and $\tilde{\nu}_t(.,.)$ guarantee the required $C^{1,2}$ smoothness of $\tilde{v}_t(.,.)$. Our goal is to show that $\tilde{v}_t(.,.)$ vanishes identically. In order to do so, we first prove that

it is the solution of a parabolic partial differential equation, and then we check that the initial condition it satisfies is identically 0. The first step is rather straightforward. By differentiation in the same way as in the first part of the proof, and using the fact that $\tilde{v}_t = \tilde{\mu}_t - \partial_\tau \tilde{C}_t$, we obtain

$$\partial_\tau \tilde{v}_t(\tau, K) = \frac{1}{2} K^2 \tilde{a}_t^2(\tau, K) \partial_{KK}^2 \tilde{v}_t(\tau, K), \qquad \tau > 0, K > 0.$$

For the remainder of the proof we show that it is possible almost surely to construct a subsequence $\tau_n \searrow 0$ such that $\tilde{v}_t(\tau_n, \cdot) \to 0$ weakly as functions of K. Uniqueness of weak solutions of the above partial differential equation guarantees that we have $\tilde{v}_t(\tau, K) = 0$ for all $\tau > 0$ and $K > 0$. This implies that $\tilde{C}_t(T-t, K)$ is a local martingale in t for any $T > 0$ and $K > 0$, and since $\tilde{C}_t \le S_t$ is square integrable we can conclude that $\{\tilde{C}_t(T-t, K)\}_{0 \le t \le T}$ is a bona fide martingale.

Monte Carlo Implementation

We now explain how the drift condition (65) can be used to set up arbitrage-free dynamic models for the local volatility surface. As we already explained, we are not able to use (65) to derive a formula in close form to express the drift surface $\{\tilde{\alpha}_t(\tau, K)\}_{\tau, K}$ as a function of the volatility surface $\{\tilde{\beta}_t(\tau, K)\}_{\tau, K}$. However, it is possible to use a discretized version, in the spirit of the Euler scheme for ordinary stochastic differential equations, to constructively derive Monte Carlo samples of the volatility surface from the mere knowledge of $\{\tilde{\beta}_t(\tau, K)\}_{\tau, K}$.

- Start from a model for $\tilde{\beta}_t(\tau, K)$ (say a stochastic differential equation);
- Get S_0 and $\tilde{C}_0(\tau, K)$ from the market and compute $\partial_{KK}^2 \tilde{C}_0$, $\tilde{a}_0$ and $\tilde{\beta}_0$ from its model;
- Loop: for $t = 0, \Delta t, 2\Delta t, \cdots$
  1. Get $\tilde{\alpha}_t(\tau, K)$ from the drift condition (65);
  2. Use Euler to get
     - $\tilde{a}_{t+\Delta t}(\tau, K)$ from the dynamics of the local volatility given by (63);
     - $S_{t+\Delta t}$ from S_t Dynamics;
     - $\tilde{\beta}_{t+\Delta t}$ from its own model;

6.6 Examples

This last subsection is devoted to the applications of the above approach to two of the most popular spot models.

Markovian Spot Models

Let us first consider the simplest case $\beta \equiv 0$. In this case

$$\tilde{a}_t(\tau, K) = \tilde{a}_0(\tau, K) + \int_0^t \tilde{\alpha}_s(\tau, K) ds$$

and in particular we have

$$\dot{\tilde{a}}_t(\tau, K) = \frac{d}{dt}\tilde{a}_t(\tau, K).$$

In the present situation, the drift condition (65) reads

$$\partial_\tau \tilde{a}_t(\tau, K) = \dot{\tilde{a}}_t(\tau, K)$$

and putting the two together we get

$$\partial_\tau \tilde{a}_t(\tau, K) = \frac{d}{dt}\tilde{a}_t(\tau, K)$$

which shows that for fixed K the function $\tilde{a}_t(\tau, K)$, as a function of t and τ, is the solution of a plain (hyperbolic) transport equation whose solution is given by:

$$\tilde{a}_t(\tau, K) = \tilde{a}_0(\tau + t, K)$$

and the consistency condition forces the special form

$$\sigma_t = a_0(t, S_t)$$

of the spot volatility. Hence we proved:

Proposition 3. *The local volatility is a process of bounded variation for each τ and K fixed if and only if it is the deterministic shift of a constant shape and the underlying spot is a Markov process.*

Stochastic Volatility Models

Next we attempt to bridge our analysis of the dynamics of the local volatility with stochastic volatility models widely used in the industry. We start with an explicit form for the dynamics of the stock and spot volatility under a risk-neutral measure, and we derive an *explicit* form for the local volatility surface together with the random fields $\tilde{\alpha}_t(.,.), \tilde{\beta}_t(.,.)$ at each fixed time t.

For the sake of illustration, we consider a simplified version of the SABR model with a stochastic volatility given by a geometric Brownian motion. To be specific we assume that

$$dS_t = S_t \sigma_t dB_t^1$$
$$d\sigma_t = \sigma_t \tilde{\sigma} dB_t^2$$

with initial conditions $S_0 = S$ and $\sigma_0 = \sigma$. Here, $\tilde{\sigma} > 0$ is a constant (usually called the vol-vol) and $\{B_t^1\}_{t\geq 0}$ and $\{B_t^2\}_{t\geq 0}$ are standard Wiener processes. If we also assume that these two Wiener processes are independent, by conditioning on $\mathcal{F}^{B^2}$ we can easily obtain a closed form formula for the call prices at time zero:

$$\tilde{C}_0(\tau, K) = \mathbb{E}\left[BS\left(S, \tau, \sqrt{\frac{1}{\tau}\int_0^\tau \sigma_u^2 du}, K\right)\right]$$

where the notation $BS(S, \tau, \sigma, K)$ for the Black-Scholes price of a European call option was introduced in (49). We can then compute the partial derivatives with respect to τ and K passing the derivatives under the expectation and get from (53) the following formula for the local volatility

$$\tilde{a}_0^2(\tau, K) = \frac{S}{K} \cdot \frac{\mathbb{E}\left[\left(2\sigma_\tau^2/\bar{\sigma}_\tau - \bar{\sigma}_\tau\right) e^{-d_1^2/2}\right]}{\mathbb{E}\left[e^{-d_2^2/2}/\bar{\sigma}_\tau\right]} \quad (70)$$

where

$$\bar{\sigma}_\tau = \sqrt{\frac{1}{\tau}\int_0^\tau \sigma_u^2 du} \quad (71)$$

and

$$d_1 = \frac{\log(S/K) + \bar{\sigma}_\tau^2 T/2}{\bar{\sigma}_\tau \sqrt{T}} \quad \text{and} \quad d_2 = d_1 - \bar{\sigma}_\tau \sqrt{\tau}. \quad (72)$$

The independence assumption is often made for the computations to be easier, but it is not necessary. Indeed, similar formula can be obtained if we assume that the two Wiener processes are correlated, say if they satisfy $dB_t^1 \, dB_t^2 = \rho \, dt$. In this case, the formula for the price of a call option becomes

$$\tilde{C}_0(\tau, K) = \mathbb{E}\left[BS\left(Se^{\frac{\rho \sigma_0}{\bar{\sigma}}(\tilde{\sigma}_T - 1) - \sigma_0^2 \frac{\rho^2}{2} T \tilde{\sigma}_T^2}, K, \tau, \sqrt{1-\rho^2}\sigma_0 \bar{\sigma}_T\right)\right]$$

where $\bar{\sigma}_\tau$ is defined as above in (71), and $\tilde{\sigma}_t = \sigma_t/\sigma_0$. It now holds

$$\tilde{a}_0^2(\tau, K) = \sigma_0^2 \sqrt{1-\rho^2} \frac{\mathbb{E}\left[\sigma_\tau^2/\bar{\sigma}_\tau e^{-d_1^2/2}\right]}{\mathbb{E}\left[e^{-d_2^2/2}/\bar{\sigma}_\tau\right]} \quad (73)$$

where d_1 is now defined by

$$d_1 = \frac{\log(S/K) + \rho\frac{\sigma_0}{\bar{\sigma}_\tau}(\tilde{\sigma}_\tau - 1) + (0.5 - \rho^2)\sigma_0^2 \tilde{\sigma}_\tau^2 \tau}{\sqrt{1-\rho^2}\sigma_0 \bar{\sigma}_\tau \sqrt{\tau}}. \quad (74)$$

Example of these local volatility surfaces are given in [7].

6.7 Factor Models and Consistency

In our discussion of the classical HJM approach to the fixed income markets in Section 3, we explained the important role played by the use of factor models based on parametric families of forward curves. Motivated by the computations of the previous subsection, we single out a simple parametric family of two-dimensional surfaces which appear to give a reasonable parametric family for local volatility surfaces. This family is given by the local volatility surfaces of stochastic volatility models where the stochastic volatility is restricted to take only three different values.

Example of a Parametric Family of Local Volatility Surfaces

Parametric families of forward curves have played a crucial role in the major developments in the econometric analysis of interest rate data. Moreover, they were also a major impetus in some of the recent formulation and solution of the consistency problem. As far as we know, parametric families of local volatility surfaces have not been introduced and systematically studied, at least with the same intensity, and at least in the academic literature. For the sake of definiteness we introduce a simple example of such a family. For each (multivariate) parameter

$$\Theta = (\sigma, \sigma_1, \sigma_2, p_1, p_2)$$

such as $\sigma > 0$, $\sigma_1 > 0$, $\sigma_2 > 0$, $p_1 > 0$, $p_2 > 0$, and also satisfying $p_1 + p_2 \leq 1$, we use formula (53) to define a surface $\tilde{a}_0(\tau, K)$ from a call function $\tilde{C}_0(\tau, K)$ obtained by randomization of the volatility assuming that it takes the values σ_1, σ and σ_2 with probabilities p_1, $1 - p_1 - p_2$ and p_2 respectively. Consequently,

$$\tilde{a}_0(\tau, K)^2 = \frac{p_1 \partial_\tau C(\sigma_1) + (1 - p_1 - p_2)\partial_\tau C(\sigma) + p_2 \partial_\tau C(\sigma_2)}{p_1 \partial_{KK}^2 C(\sigma_1) + (1 - p_1 - p_2)\partial_{KK}^2 C(\sigma) + p_2 \partial_{KK}^2 C(\sigma_2)} \quad (75)$$

where we use the notation $C(\tilde{\sigma})$ for the Black-Scholes price $\tilde{C}_0(\tau, K)$ if the volatility parameter is $\tilde{\sigma}$. Now, using the following expressions for the partial derivatives of the Black-Scholes price

$$\partial_\tau B(S, K, \tau, \sigma) = \frac{\sqrt{SK}}{\sqrt{2\pi}} \frac{\sigma}{2\sqrt{\tau}} e^{-(\log S/K)^2/2\sigma^2 \tau - \tau \sigma^2/8}$$

and

$$K^2 \partial_{KK}^2 B(S, K, \tau, \sigma) = \frac{\sqrt{SK}}{\sqrt{2\pi}} \frac{1}{\sigma \sqrt{\tau}} e^{-(\log S/K)^2/2\sigma^2 \tau - \tau \sigma^2/8}$$

we get the following formula for the definition of our local volatility parametric family:

$$a^2(\tau, x, \Theta) = \frac{\sum_{i=0}^2 p_i \sigma_i e^{-x^2/(2\tau \sigma_i^2) - \tau \sigma_i^2/8}}{\sum_{i=0}^2 (p_i/\sigma_i) e^{-x^2/(2\tau \sigma_i^2) - \tau \sigma_i^2/8}} \quad (76)$$

where we use the variable x for the log-moneyness $\log(S/K)$ and where we set $p_0 = 1 - p_1 - p_2$ and $\sigma_0 = \sigma$ to simplify the form of the formula. Figure 2 gives an example of such a surface.

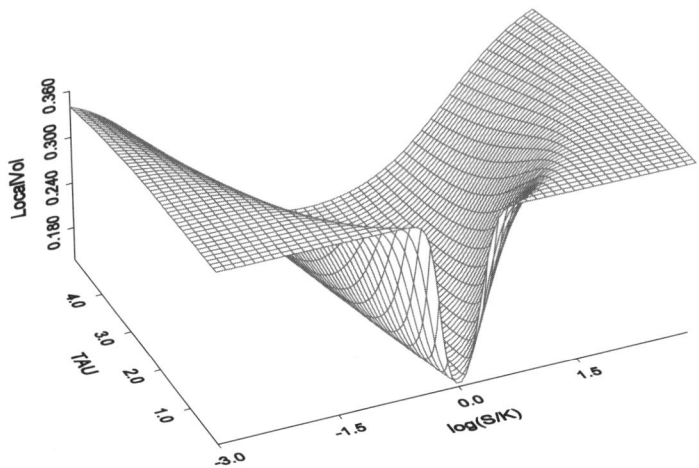

Fig. 2. Parametric local volatility surface from the family described in the text. We used the parameters $\sigma_0 = 0.4$, $\sigma_1 = 0.3$, $\sigma_2 = 0.6$, $p_0 = 0.3$, and $p_1 = 0.5$

This plot clearly hints at one of the major shortcomings of this family: the singular behavior of the surface for short time to maturity, i.e. for $\tau \searrow 0$. Indeed $a^2(\tau, x, \Theta)$ converges toward the maximum of the three σ_i when $\tau \searrow 0$ and $x \neq 0$, while the same limit is strictly smaller (a weighted average of the σ_i's) when $x = 0$. Possible fixes to this problem include the choice of time dependent volatilities σ_i, and a solution in this spirit is implemented in [7] where a different parametric family is proposed.

6.8 Local Volatility Factor Models

Studying the consistency of local volatility factor models is a very interesting problem, and as far as we know, such a problem is completely open. As explained in Section 3, factor models are based on the choice of a parametric family as defined in (76) for example. So if we assume that we are given a parametric family G as before and if we suppose that $\Theta = \{\theta_t\}_{t \geq 0}$ is a d-dimensional semi-martingale with values in the parameter space Θ, then consistency of the factor model means that the random field

$$a_t(\tau, K) = G(\theta_t, \tau, K), \qquad t \geq 0, \tau > 0, K > 0.$$

gives a local volatility model satisfying the no-arbitrage condition.

References

1. Y. Achdou and O. Pironneau, *Computational methods for option pricing*, SIAM.
2. T. Aven, *A theorem for determining the compensator of a counting process*, Scandinavian Journal of Statistics **12(1)** (1985), 69–72.
3. N. Bennani, *The forward loss model: a dynamic term structure approach for the pricing of portfolios of credit derivatives*, Tech. report, November 2005.
4. D. Blackwell, *Equivalent comparisons of experiments*, Annals of Mathematical Statistics **24** (1953), 265–272.
5. H. Buehler, *Consistent variance curve models*, Tech. report, 2007.
6. _____, *Expensive martingales*, Quantitative Finance (2007), (to appear).
7. R. Carmona and S. Nadtochiy, *HJM dynamics for equity models*, Tech. report, Princeton University, 2007.
8. R. Carmona and M. Tehranchi, *A characterization of hedging portfolios for interest rate contingent claims*, The Annals of Applied Probability **14** (2004), 1267–1294.
9. _____, *Interest rate models: an infinite dimensional stochastic analysis perspective*, Springer Verlag, 2006.
10. P. Carr and D. Madan, *Toward a theory of volatility trading*, vol. Volatility, Risk Publications, pp. 417–427.
11. A. Carter and J.P. Fouque, *Review of* SemiParametric Modeling of Implied Volatility *by matthisas fengler*, SIAM Reviews (2007), (to appear).
12. R. Cont and J. da Fonseca, *Dynamics of implied volatility surfaces*, Quantitative Finance **2** (2002), 45–60.
13. L. Cousot, *Necessary and sufficient conditions for no-static arbitrage among European calls*, Tech. report, Courant Institute, New York University, October 2004.
14. R. Jarrow D. Heath and A. Morton, *Bond pricing and the term structure of interest rates: a new methodology for contingent claims valuation*, Econometrica **60** (1992), 77–105.
15. M.H.A. Davis and D.G. Hobson, *The range of traded option prices*, Tech. report, Princeton University, July 2005.
16. E. Derman and I. Kani, *The volatility smile and its implied tree*, Tech. report, Quantitative Research Notes, Goldman Sachs, 1994.
17. _____, *Stochastic implied trees: Arbitrage pricing with stochastic term and strike structure of volatility*, International Journal of Theoretical and Applied Finance **1** (1998), 61–110.
18. D. Duffie and K. Singleton, *Credit risk*, Princeton University Press, 2003.
19. B. Dupire, *Pricing with a smile*, Risk **7** (1994), 32–39.
20. V. Durrleman, *From implied to spot volatility*, Tech. report, Stanford University, April 2005.
21. V. Durrleman and N. El Karoui, *Coupling smiles*, Tech. report, Stanford University, November 2006.
22. E.B Dynkin, *Diffusions, superdiffusions and partial differential equations*, American Mathematical Society, 2002.
23. M.R. Fengler, *Semiparametric modeling of implied volatility*, Lecture Notes in Statistics, Springer Verlag, 2005.

24. D. Filipovic, *Exponential-polynomial families and the term structure of interest rates*, Bernoulli **6(6)** (2000), 1–27.
25. _____, *Consistency problems for Heath-Jarrow-Morton interest rate models*, Lecture Notes in Mathematics, vol. 1760, Springer-Verlag, 2002.
26. V. Piterbarg J. Sidenius and L. Andersen, *A new framework for dynamic credit portfolio loss modelling*, Tech. report, October 2005.
27. J. Jacod and P. Protter, *Risk neutral compatibility with option prices*, Tech. report, Université de Paris VI and Cornell University, April 2006.
28. H. Kellerer, *Markov-komposition und eine anwendung auf martingale*, Mathematische Annalen **198** (1972), 99–122.
29. D. Lando, *Credit risk*, Princeton University Press, 2004.
30. J.P. Laurent and D. Leisen, *Building a consistent pricing model from observed option prices*, World Scientific, 2000.
31. R. Lee, *Implied volatility: Statics, dynamics, and probabilistic interpretation*, International Journal of Theoretical and Applied Finance **4** (2001), 45–89.
32. _____, *Implied and local volatilities under stochastic volatility*, Recent Advances in Applied Probability, Springer Verlag, 2004.
33. _____, *The moment formula for implied volatility at extreme strikes*, Mathematical Finance **14** (2004), 469–480.
34. J. da Fonseca R. Cont and V. Durrleman, *Stochastic models of implied volatility surfaces*, Economic Notes **31** (2002), no. 2.
35. D. Brigo R. Torresetti and A. Pallavicini, *Implied expected tranche loss surface from cdo data*, Tech. report, Banca IMI, March 2007.
36. _____, *Term structure, tranche structure, and loss distributions*, Tech. report, University of Toronto, January 2007.
37. C. Rebonato, *Volatility and correlation: the perfect hedger and the fox.*, 2nd ed., Wiley, 2004.
38. D. Revuz and M. Yor, *Continuous martingales and brownian motion*, 2nd ed., Springer-Verlag, 1990.
39. P. Schönbucher, *A market model for stochastic implied volatility*, Phil. Trans. of the Royal Society, Series A **357** (1999), 2071–2092.
40. _____, *Credit derivatives pricing models*, Wiley, 2003.
41. _____, *Portfolio losses and the term structure of loss transitition rates: a new methodology for the pricing of portfolio credit derivatives*, Tech. report, ETH Z December 2005.
42. M. Schweizer and J. Wissel, *Term structures of implied volatilities: Absence of arbitrage and existence results*.
43. S. Sherman, *On a theorem of Hardy, Littlewood, Polya and Blackwell*, Proc. National Academy of Sciences USA **37** (1951), 826–831.
44. C. Stein, *Notes on the comparison of experiments*, Tech. report, University of Chicago, 1951.

Optimal Bond Portfolios

Ivar Ekeland[1] and Erik Taflin[2]

[1] Canada Research Chair in Mathematical Economics,
University of British Columbia,
Department of Mathematics,
1984 Mathematics Road,
V6T 1Z2 Canada
email: ekeland@math.ubc.ca

[2] Chair in Mathematical Finance,
EISTI, Ecole Internationale des Sciences du Traitement de l'Information,
Avenue du Parc,
95011 Cergy, France
email: taflin@eisti.fr

Summary. We aim to construct a general framework for portfolio management in continuous time, encompassing both stocks and bonds. In these lecture notes we give an overview of the state of the art of optimal bond portfolios and we re-visit main results and mathematical constructions introduced in our previous publications (Ann. Appl. Probab. **15**, 1260–1305 (2005) and Fin. Stoch. **9**, 429–452 (2005)).

A solution of the optimal bond portfolio problem is given for general utility functions and volatility operator processes, provided that the market price of risk process has certain Malliavin differentiability properties or is finite dimensional.

The text is essentially self-contained.

Keywords: Bond portfolios, optimal portfolios, utility optimization, roll-overs, Hilbert space valued processes

JEL Classification: C61, C62, G10, G11
Mathematical Subject Classification: 91B28, 49J55, 60H07, 90C46

1 Motivation

The literature on portfolio management starts with the Markowitz portfolio and the CAPM ([19], [20], [33]). It is a one-period model, where the information on assets is minimal. Every asset is characterized by two numbers, its expected return and its covariance with respect to the market portfolio. With such poor information, one cannot hope to distinguish between stocks and bonds, and indeed part of the beauty of the CAPM lies in its generality: it applies to any type of financial assets.

On the other hand, as soon as one tries to make use of all the information available on assets, important differences appear between stocks and bonds. Bonds mature, that is they are eventually converted into cash, whereas stocks do not. The price of bonds depends on interest rates, and the price of stocks, at least in the academic literature, does not. The bond market is notoriously incomplete, much more so than the stock market, as is observed in practice. As a result, the classical results on portfolio management, such as Merton's ([21], [22]), concern stock portfolios. This paper and the papers [10] and [34] were born from a desire to extend them to bond portfolios.

More generally, we aim to construct a general framework for portfolio management in continuous time, encompassing both stocks and bonds.

The first difficulty to overcome (and, in our opinion, the main financial one) is the fact that such a theory should encompass two very different kinds of financial assets: bonds, which have a finite life, and stocks, which are permanent. We do it by introducing a new type of financial asset, the *rollovers*. A rollover of time to maturity x is a bank deposit and which can be cashed at any time, with accrued interest, provided notice be given time x in advance. Roll-overs have constant time to maturity (as opposed to zero-coupon bonds, for instance), and are similar to stocks, in the sense that their main characteristics do not change with time. By decomposing bonds into rollovers, instead of decomposing them into zero-coupons, we can hope to incorporate bonds and stocks into a unified theory of portfolio management. Rollovers were considered in [32] under the name "rolling-horizon bond".

This implies that the time to maturity x, rather than the maturity date T, becomes the relevant characteristic of bonds. Thus, we shall describe bonds using a moving maturity-time frame, where at time t, the origin is the time to maturity $x = 0$, corresponding to the maturity date $T = t$. As we shall see very soon, there will be a mathematical price to pay for that.

At any time t, denote by $p_t(x)$ the price of a unit zero-coupon with time to maturity x. The function $x \mapsto p_t(x)$ will be called the zero-coupon (price) curve at time t; note that the actual time when that zero-coupon matures is $T = t + x$, and that T, is fixed while x changes with t. The zero-coupon curve p_t will be understood to move randomly, and the second difficulty we face is to describe its motion in some reasonable way. One solution is to decide that p_t belongs to a fixed family of curves, depending on finitely many parameters, so that

$$p_t(x) = f(t, x; r_1, ..., r_d)$$

and the random motion of p_t is the image of a random motion of the r_i, which could, as in spot-rates models, be modelled, for instance by diffusions. This is the *parametric* approach, which exhibits the classical difficulty of all parametric approaches, namely that there is no theoretical reason why the p_t should be written in that way, so that the choice of the function f has to be dictated by observational fit. One then has to strike the right balance between two evils: if the number of parameters is too small, the model will be unrealistic, and if it his higher, it becomes very difficult to calibrate.

We will operate in a *non-parametric* framework: we will make no assumption on p_t, beyond some very rough ones, regarding smoothness and behavior at infinity, nothing that would much constrain their shape. Mathematically speaking, we will let the curve p_t move freely in a linear space E, which will typically be an infinite-dimensional Banach space, of functions from $[0, \infty[$ to $\mathbb{R}$.

In order to reflect adequately known financial facts, the correct definition of E must incorporate some basic constraints:

1. At any time t, the zero-coupon prices $p_t(x)$ must depend continuously on the time to maturity x. In order for forward interest rates to be well-defined, they must also have some degree of differentiability with respect to x. So E must consist of continuous curves with some degree of differentiability.
2. The degree of differentiability of functions in E will determine which basic interest rates derivatives can be modelled. If p_t is continuous, for instance, then we can introduce bonds. The price of a unit zero-coupon bond with time to maturity x is $p_t(x)$; the bond itself, i.e. the value of a portfolio including exactly one bond is represented by the linear form $p_t \mapsto p_t(x)$. Mathematically speaking, this is just the Dirac mass δ_x at x. Now other derivatives such as Call's and Put's on zero-coupon bonds can be introduced, since the pay-off for each of them is a continuous function of the zero-coupon bond price $p_T(x)$, with a given time to maturity x. If p_t is continuously differentiable, then the forward interest rate with time to maturity x, $-\frac{\partial}{\partial x}p_t(x)/p_t(x)$ is well-defined, and further contingent claims can be defined, such as caps, floors and swaps.
3. The curve p_t will be understood to move randomly in E, the randomness being driven by a Brownian motion. We will therefore need to define Brownian motions in the infinite-dimensional space E, which for all practical purposes will require E to be a Hilbert space.
4. The accepted standard in mathematical modelling of zero-coupon prices (the Heath-Jarrow-Morton model, henceforth HJM-see book [6']) is to decide that the real-valued process $t \mapsto p_t(T-t)$, the price at time t of a unit zero-coupon maturing at a given time T, is an Itô process satisfying an stochastic ordinary differential equation (SODE). As is well-known, for fixed x, the real-valued process $t \mapsto p_t(x)$, which is also an Itô process, then no longer satisfies an SODE. Indeed, if $f(t,T) \equiv p_t(T-t)$, then we have $p_t(x) = f(t, t+x)$ so that for fixed x:

$$d_t p_t(x) = [d_t f(t,T) + \frac{\partial f(t,T)}{\partial T} dT]_{T=t+x} = [d_t f(t,T)]_{T=t+x} + \frac{\partial p_t(x)}{\partial x} dt. \tag{1}$$

Here the right-hand side (r.h.s) depends, not only on $p_t(x)$, but also on its partial derivative with respect to x. So, equation (1) for p is a SPDE, stochastic partial differential equation, where the first term on the r.h.s depends only on the un-known $p_t(x)$, since $f(\cdot, T)$ satisfies an SODE. This is the well-known difficulty of the Musiela parametrization (see [25]), and the space E shall permit a simple mathematical formulation of the SPDE (1).
5. At any time t, the zero-coupon prices $p_t(x)$ should go to zero as the time to maturity x goes to ∞. To include also the trivial case, where all interest rates

vanish, and also cases where the forward rates converges rapidly to zero as $x \to \infty$, we only require that $\lim p_t(x)$ exists as $x \to \infty$. N.B. We will chose E such that the elements $f \in E$ satisfying $\lim_{x \to \infty} f(x) = 0$ form a closed sub-space of E, in order to cover easily the case where $p_t(x) \to 0$.

Formula (1) is really an infinite family of coupled equations, one for each $x \geq 0$, describing the motion of the random variable $p_t(x)$, which we write

$$dp_t(x) = p_t(x)m_t(x)dt + p_t(x)\sigma_t(x)dW_t + \frac{\partial p_t(x)}{\partial x}dt, \qquad (2)$$

where for the moment W is thought of as being a high dimensional Brownian motion. Let us rewrite it as a single stochastic evolution equation for the motion of the random curve p_t in E, i.e. as a SODE in E :

$$dp_t = p_t m_t dt + p_t \sigma_t dW_t + (\partial p_t)dt \qquad (3)$$

where ∂ is the differentiation operator with respect to *time to maturity*, i.e. it is defined by $(\partial u)(x) = \frac{du(x)}{dx}$, for differentiable $u \in E$. Since the left-hand side "belongs" to E, so must the right-hand side, and then ∂p_t must belong to E. There are ways to achieve that. One is to choose a framework where the operator ∂ is continuous over all of E. Then so is its n-th iterate ∂^n, so that the space E must consist of functions which have infinitely many derivatives. Unfortunately, the natural topology of such spaces cannot be defined by a single norm, except for very particular cases, and the mathematics become more demanding. A second more standard way to proceed is to consider ∂ as an unbounded operator in a Hilbert space E, so that ∂ is defined only on a subspace $\mathcal{D}(\partial) \subset E$, called the domain of the operator. One would then hope to define the solution of equation (3) in such a way that, if the initial condition p_0 lies in $\mathcal{D}(\partial)$, then p_t remains in $\mathcal{D}(\partial)$ for every t, so that $t \mapsto p_t$ is a trajectory in $\mathcal{D}(\partial)$. In other words, if $p_0(x)$ is differentiable with respect to x, so should the functions $x \mapsto p_t(x)$ be for all $t > 0$.

To summarize, the introduction of rollovers and a moving frame forces us to complicate the equations for price dynamics, by incorporating an additional term, ∂p_t. To be able to solve the relevant equations, we have to treat ∂ as an unbounded operator in Hilbert space. The definition of the relevant Hilbert space has to incorporate basic properties which we expect of zero-coupon curves.

This suits our purpose well, for it enables us to work in a non-parametric framework, where no particular shape is assigned to the the zero-coupon curves. On the other hand, we then have to use the theory of Brownian motion in infinite-dimensional Hilbert spaces and the corresponding stochastic integrals, which creates some additional difficulties. We do not limit the number of sources of noise, indeed in our paper there can be infinitely many. This is natural, since the already mentioned experimental fact, that even using a large number of bonds, not all interest rate derivatives can be hedged. The third difficulty to overcome, is the mathematically significant fact that such a market can not be complete in the usual sense, i.e. every (sufficiently integrable) contingent claim being hedgeable. This has important implications for the solution of the portfolio optimization problem. The now classical two-step solution, so successfully applied to the case of a finite number of stocks (cf. [17], [28]), consisting of first determining the optimal final wealth by duality methods and then determining a hedging portfolio, does not (yet at least) apply to the general infinite dimensional bond markets. In this paper (see [10] and [34]) we give, within the considered general Itô process model, the optimal final wealth for every case it exists (Proposition 7). The existence of an optimal portfolio, is then established by the construction of a hedging portfolio for two cases : The first is for deterministic E-valued drift m and volatility operator σ, where we give a necessary and sufficient condition for the existence of an optimal portfolio. Here there can exist several equivalent martingale measures (e.m.m.), so the market can clearly be incomplete in every sense of the word. The second is for certain stochastic m and σ, for which there is a unique market price of risk process γ. There is then a unique e.m.m. Q. Now, certain integrability conditions on the ℓ^2-valued Malliavin derivative of the Radon-Nikodym density dQ/dP leads to the construction of a hedging portfolio.

We have tempted to make these notes self-contained, with exception of the general hedging result in Theorem 10. The notes first recall some basic facts concerning linear operators and semi-groups in Hilbert spaces, Sobolev spaces and stochastic integration in Hilbert spaces. The theory of bond portfolios and hedging of interest rate derivatives are then introduced. Once this theory is explained, the paper proceeds to a short solution of the optimization problem, leading to the results of [10] and [34]. In particular, under the assumption that the market prices of risk are deterministic, some explicit formulas are given, very similar in spirit to those who are known in the case of stock portfolios, and a mutual fund theorem is formulated. We conclude by stating an alternative formulation, of the optimization problem, within a Hamilton-Jacobi-Bellman approach.

2 Mathematical Preliminaries

2.1 Hilbert Spaces and Bounded Maps

We shall be working with separable infinite-dimensional real Hilbert spaces. Let E be a Hilbert space with scalar product $(\ ,\ )_E$ and norm $\|\ \|_E$, simply denoted $(\ ,\ )$ and $\|\ \|$ if no risk for confusion. The topology and convergence in E is w.r.t. this norm, if not otherwise stated, i.e. the strong topology and convergence. By definition E is, *separable* if it has a countable dense subset. One shows easily that E is separable iff it has a countable orthonormal basis $e_n, n \in \mathbb{N}$, i.e. $(e_i, e_j) = 0$ for $i \neq j$ and $\|e_i\| = 1$, so that every $x \in E$ can be written:

$$x = \sum_{n=0}^{\infty} (x, e_n) e_n,$$

where the right-hand side converges in E. Since the e_n are orthonormal, we have Parseval's equality:
$$\|x\|^2 = \sum_{n=0}^{\infty} |(x, e_n)|^2.$$

A typical separable Hilbert space is ℓ^2, which is the space of all real sequences $a_n, n \in \mathbb{N}$, such that $\sum |a_n|^2 < \infty$. The scalar product in ℓ^2 is given by $(a, b) = \sum a_n b_n$. In fact, every infinite dimensional separable Hilbert space E is isomorphic to ℓ^2. The map
$$x \mapsto a_n = (x, e_n)_E, n \in \mathbb{N}, \tag{4}$$
of E into ℓ^2 is a linear bijection and it preserves norms on both sides.

A linear map $L : E_1 \to E_2$ is continuous if and only if it is *bounded*, that is if there exists a constant c such that $\|Lx\|_{E_2} \leq c\|x\|_{E_1}$ for every $x \in E_1$. The (operator) norm of L is then defined to be the infimum of all such c:
$$\|L\| = \inf\left\{c \mid \|Lx\|_{E_2} \leq c\|x\|_{E_1} \; \forall x\right\}.$$

For example, the linear map in (4) of E onto ℓ^2 as well as its inverse has norm 1. The linear space of all continuous linear maps from E_1 to E_2, $L(E_1, E_2)$, is a Banach space when given this norm. One writes $L(E)$ as a shorthand for $L(E, E)$. Linear maps are also called linear operators or just operators. A *bounded operator* on E is a bounded linear map from E into itself. The *dual* space E' of E, i.e. the space of all linear continuous functionals on E, is given by $E' = L(E, \mathbb{R})$. By the F. Riesz representation theorem,
$$F \in E' \text{ iff } \exists f \in E \text{ such that } F(x) = (f, x) \; \forall x \in E. \tag{5}$$

Also $\|F\|_{E'} = \|f\|_E$, so E' and E are isomorphic. In this paper we will often use, in the context of Sobolev spaces, other representations of the dual E'.

By duality, every operator in $L(E_1, E_2)$ corresponds to an operator in $L(E_2', E_1')$. Using the representation of the dual space given by (5), the adjoint operator A^* of $A \in L(E_1, E_2)$ is defined by $A^*y = y^*$, where for $y \in E_2$ the element $y^* \in E_1$ is defined by
$$(y^*, x)_{E_1} = (y, Ax)_{E_2} \; \forall x \in E_1. \tag{6}$$

This defines an operator $A^* \in L(E_2, E_1)$. One easily checks that $(A^*)^* = A$ and $\|A^*\| = \|A\|$. Let us consider a simple example, which will be relevant in the sequel of this paper:

Example 1 (Left-translation in L^2).
i) Let $E = L^2(\mathbb{R})$ and let a be a given real number. Define the operator A on E by $(Af)(x) = f(x+a)$. Then $\|A\| = 1$ and $(A^*f)(x) = f(x-a)$. We note that A has a bounded inverse A^{-1} given by $(A^{-1}f)(x) = f(x-a)$, so $AA^* = A^*A = I$, where I is the identity operator.
ii) Let $E = L^2([0, \infty[)$ and let $a > 0$ be a given real number. Define the operator A on E by $(Af)(x) = f(x+a)$. Here we find that $\|A\| = 1$, that a.e. $(A^*f)(x) = 0$ if

$0 \leq x < a$ and that $(A^*f)(x) = f(x-a)$ if $a \leq x$. In this case A^* is one-to-one and $AA^* = I$. But A^*A is the orthogonal projection on the (non-trivial) closed subspace of E of functions with support in $[a, \infty[$. So $A^*A \neq I$.

An operator $S \in L(E_1, E_2)$ is called unitary if $SS^* = S^*S = I$. This is the case of A in (i) of Example 1. An operator $S \in L(E_1, E_2)$ is called isometric if $S^*S = I$. This is the case of A^* in (ii) of Example 1.

We will be interested in a particular class of bounded operators on E. We begin with an easy result

Lemma 1. *Suppose $L \in L(E_1, E_2)$ and that we have:*

$$\sum_{n=0}^{\infty} \|Le_n\|^2 < \infty$$

for an orthonormal basis $e_n, n \in \mathbb{N}$ in E_1. Let $f_n, n \in \mathbb{N}$ be another orthonormal basis. Then:

$$\sum_{n=0}^{\infty} \|Le_n\|^2 = \sum_{n=0}^{\infty} \|Lf_n\|^2$$

Definition 1. *An operator L on E_1 into E_2 is Hilbert-Schmidt if $\sum_{n=0}^{\infty} \|Le_n\|^2 < \infty$ for some orthonormal basis $e_n, n \in \mathbb{N}$, in E_1. Its Hilbert-Schmidt norm is defined to be:*

$$\|L\|_{\mathcal{HS}} = \left(\sum_{n=0}^{\infty} \|Le_n\|^2 \right)^{1/2}.$$

It does not depend on the choice of the orthonormal basis $e_n, n \in \mathbb{N}$, in E. The linear space of Hilbert-Schmidt operators from E_1 into E_2 is denoted $\mathcal{HS}(E_1, E_2)$.

Hilbert-Schmidt operators are bounded (in fact, $\|L\| \leq \|L\|_{\mathcal{HS}}$) and even compact: they map bounded subsets of E_1 into relatively compact subsets of E_2. In other words, if L is Hilbert-Schmidt and $(x_n)_{n \in \mathbb{N}}$ is a bounded sequence, then one can extract from $(Lx_n)_{n \in \mathbb{N}}$ a norm-convergent subsequence. This property of a Hilbert-Schmidt operator L follows from the fact that L is the limit in the operator norm of finite rank operators. The space $\mathcal{HS}(E_1, E_2)$ endowed with the Hilbert-Schmidt norm defines a Hilbert space.

Some general references for this subsection are: [16], [18], [30], [31].

2.2 Linear Semi-Groups and Unbounded Operators

Let L be a bounded linear operator on E. For every $t \in \mathbb{R}$, define:

$$\Phi(t) = e^{tL} = \sum_{i=0}^{\infty} \frac{1}{n!} t^n L^n,$$

which converges in the operator norm. Then $\Phi(t)$ is a bounded linear operator for every t, and we have the relation:

$$\Phi(t+s) = \Phi(t)\Phi(s) \ \forall s,t \in \mathbb{R} \ \text{and} \ \Phi(0) = I, \tag{7}$$

where I is the identity operator on E, from which it follows that $\Phi(t)$ and $\Phi(s)$ commute and that $\Phi(t)$ is invertible for every t. Relation (7) states that the map $t \mapsto \Phi(t)$ is a group homomorphism. Note that it is continuous in the norm topology for operators:

$$\|\Phi(t) - I\| \to 0 \ \text{when} \ t \to 0. \tag{8}$$

The solution of the Cauchy problem:

$$\frac{dx(t)}{dt} = Lx(t), \tag{9}$$

$$x(0) = x_0 \tag{10}$$

is given by $x(t) = \Phi(t)x(0)$. In other words, $\Phi(t)$ is the flow associated with the ordinary differential equation (9). We can recover L from $\Phi(t)$ by writing:

$$Lx = \lim_{h \to 0} \frac{1}{h}[\Phi(h)x - x], \ x \in E. \tag{11}$$

The norm continuity of the mapping $t \mapsto \Phi(t)$ is exceptional and has to be replaced by a more useful weaker property (cf. Definition 1, Sect. 1, Chap. IX of [35]):

Definition 2. *A family $\Phi(t)$, $t \geq 0$, of bounded operators on E is called a one parameter semi-group if $\Phi(0) = I$, and for all $t \geq 0$ and $s \geq 0$ we have:*

$$\Phi(t+s) = \Phi(t)\Phi(s) = \Phi(s)\Phi(t). \tag{12}$$

It is said to be strongly continuous or to be of class (C_0) if, for every $x \in E$, we have:

$$\lim_{t \to 0} \Phi(t)x = x. \tag{13}$$

It is said to be a contraction semi-group if $\|\Phi(t)\| \leq 1$ for all $t \geq 0$.

Note that, since equality (12) is supposed to hold only for positive s and t, the operators $\Phi(t)$ are no longer necessarily invertible, as in the case of a group. It can be proved easily that, if the semi-group $\Phi(t)$ is strongly continuous, then $\lim_{s \to t} \Phi(s)x = \Phi(t)x$ and there are constants c and C such that $\|\Phi(t)\| \leq C \exp(ct)$. We also note that if $[0, \infty[\ni t \mapsto \Phi(t)$ is a one parameter semi-group, so is the family of adjoint operators $[0, \infty[\ni t \mapsto \Phi^*(t)$, where we define $\Phi^*(t) = (\Phi(t))^*$.

Example 2.
In the situation of (i) (resp. of (ii)) of Example 1, for given a, let $\Phi_1(a) = A$ (resp. $\Phi_2(a) = A$). Then $\mathbb{R} \ni t \mapsto \Phi_1(t)$ is a strongly continuous contraction group. However $[0, \infty[\ni t \mapsto \Phi_2(t)$ is only a strongly continuous contraction semi group, which can not be extended to a group. In fact, $\Phi_2(t)$ is not invertible for $t > 0$.

We now try to extend formula (11). It turns out that when Φ is no longer norm-continuous, but only strongly continuous, the right-hand side does not converge for every x, and if the limit exists, it does not depend continuously on x. The set of x for which the limit exists is obviously a linear subspace of E and on this subspace the limit is a linear function, let's say G of x. More formally, let $\mathcal{D}(G)$ be the subset of E of all elements $x \in E$ for which the strong limit

$$Gx = \lim_{h \to 0} \frac{1}{h} \left[\Phi(h) x - x \right] \tag{14}$$

exists.

Theorem 1. *Assume Φ is a strongly continuous semi-group. The set $\mathcal{D}(G)$ is then a dense linear subspace of E and G given by (14) defines a linear map $G : \mathcal{D}(G) \to E$. This map is closed, i.e. if x_n is a sequence in $\mathcal{D}(G)$ such that $x_n \to \bar{x} \in E$ and $Gx_n \to \bar{y} \in E$ then $\bar{x} \in \mathcal{D}(G)$ and $\bar{y} = G\bar{x}$.*
For every $x \in \mathcal{D}(G)$ and $t \geq 0$ we have $\Phi(t) x \in \mathcal{D}(G)$,

$$G\Phi(t) x = \Phi(t) Gx \tag{15}$$

and

$$\frac{d}{dt} \Phi(t) x = G\Phi(t) x. \tag{16}$$

Proof:
By definition $\mathcal{D}(G)$ is the set of x where the limit in formula (14) exists (note that this is a strong limit, meaning that we should have norm-convergence), and Gx then is the value of that limit. Clearly $G : \mathcal{D}(G) \to E$ is a linear map.

Given any $x \in E$ and $t > 0$, consider the integral:

$$X(t) = \int_0^t \Phi(s) x\, ds.$$

It is well-defined since the integrand is a continuous function from $[0, t]$ into E. Using the semi-group property, we have:

$$\begin{aligned}
\frac{1}{h} \left[\Phi(h) X(t) - X(t) \right] &= \frac{1}{h} \left[\Phi(h) \int_0^t \Phi(s) x\, ds - \int_0^t \Phi(s) x\, ds \right] \\
&= \frac{1}{h} \left[\int_0^t \Phi(s+h) x\, ds - \int_0^t \Phi(s) x\, ds \right] \\
&= \frac{1}{h} \int_0^h \Phi(s+t) x\, ds - \frac{1}{h} \int_0^h \Phi(s) x\, ds \\
&\to \Phi(t) x - x.
\end{aligned}$$

This proves that $X(t)$ belongs to $\mathcal{D}(G)$. Then so does $\frac{1}{t} X(t)$, and when $t \to 0$, we have $\frac{1}{t} X(t) \to x$, so $\mathcal{D}(G)$ is dense in H, as announced.

Now write:
$$\frac{1}{h}[\Phi(t+h) - \Phi(t)]x = \Phi(t)\frac{\Phi(h) - I}{h}x = \frac{\Phi(h) - I}{h}\Phi(t)x.$$

If $x \in \mathcal{D}(G)$, the second term converges to $\Phi(t)Gx$ and the third one to $G\Phi(t)x$. Formulas (15) and (16) now follow, since these two terms must be equal.

To prove the last condition, note that:
$$\forall x \in \mathcal{D}(G), \quad \Phi(t)x - x = \int_0^t \Phi(s)Gx\,ds. \tag{17}$$

Indeed, we have two functions of t, with values in E, which are zero for $t = 0$ and which have the same derivative, namely $\Phi(t)Gx$, for every $t > 0$. So they must be equal. Now take a sequence $x_n \to \bar{x}$, and assume that $Gx_n = y_n \to \bar{y}$ in E. Writing $x = x_n$ in formula (17), we get:
$$\Phi(t)\bar{x} - \bar{x} = \int_0^t \Phi(s)\bar{y}\,ds.$$

Dividing by t and letting $t \to 0$, we find that $\bar{x} \in \mathcal{D}(G)$ and that $\bar{y} = G\bar{x}$. □

Definition 3. *In the situation of Theorem 1, G is called the* infinitesimal generator *of the semi-group Φ.*

A linear map $L : \mathcal{D}(L) \to E_2$, where $\mathcal{D}(L)$ is a subspace of E_1, is called an operator from E_1 to E_2 with domain $\mathcal{D}(L)$. That two operators are equal, $L_1 = L_2$, means that they have the same domain $\mathcal{D}(L_1) = \mathcal{D}(L_2)$ and that $L_1 x = L_2 x$ for all x in the domain. The operator L is densely defined if $\mathcal{D}(L)$ is dense in E_1. It is called a bounded operator if there exists a finite constant $C \geq 0$ such that for all $x \in \mathcal{D}(L)$ one has $\|Lx\| \leq C\|x\|$ and it is called an *unbounded operator* if such C does not exist. It is *closed* if its graph $\{(x, Lx) \mid x \in \mathcal{D}(L)\}$ is a closed subset of $E_1 \times E_2$, which extends the definition in the preceding theorem. With these definitions, we can rephrase part of the preceding theorem by saying that every strongly continuous semi-group in E has a unique infinitesimal generator, which is a densely defined closed operator in E. The problem to determine if a given densely defined closed operator L in E is the infinitesimal generator of a strongly continuous semi-group is more difficult and we refer the interested reader to the references mentioned in the end of this subsection.

The definition of the *adjoint* of an operator can be extended to unbounded operators. Let L be a densely defined operator from E_1 to E_2. We introduce the adjoint operator L^* to L. The domain of $\mathcal{D}(L^*)$ consists of all $y \in E_2$ for which the linear functional
$$x \mapsto (y, Lx) \tag{18}$$
is continuous on $\mathcal{D}(L)$, endowed with the strong topology of E_1. For $y \in \mathcal{D}(L^*)$ we define L^*y by
$$(L^*y, x) = (y, Lx) \quad \forall x \in \mathcal{D}(L). \tag{19}$$

This defines L^*y uniquely, since $\mathcal{D}(L)$ is dense in E_1. One proves that $\mathcal{D}(L^*)$ is dense in E_2 if L is also closed.

An operator L in E is called *selfadjoint* if $L^* = L$ and skew-adjoint if $L^* = -L$. We have the following clear-cut result (Stone's theorem): L is the infinitesimal generator of a group of unitary operators iff L is skew-adjoint.

Example 3.

In the situation of Example 2, let L_1 and L_2 be the infinitesimal generators of Φ_1 and Φ_2 respectively. L_1 is given by

$$\mathcal{D}(L_1) = \{f \in L^2(\mathbb{R}) \mid f' \in L^2(\mathbb{R})\},$$

and $(L_1 f)(x) = f'(x)$, where f' is the derivative of f. L_2 is given by

$$\mathcal{D}(L_2) = \{f \in L^2([0,\infty[) \mid f' \in L^2([0,\infty[)\},$$

and $(L_2 f)(x) = f'(x)$. Since Φ_1 is a group of unitary operators, we have that $L_1^* = -L_1$. Φ_2 is not a semi-group of unitary operators, so $L_2^* \neq -L_2$. A simple calculation shows that

$$\mathcal{D}(L_2^*) = \{f \in L^2([0,\infty[) \mid f(0) = 0 \text{ and } f' \in L^2([0,\infty[)\}$$

and $(L_2^* f)(x) = -f'(x)$. So here $\mathcal{D}(L_2^*) \subset \mathcal{D}(L_2)$, with strict inclusion. One checks that Φ_2^* is a strongly continuous semi-group in $L^2([0,\infty[)$. It represents right translations of functions. Its infinitesimal generator is L_2^*.

Some general references for this subsection are: [16], [18], [31], [35].

2.3 Sobolev Spaces

For any integer $n \geq 0$, the Sobolev space $H^n(\mathbb{R})$ is defined to be the set of functions f which are square-integrable together with all their derivatives of order up to n:

$$f \in H^n(\mathbb{R}) \iff \int_{-\infty}^{\infty} \left[f^2 + \sum_{k=1}^{n} \left(\frac{d^k f}{dx^k}\right)^2 \right] dx < \infty.$$

This is a linear space, and in fact a Hilbert space with norm given by:

$$\|f\|_{H^n} = \left(\int_{-\infty}^{\infty} \left[f^2 + \sum_{k=1}^{n} (\frac{d^k f}{dx^k})^2 \right] dx \right)^{1/2}.$$

It is a standard fact that this norm of f can be expressed in terms of the Fourier transform $\hat{f}$ (appropriately normalized) of f by:

$$\|f\|_{H^n}^2 = \int_{-\infty}^{\infty} (1+y^2)^n \left|\hat{f}(y)\right|^2 dy.$$

The advantage of that new definition is that it can be extended to non-integral and non-positive values. For any real number s, not necessarily an integer nor positive, we define the Sobolev space $H^s(\mathbb{R})$ to be the Hilbert space of functions associated with the following norm:

$$\|f\|_{H^s}^2 = \int_{-\infty}^{\infty} (1+y^2)^s \left|\hat{f}(y)\right|^2 dy. \tag{20}$$

Clearly, $H^0(\mathbb{R}) = L^2(\mathbb{R})$ and $H^s(\mathbb{R}) \subset H^{s'}(\mathbb{R})$ for $s \geq s'$ and in particular $H^s(\mathbb{R}) \subset L^2(\mathbb{R}) \subset H^{-s}(\mathbb{R})$, for $s \geq 0$. $H^s(\mathbb{R})$ is, for general $s \in \mathbb{R}$, a space of (tempered) distributions. For example $\delta^{(k)}$, the k-th derivative of a delta Dirac distribution, is in $H^{-k-1/2-\epsilon}(\mathbb{R})$ for $\epsilon > 0$.

In the case when $s > 1/2$, there are two classical results.

Theorem 2 (Continuity of multiplication). *If $s > 1/2$, if f and g belong to $H^s(\mathbb{R})$, then fg belongs to $H^s(\mathbb{R})$, and the map $(f,g) \to fg$ from $H^s \times H^s$ to H^s is continuous.*

Denote by $C_b^n(\mathbb{R})$ the space of n times continuously differentiable real-valued functions which are bounded together with all their n first derivatives. Let $C_{b0}^n(\mathbb{R})$ the closed subspace of $C_b^n(\mathbb{R})$ of functions which converges to 0 at $\pm\infty$ together with all their n first derivatives. These are Banach spaces for the norm:

$$\|f\|_{C_b^n} = \max_{0 \leq k \leq n} \sup_x \left|f^{(k)}(x)\right| = \max_{0 \leq k \leq n} \left\|f^{(k)}\right\|_{C_b^0}.$$

Theorem 3 (Sobolev embedding). *If $s > n + 1/2$ and if $f \in H^s(\mathbb{R})$, then there is a function g in $C_{b0}^n(\mathbb{R})$ which is equal to f almost everywhere. In addition, there is a constant c_s, depending only on s, such that:*

$$\|g\|_{C_b^n} \leq c_s \|f\|_{H^s}.$$

From now on we shall no longer distinguish between f and g, that is, we shall always take the continuous representative of any function in $H^s(\mathbb{R})$. As a consequence of the Sobolev embedding theorem, if $s > 1/2$, then any function f in $H^s(\mathbb{R})$ is continuous and bounded on the real line and converges to zero at $\pm\infty$, so that its value is defined everywhere.

We define, for $s \in \mathbb{R}$, a continuous bilinear form on $H^{-s}(\mathbb{R}) \times H^s(\mathbb{R})$ by:

$$<f,g> = \int_{-\infty}^{\infty} \overline{\left(\hat{f}(y)\right)} \hat{g}(y) dy, \tag{21}$$

where $\bar{z}$ is the complex conjugate of z. Schwarz inequality and (20) give that

$$|<f,g>| \leq \|f\|_{H^{-s}} \|g\|_{H^s}, \tag{22}$$

which indeed shows that the bilinear form in (21) is continuous. We note that formally the bilinear form (21) can be written

$$<f,g> = \int_{-\infty}^{\infty} f(x)g(x)dx,$$

where, if $s \geq 0$, f is in a space of distributions $H^{-s}(\mathbb{R})$ and g is in a space of "test functions" $H^s(\mathbb{R})$.

Any continuous linear form $g \to u(g)$ on $H^s(\mathbb{R})$ is, due to (20), of the form $u(g) = <f,g>$ for some $f \in H^{-s}(\mathbb{R})$, with $\|f\|_{H^{-s}} = \|u\|_{(H^s)'}$, so that henceforth we can identify the dual $(H^s(\mathbb{R}))'$ of $H^s(\mathbb{R})$ with $H^{-s}(\mathbb{R})$. In particular, if $s > 1/2$ then $H^s(\mathbb{R}) \subset C_{b0}^0(\mathbb{R})$, so $H^{-s}(\mathbb{R})$ contains all bounded Radon measures.

In the sequel, we will also be interested in functions defined only on the half-line $[0, \infty[$. Let $s \geq 0$. We define the space $H^s([0, \infty[)$ to be the set of restrictions to $[0, \infty[$ of functions in $H^s(\mathbb{R})$. This is clearly a linear space. To turn it into a Hilbert space, we have to use the following norm:

$$\|f\|_{H^s([0,\infty))} = \inf \left\{ \|g\|_{H^s(\mathbb{R})} \mid g(x) = f(x) \text{ a.e. on } [0, \infty) \right\}. \tag{23}$$

This is a Hilbert space norm on $H^s([0, \infty[)$, which is the natural restriction of the norm on $H^s(\mathbb{R})$. For instance, if f is a function in $H^s(\mathbb{R})$ such that $f(x) = 0$ for $x \leq 0$, then its restriction f_0 to $[0, \infty[$ belongs to $H^s([0, \infty[)$, and we have:

$$\|f_0\|_{H^s([0,\infty[)} = \|f\|_{H^s(\mathbb{R})}$$

If $s = n$ is an integer, the norm on $H^s([0, \infty[)$ turns out to be equivalent to the following one:

$$\|\|f\|\|_{H^s}^2 = \int_0^\infty \left[f^2 + \sum_{k=1}^n \left(\frac{d^k f}{dx^k} \right)^2 \right] dx.$$

To establish properties of translations in $H^s([0, \infty[)$, we need to know if there is a continuous linear embedding of $H^s([0, \infty[)$ into $H^s(\mathbb{R})$, i.e. to know if the restriction operator has a continuous right-inverse. Fortunately, as we are in a Hilbert space setting, this problem is easy to solve. Let $s \geq 0$ and let H_-^s be the subset of functions in $H^s(\mathbb{R})$ with support in $]-\infty, 0]$, so that $f \in H_-^s$ if and only if $f \in H^s(\mathbb{R})$ and $f(x) = 0$ for all $x > 0$. H_-^s is a closed subspace of $H^s(\mathbb{R})$. Two functions $f_1, f_2 \in H^s(\mathbb{R})$ have the same restriction to $[0, \infty[$ iff $f_1 - f_2 \in H_-^s$. This means exactly that $H^s([0, \infty[)$ is a quotient space: $H^s([0, \infty[) = H^s(\mathbb{R})/H_-^s$. Introducing the notation $\oplus$ for the Hilbert space direct sum, we have the following result, which proof we omit since its trivial:

Proposition 1. *For $s \geq 0$ we have:*

i) $H^s(\mathbb{R}) = H^s([0, \infty[) \oplus H_-^s$.

ii) *Let M be the orthogonal complement of H_-^s in $H^s(\mathbb{R})$ w.r.t. the scalar product in $H^s(\mathbb{R})$, let κ be the canonical projection of $H^s(\mathbb{R})$ on $H^s([0, \infty[)$ and let ι be the canonical bijection of $H^s([0, \infty[)$ onto M. Then κ is continuous, ι is a Hilbert space isomorphism, $\kappa\iota$ is the identity map on $H^s([0, \infty[)$ and $\iota\kappa$ is the orthogonal projection map in $H^s(\mathbb{R})$ on M.*

We note that ι is a continuous operator extending functions on $[0,\infty[$ to functions on $\mathbb{R}$ and that $\|f\|_{H^s([0,\infty[)} = \|\iota f\|_{H^s(\mathbb{R})}$.

The dual space of $H^s([0,\infty[)$ can easily be characterized in terms of distributions. For $s \geq 0$, $H^s([0,\infty[) = H^s(\mathbb{R})/H^s_-$, so

$$(H^s([0,\infty[))' = \{f \in H^{-s}(\mathbb{R}) \mid <f,g> = 0 \ \forall g \in H^s_-\}. \tag{24}$$

For $s \geq 0$, we define $H^{-s}([0,\infty[)$ to be the closed subspace of all distributions in $H^{-s}(\mathbb{R})$ with support in $[0,\infty[$. It then follows that $(H^s([0,\infty[))'$ can be identified with $H^{-s}([0,\infty[)$. Since $(H^s([0,\infty[))'' = H^s([0,\infty[)$, it then follows that

$$(H^s([0,\infty[))' = H^{-s}([0,\infty[), \ s \in \mathbb{R}. \tag{25}$$

If $s \in \mathbb{R}$, then the constant function taking the value 1 is not in $H^s([0,\infty[)$. If $s > 1/2$, then even every function in $H^s([0,\infty[)$ converges to zero at ∞. For this reason, we will need a larger class of distributions containing the constant functions. Let $s \in \mathbb{R}$ and let f be a distribution with support in $[0,\infty[$ such that it admits the decomposition $f = g + a$, where $g \in H^s([0,\infty[)$ and $a \in \mathbb{R}$. This decomposition of f is then unique and the set of all such distributions is naturally given the Hilbert space structure $H^s([0,\infty[) \oplus \mathbb{R}$. The norm of $f = g + a$ is then given by

$$\|f\|^2 = \|g\|^2_{H^s([0,\infty[)} + a^2.$$

This unique decomposition property leads us to the following

Definition 4. *For $s \in \mathbb{R}$, set $E^s([0,\infty[) = H^s([0,\infty[) \oplus \mathbb{R}$ with the corresponding Hilbert space norm. If $f \in E^s([0,\infty[)$ and if $g \in H^s([0,\infty[)$ and $a \in \mathbb{R}$ are related by the unique decomposition $f = g + a$, then the norm of f is given by*

$$\|f\|^2_{E^s} = \|g\|^2_{H^s} + a^2.$$

The dual $(E^s([0,\infty[))'$, of $E^s([0,\infty[)$ is identified with $(H^s([0,\infty[))' \oplus \mathbb{R} \approx E^{-s}([0,\infty[)$ by extending the bi-linear form, defined in (21), to $E^{-s}([0,\infty[) \times E^s([0,\infty[)$:

$$<F,G> = ab + <f,g>, \tag{26}$$

where $F = a + f \in E^{-s}([0,\infty[)$, $G = b + g \in E^s([0,\infty[)$, $a,b \in \mathbb{R}$, $f \in H^{-s}([0,\infty[)$ and $g \in H^s([0,\infty[)$.

For all the Sobolev spaces H^s we have introduced, and also for the spaces E^s, there are two natural realizations of the dual space. Let us consider only the case of $E^s([0,\infty[)$, the other being similar. One possibility, the canonical one, is to identify $(E^s([0,\infty[))'$ with $E^s([0,\infty[)$ by the scalar product in $E^s([0,\infty[)$. This gives the Riesz representation in (5). Another possibility, is, as we have seen, to identify $(E^s([0,\infty[))'$ with $E^{-s}([0,\infty[)$, by the bi-linear form defined in (26). There is a linear continuous map $\mathcal{S} : E^s([0,\infty[) \to (E^s([0,\infty[))'$ with continuous inverse, relating the two realizations. It is defined by:

$$(f,g)_{E^s([0,\infty[)} = <\mathcal{S}f,g>, \ \forall f,g \in E^s([0,\infty[). \tag{27}$$

Now, different realizations of the dual space leads to different realizations of adjoint operators. Let A be a closed and densely defined operator from a Hilbert-space H to $E^s([0, \infty[)$. We have already defined in (18) its adjoint operator A^* from $E^s([0, \infty[)$ to H w.r.t. the duality defined by the scalar product. Let the dual H' of H be realized by H_1 and the continuous bi-linear form $<\,,\,>_1: H_1 \times H \to \mathbb{R}$. The adjoint A', w.r.t. the duality realized by $<\,,\,>_1$ and $<\,,\,>$ is the operator from $E^{-s}([0, \infty[)$ to H_1, defined by: The domain of $\mathcal{D}(A')$ consists of all $y \in E^{-s}([0, \infty[)$ for which the linear functional

$$x \mapsto <y, Ax> \tag{28}$$

is continuous on $\mathcal{D}(A)$. For $y \in \mathcal{D}(A')$ we define $A'y$ by

$$<A'y, x>_1 = <y, Ax> \quad \forall x \in \mathcal{D}(A). \tag{29}$$

This defines $A'y$ uniquely, since $\mathcal{D}(A)$ is dense in H.

We now study translation semi-groups in the different spaces we have introduced. It follows directly from the definition (20) of the norm in $H^s(\mathbb{R})$ and by dominated convergence that that left-translations defines a strongly continuous group of unitary operators $\tilde{\mathcal{L}}$ in $H^s(\mathbb{R})$ for $s \in \mathbb{R}$ (similarly as to the case of Φ_1 in Example 2):

$$(\tilde{\mathcal{L}}_t f)(x) = f(x + t), \quad \forall f \in H^s(\mathbb{R}) \text{ and } t \in \mathbb{R}. \tag{30}$$

Since, for $s \geq 0$, the closed subspace H^s_- of $H^s(\mathbb{R})$ is invariant under the semi-group $\tilde{\mathcal{L}}_t$, $t \geq 0$, it defines a semi-group $\mathcal{L}$ in $H^s([0, \infty[)$. Defining $\mathcal{L}$ also on constants $a \in \mathbb{R}$ by $\mathcal{L}_t a = a$ we extend the semi-group $\mathcal{L}$ to $E^s([0, \infty[)$, $s \geq 0$:

$$(\mathcal{L}_t f)(x) = f(x + t), \quad \forall f \in E^s([0, \infty[) \text{ and } t \geq 0. \tag{31}$$

Proposition 2. *If $s \geq 0$, then $\mathcal{L}$ is a strongly continuous contraction semi-group on $E^s([0, \infty[)$. Its infinitesimal generator, denoted ∂, has domain $\mathcal{D}(\partial) = E^{s+1}([0, \infty[)$. If $f \in E^{s+1}([0, \infty[)$ then $\partial f = f'$, where f' is the derivative of f.*

Proof:
We first observe that, in the canonical decomposition $E^s([0, \infty[) = H^s([0, \infty[) \oplus \mathbb{R}$ in Definition 4, $\mathcal{L}_a$ leaves the subspace $H^s([0, \infty[)$ invariant and acts trivially on $\mathbb{R}$. It is therefore sufficient to prove the statement with $E^s([0, \infty[)$ replaced by $H^s([0, \infty[)$.

We use the notations of Proposition 1 and let $P = \iota\kappa$ be the orthogonal projection on M. Since $\tilde{\mathcal{L}}_t H^s_- \subset H^s_-$, for $t \geq 0$, it follows that $P\tilde{\mathcal{L}}_t(I - P) = 0$. The group composition law $\tilde{\mathcal{L}}_t \tilde{\mathcal{L}}_u = \tilde{\mathcal{L}}_{t+u}$, then gives for $t, u \geq 0$:

$$(P\tilde{\mathcal{L}}_t P)(P\tilde{\mathcal{L}}_u P) = P\tilde{\mathcal{L}}_{t+u} P$$

So, $[0, \infty[\ni t \mapsto P\tilde{\mathcal{L}}_t P$ is a semi-group of bounded operators on M. It is a strongly continuous contraction semi-group since this is the case for $\tilde{\mathcal{L}}$ and $\|P\| = 1$.

We have that $\mathcal{L}_t = \kappa \tilde{\mathcal{L}}_t \iota$, for $t \geq 0$. Using that $\mathcal{L}_t = \kappa P \tilde{\mathcal{L}}_t P \iota$ it easily follows from the semi-group property of $P\tilde{\mathcal{L}}_t P$ that $\mathcal{L}$ is a semi-group on $H^s([0, \infty[)$. It is a strongly continuous contraction semi-group, since this is the case for $P\tilde{\mathcal{L}}P$ and since $\|\kappa\| = \|\iota\| = 1$. Let ∂ be the infinitesimal generator of $\mathcal{L}$. By the definition of $\mathcal{L}$ it follows that $\mathcal{D}(\partial) = \{f \in H^s([0,\infty[) \mid f' \in H^s([0,\infty[)\}$ and $\partial f = f'$ for $f \in \mathcal{D}(\partial)$. But $H^{s+1}([0,\infty[) = \{f \in H^s([0,\infty[) \mid f' \in H^s([0,\infty[)\}$, which proves the proposition. □

Example 4.
Let $\mathcal{L}'_t : E^{-s}([0,\infty[) \to E^{-s}([0,\infty[)$ be the adjoint of $\mathcal{L}_t$, $t \geq 0$ in Proposition 2, w.r.t. duality defined by the bilinear form $<\ ,\ >$. $\mathcal{L}'$ is then a semi-group of right-translations on the space of distributions $E^{-s}([0,\infty[)$. Loosely speaking $(\mathcal{L}'_t f)(x) = f(x-t)$. Let $s \geq 1$. Then the generator ∂' has domain $E^{-s+1}([0,\infty[)$ and $-\partial'$ is the derivative of distributions, so $(\partial' f)(x) = -df(x)/dx$ if f is a differentiable function. One is easily convinced that the expressions for $\mathcal{L}_t^*$ and ∂^* are more complicated.

Some general references for this subsection are: [1], [5], [13].

2.4 Infinite-Dimensional Brownian Motion

In this sub-section we consider a separable Hilbert space E and an index-set $\mathbb{I}$ with the cardinality equal to the dimension of E. The space E can be infinite-dimensional or finite-dimensional. There is given a family W^i, $i \in \mathbb{I}$ of standard independent Brownian motions on a complete filtered probability space $(\Omega, P, \mathcal{F}, \mathcal{A})$. The filtration $\mathcal{A} = \{\mathcal{F}_t\}_{0 \leq t \leq T}$, is generated by the W^i, and $\mathcal{F} = \mathcal{F}_T$.

Definition 5. *A standard cylindrical Brownian motion W_t, $0 \leq t \leq T$, on E is a sequence $e_i W_t^i$, $i \in \mathbb{I}$ of E-valued processes, where the e_i are the elements of an orthonormal basis of E and the W_t^i, $i \in \mathbb{I}$, are independent real-valued standard Brownian motions on a filtered probability space $(\Omega, P, \mathcal{F}, \mathcal{A})$.*

From now on, given a standard cylindrical Brownian motion W, we shall write informally $W_t = \sum_{i \in \mathbb{I}} W_t^i e_i$. If $\mathbb{I}$ is finite, we have:

$$\|W_t\|^2 = \sum_{i \in \mathbb{I}} \|W_t^i\|^2 < \infty \quad \text{a.s.}$$

and W_t is a stochastic process with values in E. If $\mathbb{I}$ is infinite, then for every t the right-hand side is the sum of infinitely many i.i.d. positive random variables, which does not converge in any reasonable way. In that case, the formula $W_t = \sum_{i \in \mathbb{I}} W_t^i e_i$ cannot be understood as an equality in E, and must be given another meaning.

Proposition 3. *If $W_t = \sum_{i \in \mathbb{I}} W_t^i e_i$ is a standard cylindrical Brownian motion, then, for every $f \in E$ with $\|f\| = 1$, the real-valued stochastic process W_t^f defined by*

$$W_t^f = \sum_{i \in \mathbb{I}} (e_i, f) W_t^i \tag{32}$$

is a standard Brownian motion on the real line.

Proof:

If $\mathbb{I}$ is finite, the result is obvious. Let us then consider the case when $\mathbb{I} = \mathbb{N}$. We first have to check if the right-hand side is well-defined. By Doob's inequality for martingales:

$$E\left[\sup_{0 \leq t \leq T}\left|\sum_{i=n}^{n+p}(e_i, f)W_t^i\right|^2\right] \leq 4E\left[\left|\sum_{i=n}^{n+p}(e_i, f)W_T^i\right|^2\right]$$

$$\leq 4T\sum_{i=n}^{n+p}(e_i, f)^2 \to 0.$$

This implies that the right-hand side of (32) converges in probability to a continuous process. Since each finite sum is Gaussian, so is the limit, and the result follows. $\square$

So, in the case when $\mathbb{I}$ is infinite, the r.h.s. of $W_t = \sum_{i \in \mathbb{I}} W_t^i e_i$ makes no sense in E, but every projection does. Equation (32) can be rewritten as:

$$\forall f \in E, \quad (W_t, f) = \sum_{i \in \mathbb{I}}(e_i, f)W_t^i.$$

We will now show that the stochastic integrals with respect to cylindrical Brownian motion make sense, provided the integrand satisfies a strong integrability condition. Consider the space $\mathcal{HS}(E, F)$ of all Hilbert-Schmidt operators from E into a Hilbert space F. Let the space $\mathcal{L}^2(\mathcal{HS}(E, F))$ consist of all progressively measurable processes A with values in the Hilbert space $\mathcal{HS}(F, F)$, such that:

$$E\left[\int_0^T \|A_t\|_{\mathcal{HS}}^2 \, dt\right] < \infty.$$

Recall that we have, according to Definition 1:

$$\|A_t\|_{\mathcal{HS}}^2 = \sum_{n=0}^{\infty} \|A_t e_n\|^2,$$

where $(e_n)_{n \in \mathbb{N}}$ is any orthonormal basis of E.

Theorem 4. *The stochastic integral:*

$$\int_0^T A_t dW_t$$

is well-defined for every process $A \in \mathcal{L}^2(\mathcal{HS}(E, F))$. It is a continuous martingale with values in E, and we have the usual isometry:

$$\left\|\int_0^T A_t dW_t\right\|_{L^2}^2 = \int_0^T E\left[\|A_t\|_{\mathcal{HS}}^2\right] dt.$$

In other words, the random variable $\int_0^T A_t dW_t$ has mean 0 and its variance is $\sum_{n=0}^{\infty} \int_0^T E\left[\|A_t e_n\|^2\right] dt$, the sum of the variances of the independent sources of Gaussian noise.

As usual, by localization the stochastic integral can be extended to a wider class of processes. Denote by $\mathcal{L}^2_{loc}(\mathcal{HS})$ the set of all progressively measurable processes with values in $\mathcal{HS}$, such that:

$$P\left[\int_0^T \|\Phi\|^2_{\mathcal{HS}} dt < \infty\right] = 1.$$

Then the stochastic integral defines a continuous local martingale.

Some general references for this subsection are: [7], [14], [23], [24], [26].

3 The Dynamics of Bond Prices

3.1 The Non-Parametric Framework

From now on, and for the rest of the paper, we are given a finite time interval of possible trading times $\mathbb{T} = [0, \bar{T}]$ and we are given a family W^i, $i \in \mathbb{I}$ of standard independent Brownian motions on a complete filtered probability space $(\Omega, P, \mathcal{F}, \mathcal{A})$, the filtration $\mathcal{A} = \{\mathcal{F}_t\}_{0 \le t \le \bar{T}}$, is generated by the W^i, and $\mathcal{F} = \mathcal{F}_{\bar{T}}$. The family $\mathbb{I}$ itself can be finite or infinite, in which case we take $\mathbb{I} = \mathbb{N}$. Let $\ell^2(\mathbb{I})$, be the Hilbert space of all real sequences $x = (x_i)_{i \in \mathbb{I}}$, such that $\|x\|_{\ell^2(\mathbb{I})} = (\sum_{i \in \mathbb{I}} (a_n)^2)^{1/2} < \infty$. So, when $\mathbb{I}$ has a finite number $\bar{m}$ of elements, then $\ell^2(\mathbb{I}) = \mathbb{R}^{\bar{m}}$. Often we write just ℓ^2 for $\ell^2(\mathbb{I})$.

Heath, Jarrow and Morton (henceforth HJM) were the first to study the term structure of interest rates in a non-parametric framework. Their basic idea (see [12]) consists of writing one equation for the price of every zero-coupon at time t. Denoting by $\hat{B}_t(T)$ the price at time t of a zero-coupon bond maturing at time $T \ge t$, the HJM equation has the following form:

$$\hat{B}_t(T) = \hat{B}_0(T) + \int_0^t \hat{B}_s(T) a_s(T) ds + \int_0^t \sum_{i \in \mathbb{I}} \hat{B}_s(T) v_s^i(T) dW_s^i, \ 0 \le t \le T \tag{33}$$

There are infinitely many such equations, one for each maturity $T \ge t$.

The trend $a_t(T)$ and the volatilities $v_t^i(T)$ are supposed to be progressively measurable processes, which means, for instance, that they could be functions of all the $\hat{B}_s(T)$, $T \ge s$ and $s \le t$. In due course, we will make further assumptions so as to ensure that equations such as (33) make mathematical sense.

Let us discount all prices to $t = 0$, by the spot interest rate r_t, which in terms of the zero-coupon bond price is given by

$$r_t = -\left.\frac{\partial \hat{B}_t(T)}{\partial T}\right|_{T=t}. \tag{34}$$

The discounted prices of zero-coupons are now:

$$B_t(T) = \hat{B}_t(T) \exp(-\int_0^t r_s ds) \qquad (35)$$

and the equations (33) become:

$$B_t(T) = B_0(T) + \int_0^t B_s(T)(a_s(T) - r_s)ds + \int_0^t \sum_{i \in \mathbb{I}} B_s(T) v_s^i(T) dW_s^i, \ 0 \le t \le T \qquad (36)$$

and, again, there is one such equation for every maturity $T \ge t$. Note the boundary condition $\hat{B}_T(T) = 1$, and hence, from (35):

$$B_t(t) = \exp(-\int_0^t r_s ds). \qquad (37)$$

3.2 The Bond Dynamics in the Moving Frame

For every $x \ge 0$, we denote by $\hat{p}_t(x)$ the price and by $p_t(x)$ the discounted price at time t of a zero-coupon maturing at time $t + x$. The stochastic processes $B_t(T)$ and $p_t(x)$ are related by:

$$p_t(x) = B_t(t+x).$$

In other words, as explained in the introduction, instead of dating events by their distance from a fixed origin, defined to be $t = 0$, we are dating them by their distance from today: we are using a time frame which moves with the observer. The equation for p_t in the moving frame, is easily obtained from (36). For every $x \ge 0$, we have:

$$\begin{aligned} p_t(x) = & p_0(t+x) + \int_0^t p_s(t-s+x) m_s(t-s+x) ds \\ & + \int_0^t \sum_{i \in \mathbb{I}} p_s(t-s+x) \sigma_s^i(t-s+x) dW_s^i, \end{aligned} \qquad (38)$$

where

$$m_t(x) = a(t, t+x) - r_t \text{ and } \sigma_t^i(x) = v^i(t, t+x), \qquad (39)$$

for all $0 \le t \le \bar{T}$ and $x \ge 0$. Here, again, the trends $t \mapsto m_t(x)$ and the volatilities $t \mapsto \sigma_t^i(x)$ are progressively measurable processes.

Instead of looking at (38) as an infinite family of coupled equations, one for each $x \ge 0$, we shall interpret it as a single equation describing the dynamics of an infinite-dimensional object, the curve $x \mapsto p_t(x)$, which will be seen as a vector p_t in the Hilbert space $E^s([0, \infty[)$, for some fixed $s > 1/2$, chosen so that the functions m_t and σ_t^i belong to $E^s([0, \infty[)$.

Let $\mathcal{L}$ be the semi-group left translations on $E^s([0, \infty[)$ (see formula (31) and Proposition 2). From now on we shall just wright E^s instead of $E^s([0, \infty[)$, when there is no risk of confusion. The equations in (38) can be rewritten as one equation in E^s:

$$p_t = \mathcal{L}_t p_0 + \int_0^t (\mathcal{L}_{t-s}(p_s m_s))ds + \int_0^t \sum_{i \in \mathbb{I}}(\mathcal{L}_{t-s}(p_s \sigma_s^i))dW_s^i. \quad (40)$$

Theorem 5. *Let $s > 1/2$. Assume that $p_0 \in E^s$ and assume that m_t and the σ_t^i, $i \in \mathbb{I}$, are progressively measurable processes in E^s satisfying:*

$$\int_0^{\bar{T}} (\|m_t\|_{E^s} + \sum_{i \in \mathbb{I}} \|\sigma_t^i\|_{E^s}^2)dt < \infty \quad a.s. \quad (41)$$

Then equation (40) defines a unique process p in E^s satisfying:

$$\int_0^{\bar{T}} (\|p_t\|_{E^s} + \|p_t m_t\|_{E^s} + \sum_{i \in \mathbb{I}} \|p_t \sigma_t^i\|_{E^s}^2)dt < \infty \quad a.s. \quad (42)$$

The process p has continuous trajectories in E^s,

$$p_t = \exp\left\{\int_0^t \mathcal{L}_{t-s}\left((m_s - \frac{1}{2}\sum_{i \in \mathbb{I}}(\sigma_s^i)^2)ds + \sum_{i \in \mathbb{I}} \sigma_s^i dW_s^i\right)\right\} \mathcal{L}_t p_0. \quad (43)$$

and if $p_0 \in H^s$ then the process p takes its values in H^s. If $p_0 \in E^s$ satisfies $p_0 \geq 0$ (resp. $p_0 > 0$), i.e. $p_0(x) \geq 0$ (resp. $p_0(x) > 0$) for all $x \geq 0$, then so does p_t.

For a proof of this theorem see Lemma A.1 of [10], which is reproduced in the appendix of this article (Lemma 6). Note that equation (40) implies that p_0 is the value of p_t for $t = 0$.

A word here about the choice of function spaces. Assuming that p_t belongs to H^s for some $s > 1/2$ is minimal: it is basically saying that the zero-coupon prices depend continuously on time to maturity and go to zero at infinity. This, however, is too strong a requirement for m_t and the σ_t^i: we cannot expect the trend and the volatilities to go to zero when the time to maturity increases to infinity. This is why we are assuming that m_t and the σ_t^i belong to E^s. To simplify the mathematical formalism and also to include interest rate models, with vanishing long term rates, we have permitted that $p_t \in E^s$. Now according to Theorem 5, p_t is in-fact in H^s if $p_0 \in H^s$.

Condition (41) implies that $\sum_{i \in \mathbb{I}} \|\sigma_t^i\|_{E^s}^2$ is finite for almost every $(t, \omega) \in \mathbb{T} \times \Omega$. This means, when $\mathbb{I} = \mathbb{N}$, that the operator σ_t from $\ell^2(\mathbb{I})$ to E^s defined by:

$$\sigma_t e_i = \sigma_t^i, \quad i \in \mathbb{I}, \quad (44)$$

where e_i are the elements of the standard basis of $\ell^2(\mathbb{I})$, is Hilbert-Schmidt a.e. (t, ω). We have

$$\|\sigma_t\|_{\mathcal{HS}(\ell^2, E^s)}^2 = \sum_{i \in \mathbb{I}} \|\sigma_t^i\|_{E^s}^2.$$

We shall refer to σ as the *volatility operator* process. It takes its values in $\mathcal{HS}(\ell^2, E^s)$ and when we say that it is progressively measurable, it is meant that all the σ^i are progressively measurable.

We can now, using the stochastic integral introduced in Theorem 4, rewrite equation (40) on a more compact form in E^s, where $s > 1/2$:

$$p_t = \mathcal{L}_t p_0 + \int_0^t \mathcal{L}_{t-s}(p_s m_s) ds + \int_0^t \mathcal{L}_{t-s}(p_s \sigma_s) dW_s. \tag{45}$$

This makes sens in E^s. Indeed, the only difference with equation (40) is the last term on the r.h.s. When condition (41) is satisfied then the volatility operator σ_u, defined by (44), from ℓ^2 to E^s, is Hilbert-Schmidt a.e. (u, ω). Since pointwise multiplication of functions in E^s is a continuous operation for $s > 1/2$ it follows that the linear operator $x \mapsto p_u \sigma_u x$, from ℓ^2 to E^s, is Hilbert-Schmidt a.e. (u, ω). $\mathcal{L}_v$ is bonded for every $v \geq 0$, so the integrand is a progressively measurable $\mathcal{HS}(\ell^2, E^s)$-valued process satisfying the conditions of Theorem 4.

A process p with values in E^s satisfying (45) (or equivalently (40)) and (42) will be called a *mild solution* of the bonds dynamics.

Note that we are not worrying about the boundary condition (37) at this time, because it does not make mathematical sense: how do we define r_t? This will be taken care of in the next section.

3.3 Smoothness of the Zero-Coupon Curve

Another way to proceed is to write (38) in differentiated form. For fixed $x \geq 0$, a formal calculation using Itô's lemma and which can be rigorously justified gives:

$$dp_t(x) - p_t(x) m_t(x) dt - \sum_{i \in \mathbb{I}} p_t(x) \sigma_t^i(x) dW_t^i$$
$$= \left(\frac{\partial}{\partial t} p_0(t+x) + \int_0^t \frac{\partial}{\partial t} (p_s(t-s+x) m_s(t-s+x)) \, ds \right.$$
$$+ \left. \int_0^t \frac{\partial}{\partial t} \sum_{i \in \mathbb{I}} p_s(t-s+x) \sigma_s^i(t-s+x) dW_s^i \right) dt.$$

In the expression on r.h.s. we can replace $\partial/\partial t$ by $\partial/\partial x$, since p_0 and the integrands on the r.h.s. are functions of $t + x$. Derivation w.r.t. x under the integral then gives:

$$dp_t(x) - p_t(x) m_t(x) dt - \sum_{i \in \mathbb{I}} p_t(x) \sigma_t^i(x) dW_t^i$$
$$= \left(\frac{\partial}{\partial x} \left(p_0(t+x) + \int_0^t p_s(t-s+x) m_s(t-s+x) \, ds \right.\right.$$
$$+ \left.\left. \int_0^t \sum_{i \in \mathbb{I}} p_s(t-s+x) \sigma_s^i(t-s+x) dW_s^i \right) \right) dt.$$

The r.h.s. is equal to $((\partial/\partial x)p_t(x))\,dt$, according to (38), so

$$dp_t(x) - p_t(x)m_t(x)\,dt - \sum_{i\in\mathbb{I}} p_t(x)\sigma_t^i(x)dW_t^i = \left(\frac{\partial}{\partial x}p_t(x)\right)dt, \qquad (46)$$

for all $x \geq 0$ and $t \in \mathbb{T}$.

Introducing the infinitesimal generator ∂ of the semi-group $\mathcal{L}$ (see Proposition 2), this can be understood as an equation in E^s:

$$dp_t = (\partial p_t + p_t m_t)dt + \sum_{i\in\mathbb{I}} p_t \sigma_t^i dW_t^i \qquad (47)$$

or equivalently:

$$p_t = p_0 + \int_0^t (\partial p_s + p_s m_s)ds + \int_0^t \sum_{i\in\mathbb{I}} p_s \sigma_s^i dW_s^i. \qquad (48)$$

Equation (40) is the integrated version of (48), w.r.t. the semi-group $\mathcal{L}$. The connection between formulas (48) and (40) is similar to the *variations of constants* formula for ODE's in finite dimension.

We now have to give some mathematical meaning to equation (48). This will require beefing up the existence conditions given in Theorem 5. The following corollary follows from applying Theorem 5 with $s+1$ instead of s:

Corollary 1. *Let $s > 1/2$. Assume that $p_0 \in \mathcal{D}(\partial) = E^{s+1}$ and assume that m_t and the σ_t^i, $i \in \mathbb{I}$ are progressively measurable processes with values in E^{s+1} satisfying*

$$\int_0^{\bar{T}} (\|m_t\|_{E^{s+1}} + \sum_{i\in\mathbb{I}} \|\sigma_t^i\|_{E^{s+1}}^2)dt < \infty \quad a.s. \qquad (49)$$

Then the mild solution p, in Theorem 5, of the bonds dynamics satisfies the following condition:

$$p_t \in E^{s+1} \text{ and } \int_0^{\bar{T}} (\|p_t\|_{E^{s+1}} + \|p_t m_t\|_{E^s} + \sum_{i\in\mathbb{I}} \|p_t \sigma_t^i\|_{E^s}^2)\,dt < \infty \quad a.s. \qquad (50)$$

Equation (48) holds for every t. In addition p has continuous paths in E^{s+1} and $p_t \in H^{s+1}$ if $p_0 \in H^{s+1}$.

By definition a solution of equation (40) is called a strong solution of the equation (48), when condition (50) is satisfied. Here we shall say that p is a *strong solution* of the bonds dynamics.

As a consequence, in the situation of Corollary 1, the term structure $x \mapsto p_t(x)$ is C^1 for every t, and interest rates are well defined. The instantaneous forward rate $R_t(x)$ contracted at $t \in \mathbb{T}$ for time to maturity x and the spot rate r_t at time t, for instance, are defined by:

$$R_t(x) = -\frac{\partial \log p_t(x)}{\partial x} = -\frac{(\partial p_t)(x)}{p_t(x)} \quad \text{and} \quad r_t = R_t(0) = -\frac{(\partial p_t)(0)}{p_t(0)}. \qquad (51)$$

By Corollary 1, p is a strong solution and the maps $t \mapsto p_t$ and $t \mapsto \partial p_t$ are continuous from $\mathbb{T}$ into E^s, and hence into $C^0([0, \infty[)$ endowed with the topology of uniform convergence. So $p_s(0)$ and $(\partial p_s)(0)$ converge to $p_t(0)$ and $(\partial p_t)(0)$, when $s \to t$. In other words, r_t is a continuous function of t, when $p_t(0) > 0$ for all $t \in \mathbb{R}$.

We are now able to make sense of the boundary condition (37), which we rewrite in terms of p:

$$p_t(0) = \exp\left(\int_0^t \frac{(\partial p_s)(0)}{p_s(0)} ds\right), \tag{52}$$

for every $t \in \mathbb{T}$.

Proposition 4. *Let $s > 1/2$. Assume that m_t and the σ_t^i are progressively measurable processes with values in E^{s+1} satisfying (49) and*

$$m_t(0) = 0, \quad \sigma_t^i(0) = 0 \quad \forall i \in \mathbb{I} \tag{53}$$

and assume that p_0 satisfies

$$p_0 \in E^{s+1}, \quad p_0(0) = 1, \quad p_0(x) > 0 \quad \forall x \geq 0. \tag{54}$$

Then the solution of the bond dynamics, given by Corollary 1, satisfies the boundary condition (52).

Proof:

Since m_t and the σ_t^i take values in E^{s+1}, they are continuous function on $[0, \infty[$, and condition (53) makes sense. As $p_0 > 0$ it follows from Proposition 5 that $p_t > 0$. We have shown that, if p_t is a strong and strictly positive solution of the bond dynamics, then r_t given by (51) is a continuous function of t. Writing conditions (53) into equation (48), we get:

$$p_t(0) = p_0(0) + \int_0^t ((\partial p_s)(0) + p_s(0)m_s(0))ds + \int_0^t \sum_{i \in \mathbb{I}} p_s(0)\sigma_s^i(0)dW_s^i$$

$$= 1 + \int_0^t (\partial p_s)(0)ds = 1 - \int_0^t r_s p_s(0)ds.$$

In other words, $\varphi(t) = p_t(0)$ must satisfy the differential equation $\varphi'(t) = -r_t \varphi(t)$, with the initial condition $\varphi(0) = 1$. The result follows. $\square$

When we get to optimizing portfolios, we will need L^p estimates on the solutions of the bond dynamics. They are provided by the following result:

Theorem 6. *Let $q(t) = p_t/\mathcal{L}_t p_0$ and $\hat{q}(t) = \hat{p}_t/\mathcal{L}_t \hat{p}_0$. If p_0, σ and m in Proposition 4 also satisfy the following additional conditions:*

$$E\left(\left(\int_0^{\bar{T}} \|\sigma_t\|^2_{\mathcal{HS}(\ell^2, E^{s+1})} dt\right)^a + \exp\left(a \int_0^{\bar{T}} \|\sigma_t\|^2_{\mathcal{HS}(\ell^2, E^s)} dt\right)\right) < \infty, \quad \forall a \in [1, \infty[\tag{55}$$

and

$$E((\int_0^{\bar{T}} \|m_t\|_{E^{s+1}} dt)^a + \exp(a \int_0^{\bar{T}} \|m_t\|_{E^s} dt)) < \infty, \forall a \in [1, \infty[, \qquad (56)$$

then the solution p in Proposition 4 has the following property:

$$p, \hat{p}, q, \hat{q}, 1/q, 1/\hat{q} \in L^u(\Omega, P, L^\infty(\mathbb{T}, E^{s+1})), \forall u \in [1, \infty[. \qquad (57)$$

Proof:

We use the notation

$$\tilde{\mathcal{E}}_t(L) = \exp(\int_0^t \mathcal{L}_{t-s}((m_s - \frac{1}{2}\sum_{i \in \mathbb{I}}(\sigma_s^i)^2)ds + \sigma_s dW_s)), \qquad (58)$$

for

$$L_t = \int_0^t (m_s ds + \sigma_s dW_s), \quad \text{if } 0 \le t \le \bar{T}. \qquad (59)$$

Conditions $(i)-(iv)$ of Lemma 7 are satisfied for p. Estimate (145) of Lemma 7 then shows that $p \in L^u(\Omega, P, L^\infty(\mathbb{T}, E^{s+1}))$ $\forall u \in [1, \infty[$. By the explicit expression (43), $q = \tilde{\mathcal{E}}(L)$, so it follows from Lemma 7 that the conclusion holds true also for q.

Let $N_t = \int_0^t ((-m_s + \sum_{i \in \mathbb{I}}(\sigma_s^i)^2)ds - \sum_{i \in \mathbb{I}}\sigma_s^i dW_s^i)$. Then $1/q = \tilde{\mathcal{E}}(N)$. According to conditions (55), (56), the conditions $(i) - (iv)$ of Lemma 7 (with N instead of L) are satisfied. We now apply estimate (145) to $1/q$, which proves that $1/q \in L^u(\Omega, P, L^\infty(\mathbb{T}, E^{s+1}))$, for all $u \geq 1$.

To prove the cases of $\hat{q}^\alpha$, $\alpha = 1$ or $\alpha = -1$, we note that $q(t) = \hat{q}(t)p_t(0)$. Using that the case of q^α is already proved and Hölders inequality, it is enough to prove that $g \in L^u(\Omega, P, L^\infty(\mathbb{T}, \mathbb{R}))$, where $g(t) = (p_t(0))^{-\alpha}$. Since $p_t(0) = (\mathcal{L}_t p_0)(0)(q(t))(0) = p_0(t)(q(t))(0)$, it follows that

$$0 \le g(t) = (p_0(t))^{-\alpha}((q(t))(0))^{-\alpha}.$$

By Sobolev embedding, p_0 is a continuous real valued function on $[0, \infty[$ and it is also strictly positive, so the function $t \mapsto (p_0(t))^{-\alpha}$ is bounded on $\mathbb{T}$. Once more by Sobolev embedding, $((q(t))(0))^{-\alpha} \le C\|(q(t))^{-\alpha}\|_{E^s}$. The result now follows, since we have already proved the case of q^α. The case of $\hat{p}$ is so similar to the previous cases that we omit it. □

Under the hypotheses of Proposition 4, $p_t(0)$ satisfies (52), so it is the discount factor (51). It has nice properties, as follows from the second part of the proof of Theorem 6

Corollary 2. *Under the hypotheses of Theorem 6, if $\alpha \in \mathbb{R}$, then the discount factor $p_t(0)$ satisfies*

$$E(\sup_{t \in \mathbb{T}}(p_t(0))^\alpha) < \infty.$$

Remark 1. It follows from Theorem 6 that for all $t \in \mathbb{T}$, p_t and p_0 have similar asymptotic behavior. In fact for some r.v. $A > 0$, $A^{-1}p_0(t+x) \leq p_t(x) \leq Ap_0(t+x)$, for all $t \in \mathbb{T}$ and $x \geq 0$, where A is independent of x and t and $A \in L^u(\Omega, P)$ for all $u \geq 1$.

In a different context, Hilbert spaces of forward rate curves were considered in [4] and [11]. The space E^s, with $s > 1/2$ sufficiently small, contains the image of these spaces, under the nonlinear map of forward rates to zero-coupons prices. Or more precisely, it contains the image of subsets of forward rate curves f with positive long term interest rate, i.e. $f(x) \geq 0$ for all x sufficiently big.

4 Portfolio Theory

In this section $s > 1/2$, $E^s = E^s[0, \infty[)$ and $\mathbb{T} = [0, \bar{T}]$, where $\bar{T}$ is the time horizon of the model. We also write E for $E^s = E^s[0, \infty[)$ and E' for $E^{-s}[0, \infty[)$.

4.1 Basic Definitions

We recall that, by the bilinear form $<\ ,\ >$, the space E^{-s} is identified with the dual of E^s, that is, the space of continuous linear functionals on E^s. It is important to note that, since $s > 1/2$, the space E^s is contained in $C_b^0([0, \infty[)$, the space of bounded continuous functions on $[0, \infty[$, so that E^{-s} contains the dual of $C_b^0([0, \infty[)$, which is the space of bounded Radon measure on $[0, \infty)$. In particular, all Dirac masses δ_x, for $x \geq 0$, belong to E^{-s}.

Definition 6. *A portfolio is progressively measurable process on the time interval* $\mathbb{T}$, *with values in* E^{-s}. *If* θ *is a portfolio, then its discounted value at time* $t \in \mathbb{T}$ *is*

$$V_t(\theta) = <\theta_t, p_t>. \tag{60}$$

The basic example is a portfolio of one zero-coupon:

Example 5.
Consider a portfolio containing exactly one zero-coupon bond with maturity date T, i.e. *time of maturity* T :
1) Let $T \geq \bar{T}$ and let T be fixed. The portfolio θ is then defined by

$$\theta_t = \delta_{T-t}, \forall t \leq \bar{T}. \tag{61}$$

Since $T \geq \bar{T}$, we have indeed that the support of the distribution θ_t is contained in $[0, \infty[$, so $\theta_t \in E^{-s}$. With this definition, the value of the zero-coupon is:

$$<\delta_{T-t}, p_t> = p_t(T-t)$$

which is precisely what we had in mind.

2) Let $T < \bar{T}$ and let T fixed. In this case we note that the process in (61) does not continue after time T: the zero-coupon is converted into cash. So the buy-and-hold strategy is not possible for zero-coupon bonds, unless the horizon $\bar{T}$ is less than the maturity T.

3) Let $T = t + x$ and $x \geq 0$ a fixed *time to maturity*. Then the portfolio is defined by

$$\theta_t = \delta_x, \quad \text{for } t \leq \bar{T}. \tag{62}$$

We note that the higher we choose s, the more portfolios can be incorporated into the model. For instance, if $s > 3/2$, all curves in E^s are C^1, so that the derivative δ'_x of the Dirac mass belongs to E^{-s}. The value of δ'_{T-t} is:

$$<\delta'_{T-t}, p_t> = p'_t(T-t) = -R_t(T-t)p_t(T-t), \tag{63}$$

where $p'_t(x) = \partial p_t(x)/\partial x$ and where $R_t(x)$, defined in (51), is the instantaneous forward rate with time to maturity x, contracted at time t. This also implies that the higher we choose s, the more interest rates derivatives can be incorporated into the model. If $s > 1/2$, then we can contract directly on the values of zero-coupon bond prices, and if $s > 3/2$, then we can contract directly on the values of interest rates.

We next introduce the notion of *self-financing* portfolio. We state a definition such that it will makes sense for mild solutions of the bonds dynamics:

Definition 7. *A portfolio is called* self-financing *if, for every* $t \in \mathbb{T}$

$$V_t(\theta) = V_0(\theta) + \int_0^t <\theta_s, p_s m_s> ds + \sum_{i \in \mathbb{I}} p_s \sigma^i_s dW^i_s>. \tag{64}$$

Given a strong solution p of the bonds dynamics, we have for a self-financing portfolio:

$$dV_t(\theta) = <\theta_t, dp_t - \partial p_t\, dt>. \tag{65}$$

Note that this is not the standard definition: this is because we are in the moving frame. Changes in portfolio value are due to two causes: changes in prices, as in the fixed frame, and also to changes in time to maturity.

For the right-hand side of (64) to make mathematical sense and to introduce later arbitrage free markets, we need a further definition.

Definition 8. *A portfolio θ is an* admissible *portfolio if $\|\theta\|_\mathsf{P} < \infty$, where*

$$\|\theta\|_\mathsf{P}^2 = E\left[(\int_0^{\bar{T}} |<\theta_t, p_t m_t>| dt)^2 + \int_0^{\bar{T}} \sum_{i \in \mathbb{I}} (<\theta_t, p_t \sigma^i_t>)^2 dt\right].$$

P *is the linear space of all admissible portfolios and* P_{sf} *the subspace of self-financing portfolios.*

The discounted gains process G, defined by

$$G(t,\theta) = \int_0^t (<\theta_s, p_s m_s> ds + <\theta_s, p_s \sigma_s dW_s>), \qquad (66)$$

is well-defined for admissible portfolios:

Proposition 5. *Assume that p_0, m and σ are as in Proposition 4. If $\theta \in \mathsf{P}$, then $G(\cdot, \theta)$ is continuous a.s. and $E(\sup_{t\in\mathsf{T}}(G(t,\theta))^2) < \infty$.*

Proof:
Let $\theta \in \mathsf{P}$ and introduce $X = \sup_{t\in\mathsf{T}} |G(t,\theta)|$, $Y(t) = \int_0^t <\theta_s, p_s m_s> ds$ and $Z(t) = \int_0^t <\theta_s, p_s \sigma_s dW_s>$. Then $G(t,\theta) = Y(t) + Z(t)$, according to formula (66). Let p be given by Proposition 4, of which the hypotheses are satisfied. We shall give estimates for Y and Z. By the definition of P:

$$E((\sup_{t\in\mathsf{T}} Y(t))^2) \leq E((\int_0^{\bar{T}} |<\theta_s, p_s m_s>| ds)^2) \leq \|\theta\|_{\mathsf{P}}^2. \qquad (67)$$

By isometry we obtain

$$\begin{aligned} E(Z(t)^2) &= E(\int_0^t <\theta_s, p_s \sum_{i\in\mathbb{I}} \sigma_s^i dW_s^i>)^2 \\ &= E(\int_0^t \sum_{i\in\mathbb{I}} (<\theta_s, p_s \sigma_s^i>)^2 ds) \leq \|\theta\|_{\mathsf{P}}^2. \end{aligned} \qquad (68)$$

Doob's L^2 inequality and inequality (68) give $E(\sup_{t\in\mathsf{T}} Z(t)^2) \leq 4\|\theta\|_{\mathsf{P}}^2$. Inequality (67) then gives $E(X^2) \leq 10\|\theta\|_{\mathsf{P}}^2$, which proves the proposition. □

Example 6.
1) The portfolio in 1) of Example 6 is self-financing and the portfolios in 2) and 3) of Example 6 are not self-financing.
2) The interest rate portfolio in formula (63) is self-financing.

4.2 Rollovers

Definition 9. *Let $S \geq 0$. A S-rollover is a self-financing portfolio θ of a number of zero-coupon bonds with constant time to maturity S and with initial price $V_0(\theta) = p_0(S)$.*

It follows directly from the definition that a S-rollover have the same initial price as a zero-coupon with maturity date S. It also follows that, if x_t is the number of zero-coupon bonds in the portfolio at t, then we must have:

$$\theta_t = x_t \delta_S,$$

where the real-valued process x makes the portfolio self-financing.

Proposition 6. *If θ_t is a S-rollover, then:*

$$x_t = \exp(\int_0^t R_s(S)\, ds). \tag{69}$$

Proof:

The portfolio θ_t only contains zero-coupons with time to maturity S, so that $V_t(\theta) = x_t p_t(S)$. Assuming the process x to be of bounded variation it follows that:

$$dV_t(\theta) = p_t(S)dx_t + x_t dp_t(S).$$

Substituting the expression for $dp_t(S)$ this becomes:

$$dV_t(\theta) = p_t(S)\frac{dx_t}{dt}dt + x_t \partial_x p_t(S)dt + x_t p_t(S)(m_t(S)dt + \sum_{i \in \mathbb{I}} \sigma_t^i(S)dW_t^i)$$

$$= (p_t(S)\frac{dx_t}{dt} + x_t \partial_x p_t(S) + x_t p_t(S)m_t(S))dt + x_t p_t(S)\sum_{i \in \mathbb{I}} \sigma_t^i(S)dW_t^i.$$

According to (64) the portfolio is then self-financing if and only if:

$$p_t(S)\frac{dx_t}{dt} + x_t(\partial p_t)(S) = 0.$$

This means that:

$$\frac{1}{x_t}\frac{dx_t}{dt} = -\frac{1}{p_t(S)}\frac{\partial p_t(S)}{\partial S} = R_t(S).$$

and the formula (69) follows by integration. This proves the proposition since x then is of bounded variation. □

In particular, if $S = 0$, then we get the usual bank account with spot rate r_t.

Henceforth, we will denote by $q_t(S)$ the value (discounted to $t = 0$) at time t of a S-rollover. In the preceding notation, $q_t(S) = V_t(\theta)$.

Introducing the price curve of the roll-over at time t, $q_t : [0, \infty[\to \mathbb{R}$, we find that the price dynamics of roll-overs is given by:

$$q_t = p_0 + \int_0^t q_s m_s ds + \int_0^t q_s \sum_{i \in \mathbb{I}} \sigma_s^i dW_s^i, \tag{70}$$

Note that, compared to the same formula for bond prices, the term in ∂ has disappeared from the right-hand side.

A S-rollover is a bank account which needs advance notice to be cashed: if notice is given at time t, the rollover will then pay x_t units of account at time $t + S$. In other words, at time t, when notice is given, the rollover is exchanged for $q_t(S)/p_t(S) = x_t$ units of a unit zero-coupon with time of maturity $t + S$.

As we noted earlier, zero-coupons do not in general allow buy-and-hold strategies. However rollovers do: a constant portfolio of rollovers is always self-financing. A general bond portfolio θ_t can be expressed in terms of a portfolio of rollovers η_t and vice versa.

4.3 Absence of Arbitrage Opportunities

Let p be a mild solution of the price dynamics. Suppose that θ_t is a self-financing portfolio such that, for almost every $(t,\omega) \in \mathbb{T} \times \Omega$, we have:

$$\forall i \in \mathbb{I}, \quad <\theta_t(\omega), p_t(\omega)\sigma_t^i(\omega)> = 0. \tag{71}$$

(We note that $p_t(\omega) \in E^s$ is a function of time to maturity, $x \mapsto p_t(\omega, x)$, and similarly for θ_t etc.) Then (64) gives $dV_t(\theta) = <\theta_t, m_t p_t> dt$, so that θ_t is risk-free. Since the spot rate is zero (after discounting values to $t = 0$), in an arbitrage free market it must follow that for almost every (t,ω):

$$<\theta_t(\omega), m_t(\omega)p_t(\omega)> = 0. \tag{72}$$

Comparing (71) and (72) for all such θ, we find that $p_t(\omega)m_t(\omega)$ must belong to the closure of the linear span of $\{p_t(\omega)\sigma_t^i(\omega) \mid i \in \mathbb{I}\}$. In fact this follows rigorously using Lemma 3, proved independently of this subsection. There are now two cases:

- $\mathbb{I}$ is finite. Then the linear span is finite-dimensional, and it coincides with its closure. So there are numbers $\gamma_t^i(\omega), i \in \mathbb{I}$ such that

$$p_t(\omega)m_t(\omega) = p_t(\omega) \sum_{i \in \mathbb{I}} \gamma_t^i(\omega)\sigma_t^i(\omega) \quad \text{(finite sum)}.$$

Since $p_t(\omega) > 0$ for almost every (t,ω), this leads to:

$$m_t(\omega) = \sum_{i \in \mathbb{I}} \gamma_t^i(\omega)\sigma_t^i(\omega)$$

and since the processes m and σ^i are progressively measurable, so can one choose the processes γ^i. Note that the preceding equation holds in E^s, and that it translates into a family of equations in $[0, \infty[$:

$$m_t(\omega, x) = \sum_{i \in \mathbb{I}} \gamma_t^i(\omega)\sigma_t^i(\omega, x) \quad \forall x \geq 0$$

or, as usual, omitting to mention the ω variable:

$$m_t(x) = \sum_{i \in \mathbb{I}} \gamma_t^i \sigma_t^i(x) \quad \forall x \geq 0.$$

The γ_t^i are the components of a market price of risk, and they do not depend on the time to maturity x. Using the volatility operator process σ the last equality reads

$$m_t = \sigma_t \gamma_t \quad \forall t \in \mathbb{T} \tag{73}$$

and any γ, progressively measurable with values in $\ell^2(\mathbb{I})$, satisfying this equation is called a market price of risk process.

- $\mathbb{I} = \mathbb{N}$. Then the linear span is not closed in general; in fact, it is closed if and only if it is finite-dimensional. In that case, we shall impose a stronger condition. To prove that the market is arbitrage-free, we shall use that $m_t(\omega)$ is in the range of the volatility operator $\sigma_t(\omega)$ which is a subset of the above closed linear span. So, once more we impose that the condition (73) should be satisfied, but for γ with values in $\ell^2(\mathbb{I})$. If the range of $\sigma_t(\omega)$ is infinite dimensional, then this condition is indeed stronger, since $\sigma_t(\omega)$ is a.e. a compact operator.

In both cases, we also need that γ satisfy some integrability condition in (ω, t). This leads us to the following

Definition 10. *We shall say that the market is strongly arbitrage-free if there exists a progressively measurable process γ with values in $\ell^2(\mathbb{I})$, such that*

$$m_t = \sigma_t \gamma_t, \quad \forall t \in \mathbb{T} \tag{74}$$

and

$$E\left[\exp\left(a \int_0^{\bar{T}} \|\gamma_t\|_{\ell^2}^2 dt\right)\right] < \infty, \quad \forall a \geq 0. \tag{75}$$

If the market is strongly arbitrage-free then, by the Girsanov theorem, a martingale measure is given by $dQ = \xi_{\bar{T}} dP$, with:

$$\xi_t = \exp\left(-\frac{1}{2} \int_0^t \|\gamma_s\|_{\ell^2}^2 ds - \sum_{i \in \mathbb{I}} \gamma_s^i dW_s^i\right). \tag{76}$$

The $\tilde{W}^i$, $i \in \mathbb{I}$, where

$$\tilde{W}_t^i = W_t^i + \int_0^t \gamma_s^i ds, \tag{77}$$

are independent Wiener process with respect to Q. The expected value of a random variable X with respect to Q is given by:

$$E_Q[X] = E[\xi_{\bar{T}} X].$$

Under a martingale measure, the discounted zero-coupon price process p satisfies the equation

$$p_t = \mathcal{L}_t p_0 + \int_0^t \mathcal{L}_{t-s}(p_s \sigma_s) d\tilde{W}_s \tag{78}$$

and also the equation

$$p_t = p_0 + \int_0^t \partial p_s ds + \int_0^t p_s \sigma_s d\tilde{W}_s. \tag{79}$$

The discounted roll-over price process q_t is given by:

$$q_t = p_0 + \int_0^t q_s \sigma_s d\tilde{W}_s. \tag{80}$$

Lemma 2. *A portfolio θ is self-financing if and only if:*

$$V_t(\theta) = V_0(\theta) + \int_0^t \sum_{i \in \mathbb{I}} <\theta_t,\, p_t \sigma_t^i> d\tilde{W}_s^i. \tag{81}$$

We note that the integrand is in fact the adjoint operator of the operator $b_t(\omega) = p_t(\omega)\sigma_t(\omega)$ from $\ell^2(\mathbb{I})$ to $E^s([0,\infty[)$:

$$(b_t(\omega)'\theta_t)^i = <\theta_t,\, p_t \sigma_t^i>,\ \forall i \in \mathbb{I}. \tag{82}$$

To see this, with $x_t^i(\omega) = <\theta_t,\, p_t \sigma_t^i>$, rewrite it as follows: for all (t,ω) and all $z \in \ell^2(\mathbb{I})$

$$\begin{aligned}(x_t(\omega), z)_{\ell^2} &= \sum_{i \in \mathbb{I}} z^i <\theta_t(\omega),\, p_t(\omega)\,\sigma_t^i(\omega)> \\ &= <\theta_t(\omega),\, p_t(\omega) \sum_{i \in \mathbb{I}} \sigma_t^i(\omega)\, z^i> \\ &= <\theta_t(\omega),\, b_t(\omega) z> = \left(b_t(\omega)'\theta_t(\omega),\, z\right).\end{aligned}$$

If the market is strongly arbitrage-free and if condition (55) of Theorem 6 is satisfied, then also condition (56) is satisfied and the Theorem 6 applies.

5 Hedging of Interest Derivatives

From now on, it will be a standing assumption that p_0 satisfies condition (54), that σ satisfy conditions (53) and (55) and that the market is strongly arbitrage-free according to Definition 10.

Before we solve the optimal portfolio problem, we shall study the problem of hedging a European interest rates derivative with payoff X at maturity $\bar{T}$. X is said to be an attainable contingent claim or derivative if $V_{\bar{T}}(\theta) = X$ for some admissible self-financing portfolio θ. Here we are only interested in payoffs, relevant for the optimal portfolio problem considered in these notes, i.e. $X \in L^p(\Omega, \mathcal{F}, P)$ for every $p \geq 1$ (see Lemma 4). We first introduce the hedging equation, the Malliavin derivative and the Clark-Ocone representation formula, which then permits the reader, if he wish, to proceed directly to the study of the optimization problem in the case of deterministic σ and γ in §6.2

Assume that $X \in L^2(\Omega, \mathcal{F}, Q)$, where Q is one equivalent martingale measure given by (76). Then, by the martingale representation theorem, X can be written as a stochastic integral:

$$X = E_Q[X] + \int_0^{\bar{T}} \sum_{i \in \mathbb{I}} x_t^i d\tilde{W}_t^i, \tag{83}$$

with:

$$E_Q[\int_0^{\bar{T}} \|x_t\|_{\ell^2}^2 dt] < \infty. \tag{84}$$

Comparing with equations (81) and (82) for a self-financing portfolio, we obtain the hedging equation

$$b_t(\omega)'\theta_t(\omega) = x_t(\omega), \text{ a.e. } (t,\omega), \tag{85}$$

where the operator $b_t(\omega) = p_t(\omega)\sigma_t(\omega)$ from $\ell^2(\mathbb{I})$ to $E^s([0,\infty[)$ was introduced in (82). Equivalently: for almost every (t,ω),

$$x_t^i(\omega) = <\theta_t(\omega), p_t(\omega)\sigma_t^i(\omega)>, \forall i \in \mathbb{I}.$$

We next introduce the Malliavin derivative (c.f. [26]), $D_t X$, with respect to $\tilde{W}$, at time $t \in \mathbb{T}$ of certain $\mathcal{F} = \mathcal{F}_{\bar{T}}$ measurable real random variables X by:

D1) $D_t X = 0$, if X is a constant,
D2) $D_t X = h_t$, if $h \in L^2(\mathbb{T}, \ell^2(\mathbb{I}))$ and $X = \int_0^{\bar{T}} \sum_{i \in \mathbb{I}} h_t^i d\tilde{W}_t^i$,
D3) $D_t(XY) = X D_t Y + Y D_t X$.

The algebra of such random variables is dense in $L^2(\Omega, \mathcal{F}, Q)$, which can be used to extend the definition to larger sets. $D_t X$ takes its values in $\ell^2(\mathbb{I})$. The partial derivative, with respect to $\tilde{W}^i$, $D_{i,t} X$, is the i-th component of $D_t X$.

We will use the following expression for the Malliavin derivative of an Itô stochastic integral:

$$D_t \int_0^{\bar{T}} \sum_{i \in \mathbb{I}} x_s^i d\tilde{W}_s^i = x_t + \int_t^{\bar{T}} \sum_{i \in \mathbb{I}} (D_t x_s^i) d\tilde{W}_s^i, \tag{86}$$

when almost all the x_s^i are Malliavin differentiable and sufficiently integrable.

In the case when X is Malliavin differentiable, the Clark-Ocone representation formula states that the integrand x_t in (83) is given by

$$x_t = E_Q[D_t X \mid \mathcal{F}_t]. \tag{87}$$

We now come back to the hedging equation (85). The fact that $\theta_t = \delta_0$ is a solution to the homogeneous equation (85) permits us to construct self-financed solutions of the in-homogeneous equation (85), from solutions, which are not self-financed:

Lemma 3. *If $\bar{\theta}$ is an admissible portfolio (not necessarily self-financed) which satisfies (85), then there is a unique self-financing admissible portfolio θ_t such that the difference $\theta_t - \bar{\theta}_t$ is risk-free. It is given by:*

$$\theta_t = a_t \delta_0 + \bar{\theta}_t, \tag{88}$$

$$a_t = \frac{1}{p_t(0)}\left[E_Q[X \mid \mathcal{F}_t] - V_t(\bar{\theta})\right]. \tag{89}$$

Proof:

We here omit the argument ω. Since the portfolio $\theta_t - \bar{\theta}_t$ is risk-free, it must have time to maturity 0, and the formula (88) is true by definition. Substituting into equation (85), and bearing in mind that $\sigma_t^i(0) = 0$:

$$((p_t\sigma_t)'\theta_t)^i = <\theta_t,\, p_t\,\sigma_t^i> \;=\; <a_t\delta_0 + \bar{\theta}_t,\, p_t\,\sigma_t^i>$$
$$= a_t\,p_t(0)\,\sigma_t^i(0) + <\bar{\theta}_t,\, p_t\,\sigma_t^i>$$
$$= x_t^i \quad \forall i \in \mathbb{I}.$$

So θ_t satisfies (85). It is then a hedging portfolio of X if $V_t(\theta) = E_Q[X\,|\,\mathcal{F}_t]$. Substituting again (88) and then (89), we get:

$$V_t(\theta) = a_t V_t(\delta_0) + V_t(\bar{\theta}) = a_t p_t(0) + V_t(\bar{\theta}) = E_Q[X\,|\,\mathcal{F}_t].$$

If $\bar{\theta}$ is an admissible portfolio, then θ is also admissible, since $\|\theta\|_\mathsf{P} = \|\bar{\theta}\|_\mathsf{P}$. □

By the lemma, the construction of a hedging portfolio for X is reduced to solve equation (85) in $\theta_t(\omega)$ for every (t, ω), in such a way that $\theta \in \mathsf{P}$, i.e. θ is admissible. Any such solution θ of this equation contains the risky part of the portfolio.

To solve equation (85), for given (t, ω), we have to know if $x_t(\omega)$ is in the range of the operator $b_t(\omega)'$. The closure of the range of $b_t(\omega)'$ is equal to the orthogonal complement $(\mathcal{K}(b_t(\omega)))^\perp$ of the kernel $\mathcal{K}(b_t(\omega))$ of $b_t(\omega)$.

Consider the cases of $\mathbb{I}$ finite: The range $\mathcal{R}((b_t(\omega))')$ is then closed, since it is finite dimensional. The kernel $\mathcal{K}(b_t(\omega))$ is trivial iff the $p_t(\omega)\,\sigma_t^i(\omega)$ are linearly independent. So $(b_t(\omega))'$ is surjective and and there is a (non-unique) solution $\theta_t(\omega)$, for every $x_t(\omega)$, iff the $p_t(\omega)\,\sigma_t^i(\omega)$ are linearly independent.

Consider the cases of $\mathbb{I}$ infinite: The map $(b_t(\omega))'$ from $E^{-s}([0,\infty[)$ to $\ell^2(\mathbb{I})$, is then never surjective. In fact, $b_t(\omega)$ is a Hilbert-Schmidt operator, so it is compact. The adjoint is then also compact and since $\ell^2(\mathbb{I})$ is infinite dimensional, its range must be a proper subspace of $\ell^2(\mathbb{I})$. This is the basic reason why there are always non-attainable contingent claims, when $\mathbb{I}$ is infinite.

We have the following result (see Th.4.1 and Th.4.2 of [34] for the case $\mathbb{I} = \mathbb{N}$):

Theorem 7. *Let* $\mathsf{D}_0 = \cap_{p \geq 1} L^p(\Omega, P, \mathcal{F})$.
i) If $\mathbb{I} = \mathbb{N}$, *then there exists* $X \in \mathsf{D}_0$ *such that* $V_{\bar{T}}(\theta) \neq X$ *for all* $\theta \in \mathsf{P}_{sf}$.
ii) D_0 *has a dense subspace of attainable contingent claims if and only if the operator* $\sigma_t(\omega)$ *has a trivial kernel a.e.* $(t, \omega) \in \mathbb{T} \times \Omega$.

Statement $ii)$ says by definition that the bond market is approximately complete (notion introduced in [2] and [3]) if and only if $\sigma_t(\omega)$ has a trivial kernel a.e.

In the sequel of this section, we are interested in the hedging problem for approximately complete markets, so we only consider the solution of the hedging equation (85) in the case when $\sigma_t(\omega)$ has a trivial kernel a.e. $(t, \omega) \in \mathbb{T} \times \Omega$.

Consider the case when $\mathbb{I} = \mathbb{N}$ is an infinite and let $\ell^2 = \ell^2(\mathbb{I})$. To derive a condition under which (85) has a solution and to derive a closed formula for one of the solutions, we rewrite the l.h.s. of (85) using the notations

$$l_t = \mathcal{L}_t p_0, \quad B_t(\omega) = l_t \sigma_t(\omega) \text{ and } \eta_t(\omega) = \mathcal{S}^{-1}(p_t(\omega)/l_t)\theta_t(\omega). \tag{90}$$

Then

$$(\sigma_t(\omega))' p_t(\omega)\theta_t(\omega) = (\sigma_t(\omega))' l_t (p_t(\omega)/l_t)\theta_t(\omega) = (l_t\sigma_t(\omega))'(p_t(\omega)/l_t)\theta_t(\omega)$$
$$= (l_t\sigma_t(\omega))^* \mathcal{S}^{-1}(p_t(\omega)/l_t)\theta_t(\omega) = (B_t(\omega))^* \eta_t(\omega).$$

The linear operator $B_t(\omega)$ is given, since p_0 and $\sigma_t(\omega)$ are supposed given. Applying Theorem 6 to the factor p/l, it follows that equation (85) is equivalent to find a progressive E^s-valued process η satisfying the equation

$$(B_t(\omega))^* \eta_t(\omega) = x_t(\omega), \text{ a.e. } (t,\omega) \in \mathbb{T} \times \Omega. \tag{91}$$

We define the self-adjoint operator $A_t(\omega)$ in ℓ^2 by

$$A_t(\omega) = (B_t(\omega))^* B_t(\omega). \tag{92}$$

It is a fact of basic Hilbert space operator theory (cf. [16]) that the range $\mathcal{R}((B_t(\omega))^*) = \mathcal{R}((A_t(\omega))^{1/2})$. The solvability of each one of equations (85) and (91) is therefore equivalent to the existence of a progressive ℓ^2-valued process z satisfying

$$(A_t(\omega))^{1/2} z_t(\omega) = x_t(\omega), \text{ a.e. } (t,\omega) \in \mathbb{T} \times \Omega. \tag{93}$$

The kernel $\mathcal{K}((A_t(\omega))^{1/2})$ is trivial since $\mathcal{K}((A_t(\omega))^{1/2}) = \mathcal{K}(A_t(\omega)) = \mathcal{K}(B_t(\omega))$ = $\{0\}$. Now, if $x_t(\omega) \in \mathcal{R}((B_t(\omega))^*)$ then the unique solution of (93) is $z_t(\omega) = (((A_t(\omega))^{1/2})^{-1} x_t(\omega)$ and a solution of (91) is given by

$$\eta_t(\omega) = S_t(\omega)(A_t(\omega))^{-1/2} x_t(\omega), \tag{94}$$

where $S_t(\omega)$, the closure of the operator $B_t(\omega)(A_t(\omega))^{-1/2}$, is isometric (cf. [16]) from ℓ^2 to E^s. Let a be as in (89) and

$$\theta = a\delta_0 + \bar{\theta} \text{ and } \bar{\theta}_t = (l_t/p_t)\mathcal{S}\eta_t. \tag{95}$$

Then θ is a hedging portfolio according to Lemma 3.

In order to ensure that $x_t(\omega)$ of (85) is in the range of $(\sigma'_t p_t)(\omega)$, we introduce spaces $\ell^{s,2}$, of vectors decreasing faster (for $s > 0$) than those of ℓ^2. For $s \in \mathbb{R}$, let $\ell^{s,2}$ be the Hilbert space of real sequences endowed with the norm

$$\|x\|_{\ell^{s,2}} = (\sum_{i \in \mathbb{N}} (1+i^2)^s |x^i|^2)^{1/2}. \tag{96}$$

Obviously $\ell^2 = \ell^{0,2}$ and $\ell^{s',2} \subset \ell^{s,2}$, if $s' \geq s$. Although $(A_t(\omega))^{-1/2}$ is an unbounded operator in ℓ^2 its restriction to $\ell^{s,2}$ can be a bounded operator for some sufficient large $s > 0$, i.e. $(A_t(\omega))^{-1/2} \ell^{s,2} \subset \ell^2$. This is the idea of our assumption, which will ensure hedgeability. However a precise formulation of this assumption must, as in the case of a finite of Bm., take care of integrability properties in (t,ω).

To consider also the case of a finite $\mathbb{I}$, we define after obvious modifications the operator $A_t(\omega)$ in $\ell^2(\mathbb{I})$ by formula (92). In this case $A_t(\omega)$ has obviously a bounded inverse.

Condition 8 *i) If $Card(\mathbb{I}) < \infty$, then there exists $k \in \mathsf{D}_0$, such that for all $x \in \ell^2(\mathbb{I})$:*

$$\|x\|_{\ell^2} \leq k(\omega)\|(A_t(\omega))^{1/2}x\|_{\ell^2} \ a.e. \ (t,\omega) \in \mathbb{T} \times \Omega. \tag{97}$$

ii) If $\mathbb{I} = \mathbb{N}$, then there exists $s > 0$ and $k \in \mathsf{D}_0$, such that for all $x \in \ell^2(\mathbb{I})$:

$$\|x\|_{\ell^2} \leq k(\omega)\|(A_t(\omega))^{1/2}x\|_{\ell^{s,2}} \ a.e. \ (t,\omega) \in \mathbb{T} \times \Omega. \tag{98}$$

In the case of a finite number of Bm. Condition 8 i) leads to a complete market and one can choose a hedging portfolio such that it is continuous in the asset to hedge. To state the result let use introduce the notation $\mathsf{D}_0(F) = \cap_{p \geq 1} L^p(\Omega, P, \mathcal{F}, F)$, where F is a Banach space. $\mathsf{D}_0 = \mathsf{D}_0(\mathbb{R})$.

Theorem 9 (Finite number of random-sources, $Card(\mathbb{I}) < \infty$).
If (i) of Condition 8 is satisfied and if $X \in \mathsf{D}_0$, then the portfolio given by equation (95) satisfies $\theta \in \mathsf{P}_{sf}$ and $V_{\bar{T}}(\theta) = X$. Moreover the linear mapping $\mathsf{D}_0 \ni X \mapsto \theta \in \mathsf{P} \cap \mathsf{D}_0(L^2(\mathbb{T}, E'))$, is continuous.

Proof:
We only outline the proof of the theorem. Here $\ell^2 = \ell^2(\mathbb{I}) = \mathbb{R}^{\bar{m}}$ is finite dimensional.
Let $X \in \mathsf{D}_0$ and let x be given by (83). First one proves (see Lemma 3.1 of [34]) that

$$\mathsf{D}_0(F) = \cap_{p \geq 1} L^p(\Omega, Q, \mathcal{F}, F). \tag{99}$$

Applying the BDG inequalities to equation (83) it follows that

$$x \in \mathsf{D}_0(L^2(\mathbb{T}, \ell^2)), \tag{100}$$

where x is progressively measurable. The definition of η in (94) and the condition (97) give

$$\|\eta_t(\omega)\|_{\ell^2} \leq k_t(\omega)\|x_t(\omega)\|_{\ell^2}.$$

Inequality (100) then leads to $\eta \in \mathsf{D}_0(L^2(\mathbb{T}, E))$. Using the definition (95) of $\bar{\theta}$ we then obtain

$$\bar{\theta} \in \mathsf{D}_0(L^2(\mathbb{T}, E')). \tag{101}$$

Since $\bar{\theta}$ satisfies equation (85) by construction and since formulas (100) and (101) shows that $\bar{\theta}$ is admissible, the hypotheses of Lemma 3 are satisfied, so $\theta \in \mathsf{P}_{sf}$. This shows that θ is a hedging portfolio of X.
All the linear maps $X \mapsto x \mapsto \eta \mapsto \theta$ are continuous in the above spaces, which also proves the claimed continuity of the map $X \mapsto \theta$. □

The solution of the hedging problem, given by Theorem 9, is highly non-unique, since when $Card(\mathbb{I}) = \bar{m} < \infty$ then the kernel $\mathcal{K}((\sigma'_t p_t)(\omega))$ has infinite dimension. For instance there is a hedging portfolio $\hat{\vartheta}$ consisting of $\bar{m} + 1$ rollovers at any time.

To state the result in the case of a infinite number of Bm., we first introduce spaces of contingent claims D_s, smaller than D_0 if $s > 0$ and corresponding to that the integrand x in (83) takes values in $\ell^{s,2}$. More precisely, for $s > 0$ let

$$D_s = \{X \in D_0 \mid x \in D_0(L^2(\mathbb{T}, \ell^{s,2})) \text{ where } x \text{ is given by (83)}\}. \tag{102}$$

Condition 8 $ii)$ leads to a D_s-*complete* market, i.e. D_s is a space of attainable contingent claims, D_s is a dense subspace of D_0 and D_s is itself a complete topological vectorspace. This concept gives a natural frame-work to study existence and continuity of hedging portfolios. We have (see Theorem 4.3 of [34]):

Theorem 10 (Infinite number of random-sources $\mathbb{I} = \mathbb{N}$).
If (ii) of Condition 8 is satisfied and if $X \in D_s$, where $s > 0$ is given by Condition 8, then the portfolio given by equation (95) satisfies $\theta \in P_{sf}$ and $V_{\bar{T}}(\theta) = X$. Moreover the linear map $D_s \ni X \mapsto \theta \in P \cap D_0(L^2(\mathbb{T}, E'))$, is continuous.

For the proof, which only uses elementary spectral properties of self-adjoint operators and compact operators, the reader is referred to [34].

A Malliavin-Clark-Ocone formalism was adapted recently in reference [6], for the construction of hedging portfolios in a Markovian context, with a Lipschitz continuous (in the bond price) volatility operator. This guarantees that the Malliavin derivative of the bond price is proportional to the volatility operator (formula (30) of [6]). Hedging is then achieved for a restricted class of claims, namely European claims being a Lipschitz continuous function in the price of the bond at maturity.

References [8] and [27] studies the hedging problem in a weaker sense of approximate hedging, which in our context simply boils down to the well-known existence of the integrand x in the decomposition (83).

6 Optimal Portfolio Management

We now consider an investor, characterized by a von-Neumann-Morgenstern utility function U, an initial wealth v, and a horizon $\bar{T}$. The money is invested in a market portfolio, and the investor seeks to maximize the terminal (discounted) value $V_{\bar{T}}(\theta)$ of the portfolio. Transaction costs and taxes are neglected. The optimal portfolio problem is then to find an admissible self-financing portfolio $\hat{\theta}$ with $V_0(\hat{\theta}) = v$, such that:

$$(P_0) \begin{cases} \sup E_P\left[U\left(V_{\bar{T}}(\theta)\right)\right] = E_P[U(V_{\bar{T}}(\hat{\theta}))] \\ V_0(\theta) = v \\ \theta \in P_{sf}. \end{cases}$$

We will follow the now classical two-step approach (cf. [17], [28]) towards solving that problem. If the portfolio is self-financing and is worth v at time 0, then, by the martingale property:

$$E_P\left[\xi_{\bar{T}} V_{\bar{T}}(\theta)\right] = v$$

where the random variable $\xi_{\bar{T}}$, arising from Girsanov's theorem, was introduced earlier in (76). In general there can be several possible $\xi_{\bar{T}}$, one for each γ satisfying the conditions of Definition 10. The first step (optimization) consists of finding for given γ, among $\mathcal{F}_{\bar{T}}$-measurable random variables X such that $E_P[\xi_{\bar{T}} X] = v$, the one(s) that maximize expected utility $E_P[U(X)]$. This problem has in our setting a general solution $\hat{X}$, given by Proposition 7. The second one (accessibility) consists in hedging one of the contingent claims $\hat{X}$, obtained for the different γ, by a self-financing portfolio $\hat{\theta}$. This portfolio is then a solution of the optimal portfolio problem (P_0). By concavity, the final optimal wealth $V_{\bar{T}}(\hat{\theta})$ is unique.

6.1 Optimization

We consider, for a given γ satisfying the conditions of Definition 10, the optimization problem:
$$\begin{cases} \sup E_P[U(X)] \\ E_P[\xi_{\bar{T}} X] = v \\ X \in L^2(\Omega, \mathcal{F}_{\bar{T}}, P) \end{cases}$$
We can rewrite it in a more geometric way, involving the scalar product in $L^2(\Omega, \mathcal{F}_{\bar{T}}, P)$:
$$(P) \begin{cases} \sup \int_\Omega U(X) \, dP \\ \int_\Omega \xi_{\bar{T}} X \, dP = (\xi_{\bar{T}}, X)_{L^2} = v \\ X \in L^2(\Omega, \mathcal{F}_{\bar{T}}, P) \end{cases}$$

Problem (P) consists of maximizing a concave function on a closed linear subspace of L^2. Assume there is a maximizer $\hat{X}$. If the usual theory of Lagrange multipliers applies, there will be some $\lambda \in \mathbb{R}$ such that $\hat{X}$ actually optimizes the functional

$$\int_\Omega [U(X) - \lambda \xi_{\bar{T}} X] \, dP$$

over all of L^2. Maximizing pointwise under the integral, and bearing in mind that U is concave, we are led to the equation:

$$U'\left(\hat{X}(\omega)\right) = \lambda \xi_{\bar{T}}(\omega) \quad P\text{-a.e.,} \tag{103}$$

which fully characterizes the solution $\hat{X}$. Unfortunately this program cannot be carried through, for the function $E_P[U(X)]$ has no point of continuity in L^2 unless U is bounded, so the constraint qualification conditions do not hold for problem (P), cf. [9]. We will therefore proceed by a roundabout way: use (103) to define $\hat{X}$, and then prove that $\hat{X}$ is optimal for a suitable choice of λ. For this, we need some conditions on U.

Definition 11. *The utility function U will be called* admissible *if it satisfies the following properties:*

1. *$U : \mathbb{R} \to \{-\infty\} \cup \mathbb{R}$ is concave and upper semi-continuous*
2. *there is some $a \in \{-\infty\} \cup]-\infty, 0]$, such that $U(x) = -\infty$ if $x < a$ and $U(x) > -\infty$ if $x > a$*

3. U is twice differentiable on the interval $A =]a, \infty[$; set $B = U'(A)$
4. $\sup B = +\infty$; $\inf B = 0$ or $\inf B = -\infty$.
5. $U' : A \to B$ is one-to-one, and there are some positive constants r, c_1, c_2 and c_3 such that its inverse $I = [U']^{-1}$ satisfies the estimate $|I(y)| + |yI'(y)| \leq c_1 + c_2 |y|^r + c_3 |y|^{-r}$ for $y \in B$.

It follows from these assumptions that I is continuous and strictly decreasing, with:
$$I(\lambda) \to +\infty \text{ when } \lambda \to \inf B$$
$$I(\lambda) \to a \text{ when } \lambda \to +\infty.$$

We note that the estimate, in point 5) of Definition 11, is satisfied iff there exist $C \geq 0$ such that
$$|I(y)| + |yI'(y)| \leq C (|y|^r + |y|^{-r}),$$
for all $y \in B$. All usual utility functions are admissible:

Example 7.
i) Quadratic utility; Set $U(x) = \mu x - \frac{1}{2}x^2$, $\mu \in \mathbb{R}$. Then $a = -\infty$, and $U'(x) = \mu - x$, so that $B = \mathbb{R}$ and $I(y) = \mu - y$. The estimate is satisfied with $r = 1$.
ii) Exponential utility; Set $U(x) = 1 - \frac{1}{\mu} \exp(-\mu x)$, $\mu > 0$. Then $a = -\infty$, and $U'(x) = \exp(-\mu x)$, so that $B =]0, \infty[$ and $I(y) = -\frac{1}{\mu} \ln(y)$. The estimate is satisfied for any $r > 0$.
iii) Power utility; Set $U(x) = \frac{1}{\mu} x^\mu$ for some $\mu < 1$ and $\mu \neq 0$ (note that μ may be negative). Then $a = 0$, and $U'(x) = x^{\mu-1}$, so that $B =]0, \infty[$ and $I(y) = y^{1/(\mu-1)}$. The estimate is satisfied with $r = \frac{1}{1-\mu}$.
iv) Logarithmic utility; Set $U(x) = \ln x$. Then $a = 0$ and $U'(x) = \frac{1}{x}$, so that $B =]0, \infty[$ and $I(y) = \frac{1}{y}$. The estimate is satisfied with $r = 1$.

Take some $\lambda \in B$ and a γ satisfying the conditions of Definition 10, and define a random variable X_λ by:
$$X_\lambda(\omega) = I(\lambda \xi_{\bar{T}}(\omega)).$$

X_λ is $\mathcal{F}_{\bar{T}}$-measurable. In addition, we have:

Lemma 4. $X_\lambda \in L^p(\Omega, \mathcal{F}_{\bar{T}}, P)$ for every $p \geq 1$.

Proof:
Since U is admissible, we know from condition 4 that, for some $r > 0$ we have:
$$|I(\lambda \xi_{\bar{T}})|^p \leq \left(c_1 + c_2 |\lambda \xi_{\bar{T}}|^r + c_3 |\lambda \xi_{\bar{T}}|^{-r} \right)^p$$
$$\leq k_1 + k_2 |\lambda|^{pr} |\xi_{\bar{T}}|^{pr} + k_3 |\lambda|^{-pr} |\xi_{\bar{T}}|^{-pr}$$

and the right-hand side is integrable, for we know that $\xi_{\bar{T}}^s \in L^1(\Omega, \mathcal{F}_{\bar{T}}, P)$ for every $s \in \mathbb{R}$. □

Lemma 5. *Let $v \in A$. There is a unique $\hat{\lambda} \in B$ such that $E_P\left[X_{\hat{\lambda}}\xi_{\bar{T}}\right] = v$.*

Proof:
Consider the map $\varphi : B \to \mathbb{R}$ defined by $\varphi(\lambda) = E_P[X_\lambda \xi_{\bar{T}}] = E_P[I(\lambda\xi_{\bar{T}})\xi_{\bar{T}}]$. Since $\xi_{\bar{T}} > 0$ P-a.e., and I is strictly decreasing, φ is strictly decreasing. Using the Lebesgue dominated convergence theorem, we find that it is continuous. Using Fatou's lemma, we find that:

- $\varphi(\lambda) \to +\infty$ when $\lambda \to \inf B$
- $\limsup \varphi(\lambda) \leq a$ when $\lambda \to +\infty$

Since $v \in A$, it follows that there is a unique $\hat{\lambda}$ such that $\varphi\left(\hat{\lambda}\right) = v$. □

Denote $X_{\hat{\lambda}}$ by $\hat{X}$. We now conclude:

Proposition 7. $\hat{X}$ *is the unique solution of problem (P).*

Proof:
Let us show that $\hat{X}$ is indeed a solution of problem (P). Uniqueness follows from the strict concavity of U.

We have shown that $\hat{X}$ is in L^2, and $E_P[\hat{X}\xi_{\bar{T}}] = v$, so $\hat{X}$ satisfies the constraints. Take another $X \in L^2$ such that $E_P[X\xi_{\bar{T}}] = v$. Since U is concave, we have:

$$U(X(\omega)) \leq U\left(\hat{X}(\omega)\right) + (X(\omega) - \hat{X}(\omega))U'\left(\hat{X}(\omega)\right) \quad P\text{-a.e.}$$

By definition, $U'\left(\hat{X}(\omega)\right) = \lambda\xi_{\bar{T}}(\omega)$. Substituting into the inequality and integrating, we get:

$$\int_\Omega U(X)\,dP \leq \int_\Omega U\left(\hat{X}\right)dP + \lambda\int_\Omega (X - \hat{X})\xi_{\bar{T}}dP$$

and the last term vanishes because it is just $\lambda(v - v)$. So $\hat{X}$ is indeed an optimizer, and the result follows. □

6.2 Hedging

Once the solution $\hat{X}$ of the optimization problem (P) is found, for a given γ, the question is whether it can be hedged by a self-financing portfolio $\hat{\theta}$, so that $V_{\bar{T}}(\hat{\theta}) = \hat{X}$. We note that, if there exists such $\hat{\theta} \in P_{sf}$, then it is a solution of (P_0). In fact, let $\theta \in P_{sf}$ and $V_0(\theta) = v$ and set $X = V_{\bar{T}}(\theta)$. It follows from (P) that

$$E_P[U(V_{\bar{T}}(\theta))] = E_P[U(X)] \leq E_P\left[U(\hat{X})\right] = E_P\left[U(V_{\bar{T}}(\hat{\theta}))\right],$$

so $\hat{\theta}$ is a solution of (P_0).

Deterministic Case

In this paragraph, we shall use the general hedging results of §5 to solve this problem, in the case when the m and σ, are deterministic (i.e. they do not depend on ω).

Under these conditions, there can be several γ that satisfy the conditions of Definition 10 and some γ can even be non-deterministic. However, as we have supposed that the market is strongly arbitrage free, so equation (74) has a solution, we can choose γ to be the unique solution with the property of being orthogonal in ℓ^2 to the kernel of the volatility operator. More precisely, we choose the unique γ such that

$$(\gamma_t, x)_{\ell^2} = 0, \quad \forall x \in \ell^2(\mathbb{I}) \text{ s.t. } \sigma_t x = 0. \tag{104}$$

The γ defined by this condition is deterministic. In the sequel of this paragraph γ is given by (104). In that case, it follows from formula (76) that $\xi_{\bar{T}}$ is Malliavin differentiable. It follows from formula (86) that the partial derivative with respect to $\tilde{W}^i$ is given by:

$$D_{i,t}\xi_{\bar{T}} = -\gamma_t^i \xi_{\bar{T}}$$

and $\hat{X} = I\left(\hat{\lambda}\xi_{\bar{T}}\right)$ is Malliavin differentiable as well, with:

$$D_{i,t}\hat{X} = -\hat{\lambda}\gamma_t^i \xi_{\bar{T}} I'(\hat{\lambda}\xi_{\bar{T}}).$$

The Clarke-Ocone formula now reads:

$$X = E_Q[X \mid \mathcal{F}_0] + \sum_{i \in \mathbb{I}} \int_0^{\bar{T}} E_Q[D_{i,t}X \mid \mathcal{F}_t] \, d\tilde{W}_t^i \tag{105}$$

$$= v - \hat{\lambda} \sum_{i \in \mathbb{I}} \int_t^{\bar{T}} \gamma_t^i E_Q\left[\xi_{\bar{T}} I'(\hat{\lambda}\xi_{\bar{T}}) \mid \mathcal{F}_t\right] d\tilde{W}_t^i \tag{106}$$

We then write the equation (85) for the hedging portfolio $\hat{\theta}$, and we substitute the Clark-Ocone formula for $x_t^i(\omega)$:

$$b_t(\omega)' \theta_t(\omega) = -\hat{\lambda} E_Q\left[\xi_{\bar{T}} I'(\hat{\lambda}\xi_{\bar{T}}) \mid \mathcal{F}_t\right] \gamma_t. \tag{107}$$

This equation has a solution iff γ_t is in the range of $b_t(\omega)'$. Since σ is deterministic, this condition simplifies. In fact, let l_t and B_t be given by (90), which here both are deterministic, and let $q(t,\omega) = p_t(\omega)/l_t$. Then the expression (82) of $b_t(\omega)'$ give:

$$(b_t(\omega)'\theta_t(\omega))^i = <\theta_t(\omega),\, p_t(\omega)\sigma_t^i> = <\theta_t(\omega)q(t,\omega),\, l_t\sigma_t^i> = (B_t' f_t(\omega))^i,$$

where $f_t(\omega) \in E^{-s}$ is given by $f_t(\omega) = q(t,\omega)\theta_t(\omega)$. So, equation (107) has a solution iff γ_t is in the range of B_t'. This is always true when $\mathbb{I}$ is finite, since then the range of B_t' is equal to the orthogonal complement of the kernel of σ_t (we remember that $p_t(\omega, x) > 0$ for $x \geq 0$). When $\mathbb{I} = \mathbb{N}$, then the range is only a strictly smaller dense subset.

We are lead to following condition

Definition 12. *We shall say that the market satisfies condition (C) if there exists a deterministic portfolio θ_t^0 which is admissible and satisfies $B_t' \theta_t^0 = \gamma_t$, i.e.*

$$< \theta_t^0, (\mathcal{L}_t p_0) \sigma_t^i > = \gamma_t^i, \tag{108}$$

for each $i \in \mathbb{I}$ and t.

Condition C is then equivalent to $\gamma_t \in \mathcal{R}(B_t')$, the range of B_t'. In the case when $\mathbb{I}$ is finite, there is never uniqueness in the choice of θ_t^0.

In the case when $\mathbb{I}$ is finite, we know that condition (C) is satisfied and it can easily be verified, with n elements say, by picking n maturities $0 < S_1 < ... < S_n$ and by seeking θ_t^0 as a linear combination of rollovers: $\theta_t^0 = \sum x_t^i \delta_{S_i}$. Condition (108) then reduces to a system of n linear equations with n unknowns which determines the x_t^i.

In the case when $\mathbb{I} = \mathbb{N}$, condition (C) may not be satisfied. We will be content with reminding that the left-hand side of equation (108) is meaningful, since $(\mathcal{L}_t p_0) \sigma_t^i$ belongs to the space E^s.

If condition (C) is satisfied, equation (107) becomes:

$$< \theta_t, p_t \sigma_t^i > = -\hat{\lambda} E_Q[\xi_{\bar{T}} I'(\hat{\lambda}\xi_{\bar{T}}) \mid \mathcal{F}_t] < \theta_t^0, \frac{\mathcal{L}_t p_0}{p_t} p_t \sigma_t^i >$$

$$= < -\hat{\lambda} E_Q[\xi_{\bar{T}} I'(\hat{\lambda}\xi_{\bar{T}}) \mid \mathcal{F}_t] \frac{\mathcal{L}_t p_0}{p_t} \theta_t^0, p_t \sigma_t^i >$$

and an obvious solution $\theta_t = \bar{\theta}_t$ (the risky part of the optimal portfolio) is given by:

$$\bar{\theta}_t = -\hat{\lambda} E_Q[\xi_{\bar{T}} I'(\hat{\lambda}\xi_{\bar{T}}) \mid \mathcal{F}_t] \frac{\mathcal{L}_t p_0}{p_t} \theta_t^0.$$

Applying Lemma 3, with x defined by (89), we obtain a hedging portfolio $\hat{\theta} = x\delta_0 + \bar{\theta}$ of $\hat{X}$, where $\bar{\theta}$ is as above, and:

$$x_t = \frac{1}{p_t(0)} \left(E_Q \left[I\left(\hat{\lambda}\xi_{\bar{T}}\right) \mid \mathcal{F}_t \right] - < \bar{\theta}_t, p_t > \right).$$

To sum up, in the case when the m_s and the $\sigma_s^i, i \in \mathbb{I}$, are deterministic, with $\sigma_t^i(0) = 0$, with condition (C) and equation (74) satisfied, an optimal admissible and self-financing portfolio is given by

$$\hat{\theta}_t = x_t \delta_0 + \bar{\theta}_t, \quad \text{where} \quad \bar{\theta}_t = y_t \frac{(\mathcal{L}_t p_0)}{p_t} \theta_t^0 \tag{109}$$

and where the coefficients x_t and y_t are real-valued progressively measurable processes given by

$$y_t = -E_Q[\hat{\lambda}\xi_{\bar{T}} I'(\hat{\lambda}\xi_{\bar{T}}) \mid \mathcal{F}_t] \tag{110}$$

$$x_t = (p_t(0))^{-1} \left(E_Q[I(\hat{\lambda}\xi_{\bar{T}}) \mid \mathcal{F}_t] - y_t < \theta_t^0, \mathcal{L}_t p_0 > \right). \tag{111}$$

This leads immediately to a mutual fund theorem: whatever the utility function and the initial wealth, the optimal portfolio at time t is a linear combination of the current account δ_0 and the portfolio $f \mapsto <\theta_t^0, \frac{\mathcal{L}_t p_0}{p_t} f>$, i.e. the portfolio $\frac{\mathcal{L}_t p_0}{p_t} \theta_t^0$. This portfolio is in general not self-financed, so it can not be given the status of a *market portfolio*. However we can easily reformulate the result with a self-financed portfolio. In fact, chose an admissible utility function, with $a = 0$, according to Definition 11. For this utility function, let Θ be the optimal portfolio given by (109), with unit initial wealth. Obviously $\frac{\mathcal{L}_t p_0}{p_t} \theta_t^0$ is a linear combination of δ_0 and Θ_t. This gives us:

Theorem 11 (Mutual fund theorem). *The optimal portfolio Θ has the following properties:*

i) Θ is an admissible self-financing portfolio, with unit initial value, i.e. $<\Theta_0, p_0> = 1$, and the value at each time $t \in \mathbb{T}$ is strictly positive, i.e. $<\Theta_t, p_t> >> 0$.

ii) For each utility function U, admissible according to Definition 11 and each initial wealth $v \in]a, \infty[$, there exist two real valued processes c and d such that if $\hat\theta_t = c_t \delta_0 + d_t \Theta_t$, then $\hat\theta$ is an optimal self financing portfolio for U, i.e. a solution of problem (P_0).

Stochastic m and σ

We shall here concentrate on the case of an approximately complete market, which is equivalent to that the volatility operator is non-degenerated. In fact, according to $iii)$ of Theorem 7, the market is approximately complete if and only if $\sigma_t(\omega)$ has a trivial kernel a.e. $(t, \omega) \in \mathbb{T} \times \Omega$. We remind that the market of price process γ is unique in this case.

In the case of a finite number of Bm. we obtain easily from Lemma 4 and Theorem 9 the following result (see Theorem 3.6 of [10]):

Theorem 12. *Let $\mathbb{I}$ be a finite set, let U be admissible in the sens of Definition 11 and let $i)$ of Condition 8 be satisfied. The problem (P_0) then has a solution $\hat\theta$. One solution $\hat\theta = a\delta_0 + \bar\theta \in P_{sf}$ is given by (95).*

In the case of an infinite number of Bm. we shall impose Malliavin differentiability properties on the market price of risk γ. To this end we introduce the space D_s^1, for $s > 0$ by

$$D_s^1 = \{X \in D_0 \mid DX \in D_0(L^2(\mathbb{T}, \ell^{s,2}))\}. \tag{112}$$

We can now state a result in the case of an infinite number of Bm., quite analog to the case of a finite number of Bm. (see Theorem 4.5 of [34]):

Theorem 13. *Let $\mathbb{I} = \mathbb{N}$, let U be admissible in the sens of Definition 11, let $ii)$ of Condition 8 be satisfied and let $\ln(\xi_{\bar T}) \in D_s^1$, where $s > 0$ is given by $ii)$ of*

Condition 8. The problem (P_0) then has a solution $\hat{\theta}$. One solution $\hat{\theta} = a\delta_0 + \bar{\theta} \in \mathsf{P}_{sf}$ is given by (95).

Proof:

We only consider the case of $U' > 0$, since the case of $U'(x) = 0$ for some x is so similar. Let the hypotheses of the theorem be satisfied. The portfolio $\hat{\theta}$ is a solution of equation (P_0), if $\hat{\theta} \in \mathsf{P}_{sf}$ and if it hedges $\hat{X}$ given by Proposition 7. (See Corollary 3.4 of [10]). It is enough to verify that Theorem 10 applies to $\hat{X} = I(\hat{\lambda}\xi_{\bar{T}})$ for a certain given $\hat{\lambda} > 0$.

I is C^1, so $D_t \hat{X} = \lambda \xi_{\bar{T}} \varphi'(\lambda \xi_{\bar{T}}) D_t \ln(\xi_{\bar{T}})$. Since $\ln(\xi_{\bar{T}}) \in \mathsf{D}_s^1$, this gives $\|D\hat{X}\|_{L^2(\mathbb{T}, \ell^{s,2})} = |\lambda \xi_{\bar{T}} \varphi'(\lambda \xi_{\bar{T}})| \|D\ln(\xi_{\bar{T}})\|_{L^2(\mathbb{T}, \ell^{s,2})}$. The inequality in 5) of Definition 11 gives $\|D\hat{X}\|_{L^2(\mathbb{T}, \ell^{s,2})} \le C((\lambda \xi_{\bar{T}})^p + (\lambda \xi_{\bar{T}})^{-p})\|D\ln(\xi_{\bar{T}})\|_{L^2(\mathbb{T}, \ell^{s,2})}$, for some $p \ge 1$. Condition (75) of Definition 10 shows that $(\lambda \xi_{\bar{T}})^p + (\lambda \xi_{\bar{T}})^{-p} \in L^q(\Omega, P)$, for all $q \ge 1$. By hypothesis $\|D\ln(\xi_{\bar{T}})\|_{L^2(\mathbb{T}, \ell^{s,2})} \in \mathsf{D}_0$, so Hölder's inequality now gives that $\|D\hat{X}\|_{L^2(\mathbb{T}, \ell^{s,2})} \in \mathsf{D}_0$, i.e. $D\hat{X} \in \mathsf{D}_0(L^2(\mathbb{T}, \ell^{s,2}))$. By Lemma 4, $\hat{X} \in \mathsf{D}_0$. It follows that $\hat{X} \in \mathsf{D}_s^1$. We can now apply Theorem 10, which proves the existence of $\hat{\theta}$. □

Examples

We now give some examples of optimal bond portfolios for logarithmic and quadratic utility functions U. Other examples can be found in [10].

First we assume the drift function m_t and the volatility operator σ_t to be deterministic. We shall therefore suppose that the market satisfy condition (C), of Definition 12, so the market prices of risk γ is deterministic and satisfy condition (108). We shall derive the optimal portfolio directly, going through the steps leading to the general solution (109).

Secondly we study the general case of stochastic drift function m_t and volatility operator σ_t for the logarithmic utility function.

The final optimal discounted wealth is $\hat{X} = I(\hat{\lambda}\xi_{\bar{T}})$. The corresponding optimal discounted wealth process Y is given by $Y_t = E_Q[I(\hat{\lambda}\xi_{\bar{T}}) \mid \mathcal{F}_t]$. The initial wealth $Y_0 = v$ determines $\hat{\lambda}$ by the equation

$$v = Y_0 = E_Q[I(\hat{\lambda}\xi_{\bar{T}})]. \tag{113}$$

We recall that $(p_t)^{-1}\mathcal{L}_t p_0 \in E^s$ a.s and that $p_t(0) > 0$ a.s.

Logarithmic Utility (Deterministic m and σ)

Let

$$U(x) = \ln(x). \tag{114}$$

We have $I(x) = 1/x$, and $\hat{X} = (\hat{\lambda}\xi_{\bar{T}})^{-1}$, so that equation (113) gives:

$$v = E_Q[1/(\hat{\lambda}\xi_{\bar{T}})] = E_P[\xi_{\bar{T}}/(\hat{\lambda}\xi_{\bar{T}})] = 1/\hat{\lambda}.$$

Then using the expression (76) for ξ_t and $\tilde{W}_t^i$ we have:

$$\frac{1}{\xi_t} = \exp\left(-\frac{1}{2}\int_0^t \sum_{i\in\mathbb{I}} (\gamma_s^i)^2 \, ds + \int_0^t \sum_{i\in\mathbb{I}} \gamma_s^i d\tilde{W}_s^i\right). \tag{115}$$

The right-hand side is a Q-martingale, then so is $1/\xi_t$. It follows that the optimal discounted wealth at t is

$$Y_t = E_Q[I(\hat{\lambda}\xi_{\bar{T}}) \mid \mathcal{F}_t] = \frac{1}{\hat{\lambda}\xi_t} = \frac{v}{\xi_t}.$$

Since $d(1/\xi_t) = \sum_{i\in\mathbb{I}} (\gamma_t^i/\xi_t) d\tilde{W}_t^i$ and $\hat{X} = Y_{\bar{T}}$, it then follows that:

$$\hat{X} = v\left(1 + \sum_{i\in\mathbb{I}} \int_0^{\bar{T}} \gamma_t^i \frac{1}{\xi_t} d\tilde{W}_t^i\right). \tag{116}$$

The hedging equation (85) and the above formula give:

$$\forall i \in \mathbb{I}, \quad <\theta_t(\omega), p_t(\omega) \sigma_t^i(\omega)> = \frac{v}{\xi_t(\omega)} \gamma_t^i \tag{117}$$

By condition (C) we find a portfolio θ^0 satisfying $\gamma_t^i = <\theta_t^0, (\mathcal{L}_t p_0) \sigma_t^i>$, so

$$\gamma_t^i = <(\mathcal{L}_t p_0) \theta_t^0, \sigma_t^i>. \tag{118}$$

Substituting this expression of γ into (117) we obtain:

$$\forall i \in \mathbb{I}, \quad <p_t(\omega)\theta_t(\omega) - \frac{v}{\xi_t(\omega)}(\mathcal{L}_t p_0)\theta_t^0, \sigma_t^i(\omega)> = 0. \tag{119}$$

One solution of this equation is obviously given by $\theta = \bar{\theta}$, where

$$\bar{\theta}_t(\omega) = y_t(\omega) \frac{(\mathcal{L}_t p_0)}{p_t(\omega)} \theta_t^0, \quad y_t(\omega) = \frac{v}{\xi_t(\omega)}. \tag{120}$$

The discounted value of $\bar{\theta}$ at time t in state ω is then

$$(V_t(\bar{\theta}))(\omega) = <\bar{\theta}_t, p_t> = \frac{v}{\xi_t(\omega)} <\theta_t^0, \mathcal{L}_t p_0>. \tag{121}$$

The optimal portfolio $\hat{\theta}$ is now obtained by using Lemma 3: $\hat{\theta}_t = x_t \delta_0 + \bar{\theta}_t$, where

$$x_t = \frac{1}{p_t(0)} \frac{v}{\xi_t}(1 - <\theta_t^0, \hat{\theta}_t(\omega) \mathcal{L}_t p_0>). \tag{122}$$

As it should, the discounted value of $\hat{\theta}$ is then $V_t(\hat{\theta}) = Y_t = v/\xi_t$.

We note the following useful property: the ratio of the investment in bonds with time to maturity $S > 0$ to the total investment is deterministic. In fact this ratio is simply price at $t = 0$, of a zero-coupon bond with time to maturity $S + t$:

$$\frac{\bar{\theta}_t(S,\omega) \, p_t(S,\omega)}{(V_t(\bar{\theta}))(\omega)} = p_0(S+t). \tag{123}$$

Quadratic Utility (deterministic m and σ)

Let the utility function be:
$$U(x) = \mu x - \frac{1}{2}x^2$$

As in i) of Example 7, we find that
$$I(y) = \mu - y.$$

The final discounted optimal wealth is $\hat{X} = I(\hat{\lambda}\xi_{\bar{T}})$, so
$$\hat{X} = \mu - \hat{\lambda}\xi_{\bar{T}}.$$

We determine $\hat{\lambda}$ by the condition:
$$v = E_Q\left[\hat{X}\right] = E_Q\left[\mu - \hat{\lambda}\xi_{\bar{T}}\right] = \mu - \hat{\lambda}E_Q\left[\xi_{\bar{T}}\right]. \tag{124}$$

Set
$$Z_t = \exp\left(-\frac{1}{2}\int_0^t \sum_{i\in\mathbb{I}}(\gamma_s^i)^2 ds - \int_0^t \sum_{i\in\mathbb{I}}\gamma_s^i d\tilde{W}_s^i\right).$$

Then Z is a martingale with respect to Q and formula (77) gives
$$\xi_t = Z_t \exp\left(\int_0^t \sum_{i\in\mathbb{I}}(\gamma_s^i)^2 ds\right). \tag{125}$$

We have, by substitution into (124):
$$v = \mu - \hat{\lambda}E_Q[\xi_{\bar{T}}] = \mu - \hat{\lambda}\exp\left(\int_0^{\bar{T}} \sum_{i\in\mathbb{I}}(\gamma_s^i)^2 ds\right).$$

This gives
$$\hat{\lambda} = (\mu - v)\exp\left(-\int_0^{\bar{T}} \sum_{i\in\mathbb{I}}(\gamma_s^i)^2 ds\right). \tag{126}$$

It now follows from (125) that
$$\hat{X} = \mu - \hat{\lambda}\xi_{\bar{T}} = \mu + (v - \mu) Z_{\bar{T}} \tag{127}$$

and the optimal discounted wealth at t is
$$Y_t = E_Q[I(\hat{\lambda}\xi_{\bar{T}})\,|\,\mathcal{F}_t] = \mu + (v - \mu) Z_t.$$

Since $dZ_t = -Z_t \sum_{i\in\mathbb{I}} \gamma_t^i d\tilde{W}_t^i$, we have that
$$\hat{X} = \mu - (v - \mu)\int_0^{\bar{T}} \sum_{i\in\mathbb{I}} Z_t \gamma_t^i d\tilde{W}_t^i = \mu + \int_0^{\bar{T}} \sum_{i\in\mathbb{I}} (\mu - Y_t)\gamma_t^i d\tilde{W}_t^i,$$

so the hedging equation reads (see (85)):

$$\forall i \in \mathbb{I}, \ < \theta_t(\omega), \ p_t(\omega) \ \sigma_t^i(\omega) > \ = \ -(\mu - Y_t(\omega)) \ \gamma_t^i. \qquad (128)$$

As usually, condition (C) gives a portfolio θ^0 satisfying $\gamma_t^i = \ < \theta_t^0, (\mathcal{L}_t p_0) \sigma_t^i >$, which together with (128) gives:

$$\forall i \in \mathbb{I}, \ < p_t(\omega) \theta_t(\omega) + (Y_t(\omega) - \mu) \ (\mathcal{L}_t p_0) \theta_t^0, \ \sigma_t^i(\omega) > \ = 0.$$

One solution of this equation is $\theta = \bar{\theta}$, where

$$\bar{\theta}_t(\omega) = y_t(\omega) \frac{(\mathcal{L}_t p_0)}{p_t(\omega)} \theta_t^0, \ y_t(\omega) = \mu - Y_t(\omega).$$

$\bar{\theta}$ gives the risky part of the optimal portfolio.

Applying Lemma 3 we obtain the optimal portfolio $\hat{\theta}_t = x_t \delta_0 + \bar{\theta}_t$, where

$$x_t = (p_t(0))^{-1}(Y(t) - (\mu - Y(t)) < \theta_t^0, \mathcal{L}_t \hat{p}_0 >). \qquad (129)$$

Logarithmic Utility (stochastic m and σ)

We assume that the conditions of Definition 10 are satisfied. We chose $\gamma_t(\omega)$ to be orthogonal to the kernel of $\sigma_t(\omega)$, a.e. (t, ω). This γ satisfies the conditions of Definition 10. Formulas (114)–(117) then still hold true. As in the discussion preceding the condition (C), of Definition 12 it follows that $\gamma_t(\omega)$ is a.s. in the closure of the range of $B_t'(\omega)$. Therefore, in this example, the natural generalization of the condition (C) to the stochastic case is simply to impose the same condition (108) of Definition 12 to be satisfied with a stochastic portfolio $\theta^0 \in \mathsf{P}$. Formulas (118)–(123) are then also true statements and it follows using Theorem 6 that $\hat{\theta} \in \mathsf{P}_{sf}$. In particular the ratio of the investment in bonds with time to maturity $S > 0$ to the total investment is deterministic.

6.3 The H-J-B Approach

When m_t and σ_t^i are given functions $m_t(p_t)$ and $\sigma_t^i(p_t)$ of the price p_t, for every t, then the optimal portfolio problem (P_0) can be considered within a Hamilton-Jacobi-Bellman approach. In this subsection we illustrate this approach, without being rigorous and we suppose that the utility function U satisfies the conditions of Definition 11. For notational simplicity we exclude the price argument in m_t and σ_t^i.

The optimal value function, here denoted by F, then only depends of time t, of the value of the discounted wealth w and the discounted price function $f \in E^s$ of Zero-Coupons at time t:

$$F(t, w, f) = \sup\{E[U(V_{\bar{T}}(\theta))] \mid V_t(\theta) = w, \ p_t = f] \mid \theta \in \mathsf{P}_{sf}\}.$$

The derivative $DG(f; g)$ of a function $E^s \ni f \mapsto G(f)$ in the direction $g \in E^s$, is as usually defined by

$$DG(f;g) = \lim_{\epsilon \to 0} \frac{G(f+\epsilon g) - G(f)}{\epsilon}.$$

Suppose that G is C^2. Writing $DG(f)$ for the map $g \mapsto DG(f;g)$ and $D^2G(f)$ for the map $g_1 \times g_2 \mapsto DG(f;g_1,g_2)$, we have that $DG(f)$ is a linear continuous form on E^s and $D^2G(f)$ is a bi-linear continuous form.

Let us first consider the case of a volatility operator σ with trivial kernel, i.e. for every strictly positive price (function) $f \in E^s$, the kernel of the linear map $\sigma_t : \ell^2(\mathbb{I}) \to E^s$ is trivial a.s. According to Definition 10 there is then a unique market of price process γ. Define the Hamiltonian $H(t,w,f,x)$ by:

$$H(t,w,f,x) = \sum_{i \in \mathbb{I}} x^i(t,w,f) \gamma_t^i \frac{\partial F}{\partial w}(t,w,f) + DF(t,w,f;\partial f + \sum_{i \in \mathbb{I}} \gamma_t^i \sigma_t^i f)$$
$$+ \sum_{i \in \mathbb{I}} \left(\frac{1}{2}(x^i(t,w,f))^2 \frac{\partial^2 F}{\partial w^2}(t,w,f) + x^i(t,w,f) \right.$$
$$\left. \times \frac{\partial}{\partial w} DF(t,w,f;\sigma_t^i f) + \frac{1}{2} D^2 F(t,w,f;\sigma_t^i f, \sigma_t^i f) \right). \tag{130}$$

In that formula, $x = (x^i)_{i \in \mathbb{I}} \in \ell^2$ is the control, which is related to the optimal terminal wealth by formula (83). A control x is called *admissible* if

$$x^i(t, V_t(\theta), p_t) = <\theta_t, p_t \sigma_t^i> \tag{131}$$

for all $\theta \in P_{sf}$. In other words, x^i can be interpreted as the value invested in the i-th source of noise. Using the Ito formula, one derives the (formal) HJB equation:

$$\frac{\partial F}{\partial t}(t,w,f) + \sup_x H(t,w,f,x) = 0, \tag{132}$$

with the boundary condition

$$F(\bar{T},w) = U(w). \tag{133}$$

The optimal control $\hat{x}$, solution of the optimization problem

$$\sup_x H(t,w,f,x),$$

is given by

$$\hat{x}^i(t,w,f) = -\left(\frac{\partial^2 F}{\partial w^2}\right)^{-1} \left(\gamma_t^i \frac{\partial F}{\partial w} + (D\frac{\partial F}{\partial w})(t,w,f;\sigma_t^i f) \right), \; i \in \mathbb{I}. \tag{134}$$

Now, substitution of $H(t,w,f,\hat{x}(t,w,f))$ into equation (132) gives:

$$\frac{\partial^2 F}{\partial w^2}(t,w,f)\left(\frac{\partial F}{\partial t}(t,w,f) + DF(t,w,f;\partial f + m_t f) \right.$$
$$\left. + \frac{1}{2} \sum_{i \in \mathbb{I}} D^2 F(t,w,f;\sigma_t^i f, \sigma_t^i f) \right) = \frac{1}{2} \sum_{i \in \mathbb{I}} \left(\gamma_t^i \frac{\partial F}{\partial w} + (D\frac{\partial F}{\partial w})(t,w,f;\sigma_t^i f) \right)^2. \tag{135}$$

Once the solution F of (135), with boundary condition (133), is found, the optimal control $\hat{x}$ is given by (134). Any optimal portfolio $\hat{\theta}$ is then a solution of the equation:

$$\hat{x}^i(t, V_t(\hat{\theta}), p_t) = <\hat{\theta}_t, p_t \sigma_t^i>, \quad \forall\ i \in \mathbb{I},\ t \in \mathbb{T}.$$

Next we consider the case of a volatility operator, which does not necessarily have a trivial kernel. Once more we define the Hamiltonian $H(t, w, f, x, \gamma)$ by formula (130), which now also depends on the control γ, a $\ell^2(\mathbb{I})$ valued function of (t, w, f). A control (x, γ) is admissible if condition (131) is satisfied and if the conditions of Definition 10 are satisfied, so writing out the price argument $f \in E^s$ in m_t and σ_t^i:

$$m_t(f) = \sigma_t(f)\gamma_t(w, f). \tag{136}$$

The optimal control $\hat{\gamma}$ is determined by conditions (131) and (136). This can be seen as follows. Let $\gamma^\perp(f)$ be the unique solution of (136) such that $\gamma^\perp(f)$ is in the orthogonal complement $(\mathcal{K}(\sigma_t(f)))^\perp$ of the kernel $\mathcal{K}(\sigma_t(f))$, let $\hat{\alpha} = \hat{\gamma} - \gamma^\perp$ and let $P_t(f)$ be the orthogonal projection on $\mathcal{K}(\sigma_t(f))$. Condition (131) implies that $\hat{x} \in (\mathcal{K}(\sigma_t(f)))^\perp$. According to (134), this can only be satisfied if

$$\hat{\gamma} = \gamma^\perp + \hat{\alpha} \text{ and } \hat{\alpha}_t(w, f)\frac{\partial F}{\partial w} = P_t(f)\nu_t(w, f), \tag{137}$$

where $\nu_t^i(w, f) = (D\frac{\partial F}{\partial w})(t, w, f; \sigma_t^i f)$. So in the general the case of a volatility operator, which does not necessarily have a trivial kernel, the H-J-B approach leads to the equation (135), with γ replaced by $\hat{\gamma}$ defined by formula (137).

In the case when m_t and σ_t^i are independent of p_t, then the $\hat{x}^i$ are independent of f, $\gamma = \gamma^\perp$ and the above equations simplify:

$$\frac{\partial F}{\partial t}\frac{\partial^2 F}{\partial w^2} = \frac{1}{2}\left(\sum_{i\in \mathbb{I}}\|\gamma_t^i\|^2\right)(\frac{\partial F}{\partial w})^2,$$

with the boundary condition

$$F(\bar{T}, w) = U(w),\ w \in \mathbb{R}.$$

Each self financing portfolio $\hat{\theta} \in P_{sf}$, such that

$$<\hat{\theta}_t, p_t\sigma_t^i> = -\gamma_t^i\left(\frac{\partial F}{\partial w}\right)\left(\frac{\partial^2 F}{\partial w^2}\right)^{-1},\ \forall\ i \in \mathbb{I},\ t \in \mathbb{T},$$

where $w = V_t(\hat{\theta})$, is then a solution of problem (P_0). The solutions in the examples in §6.2, as well as the general solution (109) for deterministic m and σ, are easily obtained by solving these equations.

A Appendix

In this appendix, we reproduce results (proved in the appendix of [10]), used in this article, concerning existence of solutions of some SDE's and L^p estimates of these solutions. The notations $\mathbb{T} = [0, \bar{T}]$, W^i, $\mathbb{I}$ and $(\Omega, P, \mathcal{F}, \mathcal{A})$ are defined in §3.1. Through the appendix m and σ^i, $i \in \mathbb{I}$, are $\mathcal{A}$-progressively measurable E^s-valued processes satisfying

$$\int_0^{\bar{T}} (\|m_t\|_{E^s} + \sum_{i \in \mathbb{I}} \|\sigma_t^i\|_{E^s}^2) dt < \infty, a.s. \tag{138}$$

The E^s-valued semi-martingale L is given by

$$L(t) = \int_0^t (m_s ds + \sum_{i \in \mathbb{I}} \sigma_s^i dW_s^i), \quad \text{if } 0 \le t \le \bar{T} \tag{139}$$

and by $L(t) = L(\bar{T})$, if $t > \bar{T}$. We introduce, for $t \ge 0$, the random variable

$$\mu(t) = t + \int_0^t (\|m_s\|_{E^s} + \sum_{i \in \mathbb{I}} \|\sigma_s^i\|_{E^s}^2) ds, \quad \text{if } 0 \le t \le \bar{T} \tag{140}$$

and $\mu(t) = t - \bar{T} + \mu(\bar{T})$ if $t > \bar{T}$. μ is a.s. strictly increasing, absolutely continuous and on-to $[0, \infty[$. The inverse τ of μ also have these properties and $\tau(t) \le t$. For a continuous E^s-valued processes Y on $[0, \bar{T}]$ we introduce

$$\rho_t(Y) = (E[\sup_{s \in [0,t]} \|Y(\tau(s))\|_{E^s}^2])^{1/2}, \tag{141}$$

for $t \in [0, \infty[$, where we have defined $Y(t)$ for $t > \bar{T}$ by $Y(t) = Y(\bar{T})$. We note that $\rho_t(Y) \le (E[\sup_{s \in [0,t]} \|Y(s)\|_{E^s}^2])^{1/2}$, since $\tau(t) \le t$.

Lemma 6. *If condition (138) is satisfied and if Y is an $\mathcal{A}$-progressively measurable E^s-valued continuous process on $[0, \bar{T}]$, satisfying $\rho_t(Y) < \infty$, for all $t \ge 0$, then the equation*

$$X(t) = Y(t) + \int_0^t \mathcal{L}_{t-s} X(s) (m_s ds + \sum_{i \in \mathbb{I}} \sigma_s^i dW_s^i), \tag{142}$$

$t \in [0, \bar{T}]$, *has a unique solution X, in the set of $\mathcal{A}$-progressively measurable E^s-valued continuous process satisfying:*

$$\int_0^{\bar{T}} (\|X(t)\|_{E^s} + \|X(t) m_t\|_{E^s} + \sum_{i \in \mathbb{I}} \|X(t) \sigma_t^i\|_{E^s}^2) dt < \infty \text{ a.s.} \tag{143}$$

Moreover this solution satisfies:
i) If $\int_0^{\bar{T}} (\|m_t\|_{E^{s+1}} + \sum_{i \in \mathbb{I}} \|\sigma_t^i\|_{E^{s+1}}^2) dt < \infty$ and Y is a continuous E^{s+1}-valued process with $\rho_t(\partial Y) < \infty$, for all $t \ge 0$, then X is a continuous E^{s+1}-valued process.
ii) If (i) is satisfied and if Y is a semi-martingale, then X is a semi-martingale.
iii) If Y is H^s-valued, then X is H^s-valued.

The next lemma establish conditions under which the solution of equation (142) is in L^p, $p \in [0, \infty[$. The notation $\tilde{\mathcal{E}}$ was introduced in (58).

Lemma 7. *Let condition (138) be satisfied and let* (i)

$$E[\exp(p \int_0^{\bar{T}} (\|m_t\|_{E^s} + \sum_{i \in \mathbb{I}} \|\sigma_t^i\|_{E^s}^2) dt)] < \infty,$$

for each $p \in [1, \infty[$*. Suppose that* Y *in Lemma 6 satisfies* (ii)

$$E[\sup_{t \in \mathbb{T}} \|Y(t)\|_{E^s}^p] < \infty,$$

for each $p \in [1, \infty[$*. Then the unique solution* X *of equation (142) in Lemma 6 satisfies*

$$E[\sup_{t \in \mathbb{T}} \|X(t)\|_{E^s}^p] < \infty, \ \forall p \in [1, \infty[. \tag{144}$$

Moreover if (iii)

$$E[(\int_0^{\bar{T}} (\|m_t\|_{E^{s+1}} + \sum_{i \in \mathbb{I}} \|\sigma_t^i\|_{E^{s+1}}^2) dt)^p] < \infty$$

and (iv)

$$E[\sup_{t \in \mathbb{T}} \|Y(t)\|_{E^{s+1}}^p] < \infty,$$

for each $p \in [1, \infty[$*, then also*

$$E[\sup_{t \in \mathbb{T}} \|X(t)\|_{E^{s+1}}^p] < \infty, \ \forall p \in [1, \infty[. \tag{145}$$

In particular, estimates (144) and (145) applies to $X = \tilde{\mathcal{E}}(L)$.

References

1. Adams, R.A. and Fournier, J.J.F.: *Sobolev Spaces*, Academic Press 2003.
2. Björk, T., Kabanov, Y. and Runggaldier, W.: *Bond market structure in the presence of marked point processes*, Mathematical Finance, **7**, 211–239 (1997).
3. Björk, T., Masi, G., Kabanov, Y. and Runggaldier, W.: *Toward a general theory of bond markets*, Finance and Stochastics, **1**, 141–174 (1997).
4. Björk, T. and Svensson, L.: On the Existence of Finite Dimensional Realizations for Nonlinear Forward Rate Models, Mathematical Finance, **11**, 205–243 (2001).
5. Calderon, A.P.: *Lebesgue spaces of differentiable functions and distributions*, Proc. Symp. Pure Math. IV, AMS 1961, 33–49.
6. Carmona, R. and Tehranchi, M.: *A Characterization of Hedging Portfolios for Interest Rate Contingent Claims*, Ann. Appl. Probab. **14**, 1267–1294 (2004).

6'. Carmona, R. and Tehranchi, M.: *"Interest rate models: an infinite dimensional stochastic approach"*, Springer 2006.
7. Da Prato, G. and Zabczyk, J.: *Stochastic Equations in Infinite Dimensions*, Encyclopedia of Mathematics and its Applications, Cambridge University Press, 1992.
8. De Donno, M. and Pratelli, M.: *On the use of measure-valued strategies in bond markets*, Finance and Stochastics, **8**, 87–109 (2004).
9. Ekeland, I. and Témam, R.: *Convex Analysis and Variational Problems*, Classics in Applied Mathematics 28, SIAM 1999.
10. Ekeland, I. and Taflin, E.: *A Theory of Bond Portfolios*, Ann. Appl. Probab. **15**, 1260–1305 (2005). Also http://arxiv.org/abs/math.OC/0301278
11. Filipović, D.: *Consistency Problems for HJM Interest Rate Models*, Phd thesis, Dep. Math. ETH, Zürch 2000 Preprint 2001.
12. Heath, D.C., Jarrow, R.A. and Morton, A.: *Bond pricing and the term structure of interest rates: a new methodology for contingent claim valuation*, Econometrica, **60**, 77–105 (1992).
13. Hörmander, L.: *The analysis of linear partial differential operators*, Vol. I, Springer-Verlag 1985.
14. Kallianpur, G., and J. Xiong, Stochastic Differential Equations in Infinite Dimensional Spaces, Lecture Notes-Monograph Series, Institute of Mathematical Statistics, 1995.
15. Karatzas, I. and Shreve, S.E.: *Methods of Mathematical Finance*, Applications of Mathematics, Volume 9, Springer-Verlag 1999.
16. Kato, T. *Perturbation Theory for Linear Operators*, Die Grundlehren der mathematischen Wissenschaften, Volume 132, Springer-Verlag, New York 1966.
17. Kramkov, D. and Schachermayer W.: *The Asymptotic Elasticity of Utility Functions and Optimal Investment in Incomplete Markets*, Annals Appl. Probability, **9**, 904–950 (1999).
18. Lax, P.D.: *Functional Analysis*, Wiley-Interscience 2002.
19. Lintner, J.: *The Valuation of Risk Assets and the Selection of Risky Investments in Stock Portfolios and Capital Budgets*, The Review of Economics and Statistics, **47**, 13–37 (1965).
20. Markowitz, H.: *Portfolio Selection*, Jour. Finance, **7**, 77–91 (1952).
21. Merton, R.: *Lifetime Portfolio Selection Under Uncertainty: The Continuous-Time case*, Rev. Economics and Stat. **51**, 247–257 (1969).
22. Merton, R.: *Optimum Consumption and Portfolio Rules in a Continuous Time Model*, Jour. Economic Theory, **3**, 373–413 (1971).
23. Mikulevicius, R. and Rozovskii, B.L.: *Normalized stochastic integrals in topological vector spaces*, Seminaire de Probabilites XXXII, LNM, Springer-Verlag, 1998.
24. Mikulevicius, R. and Rozovskii, B.L.: *Martingale problems for SPDE's*, Stochastic Partial Differential Equations: Six Perspectives, Ed: R. Carmona and BL Rozovskii, AMS, Mathematical Surveys and Monographs, 1999.
25. Musiela, M., *Stochastic PDEs and term structure models*, Journées Internationales de Finance, IGR-AFFI, La Baule, 1993.
26. Nualart D.: *The Malliavin Calculus and Related Topics*, Probability and its Applications, Springer-Verlag, 1991.
27. Pham, H.: *A predictable decomposition in infinite asset model with jumps. Application to hedging and optimal investment*, Stochastics and Stochastic Reports, **5**, 343–368 (2003).

28. Pliska, S.R.: *A stochastic calculus model of continuous trading: optimal portfolios*, Math. Operations Research **11**, 371–382 (1986).
29. Revuz, D. and Yor, M.: *Continuous Martingales and Brownian Motion*, Grundlehren der mathematischen Wissenschaften, Band 293, Sprinter-Verlag 1994.
30. Rudin, W.: *Real and Complex Analysis*, 3rd edition, McGraw-Hill, 1986.
31. Rudin, W.: *Functional Analysis*, 2nd edition, McGraw-Hill, 1991.
32. Rutkowski, R.: *Self-financing Trading Strategies for Sliding, Rolling-horizon, and Consol Bonds*, Math. Finance **5**, 361–385 (1999).
33. Sharp, W.F.: *Capital Asset Prices: A Theory of Market Equilibrium under Conditions of Risk*, The Journal of Finance, **19**, 425–442 (1964).
34. Taflin, E.: *Bond Market Completeness and Attainable Contingent Claims*, Fin. Stoch. **9**, 429–452 (2005). Preprint
    http://arxiv.org/abs/math.OC/0402364
35. Yosida, K.: *Functional Analysis*, Grundlehren der mathematischen Wissenschaften, Band 123, Springer-Verlag 1974.

Note added in the proofs: Since the preparation of this paper, the optimal bond portfolio problem has further been studied in various directions:

1 The reference De Donno, M. and Pratelli, M.: *A theory of stochastic integration for bond markets*, Ann. Appl. Probab. **15**, 2773–2791 (2005) considers the optimal bond portfolio problem in a more general semi martingale bond market. Existence of optimal wealth strategies is established and existence of optimal portfolios is studied.

2 The reference Ringer, N. and Tehranchi, M.: *Optimal portfolio choice in the bond market*, Finance Stoch. **10**, 553–573 (2006) considers the optimal bond portfolio problem in a Markovien setting of local volatility operators with full range and which are globally Lipschitzien. More precisely it is assumed, with our notations and limiting us to the time homogeneous case, that the function $C : E \to \mathcal{HS}(\ell^2, E)$, where $C(f) = f\sigma(f)$, is globally Lipschitzien and that for all strictly positive $f \in E$ the closure of the range $\mathcal{R}(C(f))$ is the subset of elements $g \in E$ such that $g(0) = 0$. If moreover (the unique) market price of risk is globally Lipschitzien then they establish the existence of a solution to the optimal portfolio problem. We note that the proof of this boils down to the verification of properties of the Malliavin derivative of $\ln(\xi_{\bar{T}})$ as was already the case in Theorem 4.5 of [34] (see Theorem 13). We also note that their Gaussian example, of course satisfies our condition (C) of Definition 12, so it is covered by our treatment.

Models for Insider Trading with Finite Utility*

Arturo Kohatsu-Higa

Graduate School of Engineering Sciences,
Osaka University
Osaka 560-8531, Japan
email: kohatsu@sigmath.es.osaka-u.ac.jp

Summary. In these lecture notes we review through simple examples recent results on models for insider trading based on the theory of enlargement of filtrations and on anticipating calculus. In particular, we concentrate on the case of strong type of insiders. That is, insiders that have additional information in the a.s. sense. We explain how to treat the utility maximization problem for insiders in order to obtain models where the utility is finite. In the anticipating framework, we introduce models where the signal of the insider can not be revealed to the small trader even though the insider has an effect on the price (large trader effect).

MSC 2000: primary 91B28; 91B70; 60G48; secondary 60H07

Keywords: Insider trading, Asymmetric information, Enlargement of filtrations, Large traders, Utility maximization

1 Introduction

These lecture notes are the result of a short course delivered as part of a *Cours de formation pour la recherche" at CREST during November-December of 2004*. The goal of these lectures was to introduce young researchers and Ph D students to this area of research. I have decided to keep the spirit of the lectures in these notes although some of the concepts and results may seem to be repetitive through the text or may be well known to the advanced researcher.

The format throughout the text uses examples with concrete answers, in order to give the basic ideas behind the general results and the proofs, rather than the full generality and (sometimes non-trivial) technicalities that can be seen in the research papers mentioned in the references. We do this at the risk of becoming overly trivial but hopefully very clear in what the goals and the techniques are.

These notes are largely self-contained. They do not assume any previous knowledge of mathematical finance. They do assume knowledge of basic stochastic calculus. In particular, I have decided to introduce discrete time models as an

---

* This research was partially supported with grants BFM 2003-03324 and BFM 2003-04294.

approximation of their continuous counterparts for two reasons: The first is to introduce various concepts through their discrete time counterparts. Most importantly, the second reason is to introduce the anticipating type models where the discrete counterpart becomes essential to understand the meaning and the difficulties in this setup.

These notes are designed to be read at two levels. The objective of the main line and first level of discussion is to give a brief overview of the theory. In a second level, we give more details and properties through exercises which may assume more knowledge of stochastic calculus. Most of them are solved at the end of this survey. While I do not considered all of them to be trivial, I think they are useful for the person that wants to deepen his/her knowledge about this area of study. Still, part of the material appears here for the first time.

I have not tried my outmost to add historical notes or to assign credit to each statement appearing in the text and probably many references may be missing. I apologize for all historical imprecisions: my ignorance is only to be blamed.

I would like to thank Nizar Touzi for inviting me to deliver these lectures and for his trust in my completing these notes. Also thanks to all the people who participated in the course with interesting comments, remarks and questions. In particular to Monique Jeanblanc for her careful review of a previous version of these notes.

This article deals with the modelling of financial markets where agents may have different information. That is, agents observe the same prices but their filtration may differ due to some extra information that has been obtained in some other fashion. In such a setting the natural mathematical question is what are the process properties which need to be monitored when such change of filtration takes place. This will take us naturally to the enlargement of filtration problem: That is, does a Wiener process remain a semi-martingale in a larger filtration than the one generated by the process itself? In general, the answer to this question is negative but Jacod's theorem will tell us some cases when the answer is affirmative. The restriction of this theorem is that the only information allowed is the one given by a single random variable. So one of the goals of the present lectures is to try to show a possible route leading to the introduction of flows of information as possible differences between agents.

Another issue that we want to address is that from the financial modeling point of view, there are insiders of two types. The first class of insiders, called unlawful insiders are the agents that trade using the information they possess without any risk. In our theory these insiders will become infinitely rich. This situation corresponds to the typical application of Jacod's theorem.

The second, are insiders which try to hide the fact that they possess this perfect information or either their information does have some risks and therefore our search will be for models where utility is finite. We will try to concentrate on this type of insiders. Nevertheless not only for historical, but also for educational reasons we will have to review the theory of unlawful insiders.

We first introduce the logarithmic utility optimization problem for the classical investor (the so called Merton problem), then we introduce a first example of insider which knows exactly the price at the end of a time interval. This example is a first example of insider where the agent will make an infinite amount of money and there is arbitrage in the model.

Then we present various alternatives in order to obtain finite utilities in models based on the Wiener process. Next we will show that without these modifications the optimal logarithmic utility of the insider is finite in the case of markets with jumps. This effect is due to the high risk imbedded in such models.

Finally, we will move into the study of insiders as large traders. This will require in a natural manner the introduction of anticipating integrals. In particular, we will use the notion of forward integrals as defined by Russo-Vallois. With these models we will show that there is the possibility that the insider influences the prices but still the price information not reveal its information to the traders.

2 The Small Investor Problem

Unless stated otherwise, We will mainly consider the one dimensional setting to simplify the notation. The set-up will be the same for the discrete and continuous time cases. For this, we start considering a one dimensional Wiener process $W = \{W(t), 0 \leq t \leq T\}$ and a compound Poisson process $Z(t) = \sum_{i=0}^{N(t)} X_i$ where N is a simple Poisson process with intensity λ and X_i, $i = 1, \ldots$ are i.i.d. random variable's with density function f such that $\mathbb{E}(e^X) < \infty$. These processes are defined on a complete probability space $(\Omega, \mathcal{F}, \mathbb{P})$. We denote by $(\mathcal{F}_t)_{t \in [0,T]}$ the natural filtration generated by the Wiener and the compound Poisson process and the sets of $\mathbb{P}$-measure zero.

Suppose that $\{S(t); t \in [0, T]\}$ is a positive stochastic process defined on $(\Omega, \mathcal{F}, P)$. This process models the stock price. In the discrete time setting we suppose that trading only takes place at discrete times $0 = t_0 < \ldots < t_n = T$. Then the model for the price will be

$$\log \frac{S(t_{i+1})}{S(t_i)} = \mu(t_i)(t_{i+1} - t_i) + \sigma(t_i)(W(t_{i+1}) - W(t_i)) + (Z(t_{i+1}) - Z(t_i)) \quad (1)$$

where $\mu(t)$ and $\sigma(t) > c > 0$ are two $(\mathcal{F}_t)_{t \in [0,T]}$-adapted *cádlág* bounded processes.

The small agent who has as information at time t all the $\mathcal{F}_t$-measurable random variable's makes his investment decisions based on these random variable's. Another agent, called the insider, observes the same process S but he/she holds as information a bigger filtration $\mathcal{G}$ than the small agent (that is, $\mathcal{G} \supseteq \mathcal{F}$). Both investors are price takers. That is, none of them can change the value of S by using his/her trading strategies.

Nevertheless, insiders are usually large traders and they can change the price dynamics with their trades. This is another aspect of this area of research which is of interest. We neglect this aspect in the first part of our exposition (for more on this, see Section 11)

We start giving a brief exposition of the classical Merton problem for the small investor. That is, what is the optimal portfolio policy that a small investor should choose in order to maximize his/her utility.

Suppose that the small investor starts with a wealth of V_0 units of money and chooses a $\mathcal{F}$-adapted policy $p(t) = (p_0(t), p_1(t))$, where $p_0(t)$ denotes the money invested in his/her bank account at time t that gives an interest rate of r, and $p_1(t)$ denotes the number of shares held at time t. We also suppose that we only allow trading at time points $\{0 = t_0 < t_1 < ... < t_n = T\}$. That is, p is a step process

$$p(t) = \sum_{i=0}^{n-1} p(t_i) 1_{(t_i, t_{i+1}]}(t)$$

An investment policy p is allowed for this investor, besides other technical conditions to be established later, if it is self-financing. That is, if the investor can change $p(t_i)$ to $p(t_{i+1})$ by using the proceeds of selling all his assets. Therefore, the self-financing condition is $p_0(t_0) + p_1(t_0)S(t_0) = V_0$ and for $i = 0, ..., n-2$

$$p_0(t_{i+1}) + p_1(t_{i+1})S(t_{i+1}) = p_0(t_i)e^{r(t_{i+1}-t_i)} + p_1(t_i)S(t_{i+1}). \tag{2}$$

That is, every sequence of possible vectors $p(t_i); i = 0, ..., n-1$ have to satisfy the above $n-1$ equations. In fact with these restrictions the investor has only the freedom to choose $p_1(t_i), i = 0, ..., n-1$. All the other variables $p_0(t_i), i = 0, ..., n-1$ are determined by the self financing equations (2). In fact, we have the following result which is easy to prove.

Lemma 1. *Given an initial wealth V_0, then for every sequence of real values $p_1(t_i)$, $i = 0, ..., n-1$ there exists a unique sequence of values $p_0(t_i), i = 0, ..., n-1$, such that the vectors $p(t_i)$ for $i = 0, ..., n-1$ form a self-financing policy.*

Now we can define the wealth of the investor as $\bar{V}(t) \equiv V^n(t) = p_0(t) + p_1(t)S(t)$ where $p_0(t) = p_0(t_i)e^{r(t-t_i)}$ for $t_i \leq t < t_{i+1}$. With this condition one can rewrite the wealth as the sum of earnings/losses for each interval $[t_i, t_{i+1}]$. That is, if the initial wealth of the small investor is V_0 then

$$\bar{V}(t_i) = V_0 + \sum_{j=0}^{i-1} (\bar{V}(t_{j+1}) - \bar{V}(t_j)) \tag{3}$$

$$= V_0 + \sum_{j=0}^{i-1} \left(p_0(t_j) \left(e^{r(t_{j+1}-t_j)} - 1 \right) + p_1(t_j)(S(t_{j+1}) - S(t_j)) \right)$$

$$= \bar{V}(t_{i-1}) + p_0(t_{i-1}) \left(e^{r(t_i-t_{i-1})} - 1 \right) + p_1(t_{i-1})(S(t_i) - S(t_{i-1}))$$

Note that the above formula is valid without any assumption on the model for S. The next restriction we will impose on the actions of the investor is that we will not allow uncovered losses (sometimes called the no-borrowing condition or tameness condition when written in a more general form). That is, p should satisfy that $\bar{V}(t) \geq 0$ for all $t \in [0, T]$.

Exercise 1. Suppose that the support of the law of $S(t)$ is $(0, \infty)$ for all $t \in [0, T]$. Prove that $\bar{V}(t) \geq 0$ for all $t \in [0, T]$ if and only if $p_0(t) \geq 0$ and $p_1(t) \geq 0$ for all $t \in [0, T]$.

This exercise proves that the no-borrowing restriction is quite strong for markets with transactions in discrete time. Nevertheless this will not be so in continuous time.

Now we have to discuss the possibility for a small trader to make money without any risk. For this, we define discounted wealth process $\hat{V}(t_i) = e^{-rt_i}\bar{V}(t_i)$, $i = 0, ..., n$ and the discounted stock price process $\hat{S}(t) = e^{-rt}S(t)$.

Definition 1. *We say that there is the possibility of arbitrage for the small trader is there exists a policy* $p = (p_0, p_1)$ *such that* $\mathbb{P}(\hat{V}(T) \geq V_0) = 1$ *and* $\mathbb{P}(\hat{V}(T) > V_0) > 0$.

So far we have not made any use of the particular form of S. Now we will use equation (1) to prove that there is no possibility for arbitrage.

Proposition 1. *There is no possibility of arbitrage for the small trader in the space of admissible strategies*

$$\mathcal{A}(T) = \left\{ p(t) = \sum_{i=0}^{n-1} p(t_i)1(t_i \leq t < t_{i+1}); \{p(t)\}_t \text{ adapted and} \right.$$
$$\left. \mathbb{E}\left[|p_0(t_i)| + |p_1(t_i)|\right] < \infty, \ i = 0, ..., n-1 \right\}.$$

Proof:

First note that for $p \in \mathcal{A}(T)$, we have that $\mathbb{E}\left|\widehat{V}(t_i)\right| < \infty$ for all $i = 1, ..., n$. Given that $(W(t_{i+1}) - W(t_i))_{i=0}^{n-1}$ is a Gaussian vector we can perform the following change of measure

$$\frac{d\mathbb{Q}}{d\mathbb{P}} = \exp\left(-\sum_{i=0}^{n-1} \frac{2\theta(t_i)(W(t_{i+1}) - W(t_i)) + \theta(t_i)^2(t_{i+1} - t_i)}{2}\right).$$

with

$$\theta(t_i) = \sigma(t_i)^{-1}\left(\mu(t_i) - r + \lambda(\mathbb{E}(e^X) - 1)\right).$$

With this change we have that $\hat{V}$ is a discrete time martingale in $(\Omega, \mathcal{F}, \mathbb{Q})$. That is,

$$\mathbb{E}_\mathbb{Q}[\hat{V}(t_{i+1})|\mathcal{F}_{t_i}] = \hat{V}(t_i).$$

In fact,

$$\mathbb{E}_\mathbb{Q}[\hat{V}(t_{i+1})|\mathcal{F}_{t_i}] = \hat{V}(t_i)\left(1 + \pi(t_i)\left(\mathbb{E}_\mathbb{Q}\left[\frac{\hat{S}(t_{i+1})}{\hat{S}(t_i)}\bigg|\mathcal{F}_{t_i}\right] - 1\right)\right).$$

Furthermore using the change of variables theorem and the moment function for a Gaussian random variable and a compound Poisson random variable we have that

$$\mathbb{E}_{\mathbb{Q}}\left[\left.\frac{\hat{S}(t_{i+1})}{\hat{S}(t_i)}\right|\mathcal{F}_{t_i}\right] = \mathbb{E}\left[\exp\left(-\frac{\sigma(t_i)^2(t_{i+1}-t_i)}{2} + \sigma(t_i)(\tilde{W}(t_{i+1}) - \tilde{W}(t_i))\right.\right.$$
$$\left.\left. - \lambda(\mathbb{E}(e^X) - 1)(t_{i+1} - t_1) + Z(t_{i+1}) - Z(t_i)\right)\right]$$
$$= 1.$$

Here $\tilde{W}$ denotes a new Gaussian random variable obtained after the change of variables. Therefore $e^{-rT}\mathbb{E}_{\mathbb{Q}}[\bar{V}(T)] = \mathbb{E}_{\mathbb{Q}}[\hat{V}(t_n)] = e^{-rT}V_0$. Therefore arbitrage is not possible because otherwise we will have that $\mathbb{E}_{\mathbb{Q}}(\bar{V}(T)) > V_0 e^{rT}$.

The above proof is just a discrete version of a similar proposition in continuous time. In fact, we propose the following alternative proof.

Exercise 2. Write the limit of the above measure $\mathbb{Q}$ when the partition mesh size $\max\{t_{i+1} - t_i; i = 0, ..., n-1\}$ goes to zero. Use Girsanov's theorem to find the equation satisfied by $\log(\hat{S}(t))$. Prove that $\hat{S}$ is a martingale under this measure.

Exercise 3. Write an argument using stopping times to prove the above proposition without the assumption $\mathbb{E}\left[|p_0(t_i)| + |p_1(t_i)|\right] < \infty$, $i = 0, ..., n-1$.

Exercise 4. Show that the change of measure used in the proof of Proposition 1 is not unique.

Define $T_0 = \inf\{t \in [0, T]; \bar{V}(t) = 0\}$. Then $\bar{V}(s) = 0$ for all $s \geq T_0$. Suppose that $\bar{V}(s) > 0$ for $s \leq t_{i-1}$ then we introduce the change of variables

$$\pi(t) = \frac{p_1(t)S(t)}{\bar{V}(t)}$$

for $t \leq t_{i-1}$. The variable π represents the fraction of wealth invested in stocks. $1 - \pi$ represents the proportion invested in the bank account. Negative values of π are in general interpreted as loans of shares to the investor to be invested in the bank account.

This reduction of variables means that the investor can choose the values of $\pi(t_i)$ for $i = 0, ..., n-1$ so as to maximize his wealth. With these changes of variables one has that

$$\bar{V}(t_i) = \bar{V}(t_{i-1})\left(1 + (1-\pi(t_{i-1}))\left(e^{r(t_i-t_{i-1})}-1\right) + \pi(t_{i-1})\left(\frac{S(t_i)}{S(t_{i-1})}-1\right)\right). \tag{4}$$

Therefore the no-borrowing condition takes the form

$$\pi(t_{i-1})\left(\frac{S(t_i)}{S(t_{i-1})} - e^{r(t_i-t_{i-1})}\right) \geq -e^{r(t_i-t_{i-1})}.$$

One has trivially that if $\pi(t) \in [0, 1]$ for all $t \in [0, T]$ (no borrowing of stocks or from the bank is allowed. Try to link this with Exercise 1) then the no-borrowing condition is always satisfied.

Equation (4) leads to a linear difference equation that can be solved by induction. One obtains that

$$\bar{V}(t_i) = V_0 \prod_{j=0}^{i-1} \left(1 + (1 - \pi(t_j))\left(e^{r(t_{j+1}-t_j)} - 1\right) + \pi(t_j)\left(\frac{S(t_{j+1})}{S(t_j)} - 1\right)\right).$$

$\hat{V}$ satisfies the following linear difference equation which can also be solved by induction:

$$\hat{V}(t_i) = V_0 + \sum_{j=0}^{i-1} \hat{V}(t_j)\pi(t_j)\left(\frac{\hat{S}(t_{j+1})}{\hat{S}(t_j)} - 1\right) \tag{5}$$

$$\hat{V}(t_i) = V_0 \prod_{j=0}^{i-1} \left(1 + \pi(t_j)\left(\frac{\hat{S}(t_{j+1})}{\hat{S}(t_j)} - 1\right)\right).$$

We will now take limits in the above arguments and in the same way as it is done in any introductory course on stochastic calculus, one obtains the Itô stochastic integral and the above arguments about the non-existence of arbitrage can be carried out similarly.

In fact, one has that S is a semi-martingale and that, if we consider π to be a predictable process then one can consider the limit of the equation (5). We obtain the wealth equation for continuous time trading (we will return to this issue in sections 9 and 12). This gives

$$\hat{V}(t) = V_0 + \int_0^t \frac{\pi(s-)\hat{V}(s-)}{\hat{S}(s-)} d\hat{S}(s). \tag{6}$$

Note that there is a slight abuse of notation as we are using the same notation for the approximative wealth in equation (5) but it should be clear from the context if we are talking about the discrete time approximation or the continuous one above.

Exercise 5. Prove that the limit of the sequence $\hat{V}$ is the solution of the above linear equations under mild regularity conditions on S and π.

This is a linear equation in V which can be explicitly solved. In the case that S has jumps the next calculation is a bit different (see Section 9). For this reason and to compute explicitly an optimal portfolio we take a particular continuous model for the stock price S. In what follows we further simplify our model and assume that the stock price is the geometric Brownian motion given by

$$S(t) = S(0)\exp\left(\int_0^t \left(\mu(s) - \frac{1}{2}\sigma^2(s)\right)ds + \int_0^t \sigma(s)dW_s\right)$$

where μ is the mean rate of return and $\sigma \geq c > 0$ is the volatility of the stock price which are uniformly bounded and adapted to the filtration generated by the

Wiener process completed with respected to P. The process S also satisfies the linear equation

$$S(t) = S(0) + \int_0^t \mu(s)S(s)ds + \int_0^t \sigma(s)S(s)dW(s).$$

In such a case, equation (5) becomes

$$\hat{V}(t) = V_0 + \int_0^t \hat{V}(s)(\mu(s) - r)\pi(s)ds + \int_0^t \sigma(s)\pi(s)\hat{V}(s)dW(s). \quad (7)$$

This stochastic linear equation has an explicit solution given by

$$\hat{V}(t) = V_0 \exp\left(\int_0^t \left((\mu(s) - r)\pi(s) - \frac{1}{2}\sigma^2(s)\pi(s)^2 \right) ds + \int_0^t \sigma(s)\pi(s)dW(s) \right). \quad (8)$$

Note that the restriction on the policy $\pi \in [0,1]$ has disappeared. For any value of π, $V_t > 0$ for any $t \in [0,T]$. This will be further explained in section 9. The small investor desires to optimize his/her portfolio policy by considering the problem

$$\max_{\pi \in \mathcal{F}} \mathbb{E}\left[\log(\hat{V}(t)) \right] \quad (9)$$

where the logarithmic function is used as a utility function. This particular utility function is used because the calculations to follow will become explicit. Note the inverse relationship between the logarithmic and the exponential function in (8) and (9). In what follows $\pi \in \mathcal{F}$ is a shorthand for π is $\mathcal{F}$-predictable process.

Other utility functions can also be used (see Exercise 9). Also note that if in the discrete time model we had $T_0 < t$ with positive probability then the $\mathbb{E}(\log(\hat{V}(t))$ is defined as $-\infty$ therefore not giving the optimal value (Exercise: propose a portfolio with bigger value than $-\infty$ and compute its expected utility explicitly.). We now give an informal argument to obtain the optimal portfolio. A formal approach is given in the exercises.

To find the optimal portfolio solving the Merton problem (9) we note that if $\mathbb{E} \int_0^t \pi(s)^2 ds < \infty$ then

$$\mathbb{E}\left[\log(\hat{V}(t)) \right] = \log(V_0) + \mathbb{E}\left[\int_0^t \left((\mu(s) - r)\pi(s) - \frac{1}{2}\sigma^2(s)\pi(s)^2 \right) ds \right.$$
$$\left. + \int_0^t \sigma(s)\pi(s)dW(s) \right]$$
$$= \log(V_0) + \mathbb{E}\left[\int_0^t \left((\mu(s) - r)\pi(s) - \frac{1}{2}\sigma^2(s)\pi(s)^2 \right) ds \right],$$

as $\mathbb{E}\left[\int_0^t \sigma(s)\pi(s)dW(s) \right] = 0$. This expression says that the utility of the portfolio π is determined by the value of $f_s(\pi) = (\mu(s) - r)\pi - \frac{1}{2}\sigma^2(s)\pi^2$. This is a strictly concave function with domain $\pi \in \mathbb{R}$. Therefore its maximum value is obtained by differentiation. That is, $f_s'(\pi) = (\mu(s) - r) - \sigma^2(s)\pi = 0$ and therefore the maximum value is given by $\pi^*(s) = \frac{\mu(s)-r}{\sigma^2(s)}$. That is, the solution of Merton's problem says that the small investor has to keep the Sharpe ratio of his investments in stocks constant through the life of his investment.

Exercise 6. Compute the maximum utility given by the optimal portfolio $\pi^*(s) = \frac{\mu(s)-r}{\sigma^2(s)}$ and write the explicit expression in the case that μ and σ are constants.

Exercise 7. Prove that within the class

$$\mathcal{A}(t) = \{\pi;\ \pi \text{ is } \mathcal{F} - \text{adapted and, } \mathbb{E}\left[\int_0^t \pi(s)^2 ds\right] < \infty\},$$

the portfolio $\pi(s) = \frac{\mu(s)-r}{\sigma^2(s)}$ is the optimal portfolio for the problem

$$\max_{\pi \in \mathcal{A}} \mathbb{E}\log(\hat{V}(t)).$$

Exercise 8. : Prove that the above portfolio is also optimal in the class $\{\pi :\ \pi$ is $\mathcal{F}$-adapted, $\int_0^t \pi(s)^2 ds < \infty$ a.s. and $\mathbb{E}\left[\log(\hat{V}(t))\right] < \infty\}$.

Exercise 9. Use Girsanov's theorem to find the optimal portfolio for the problem

$$\max_{\pi \in \mathcal{A}_\theta} \mathbb{E}\left[\hat{V}(t)^\theta\right] \text{ for } \theta \in (0,1).$$

Show that the problem has no finite solution if $\theta \notin (0,1)$. Define the set $\mathcal{A}_\theta$.

An issue that is important to discuss before continuing with the optimization problem for the insider is the issue of arbitrage: Can the investor make money without risking any loss?

Definition 2. *We say that there is arbitrage in the market if there exists a self-financing portfolio π with $\mathbb{P}(\hat{V}(T) \geq V_0) = 1$ and $\mathbb{P}(\hat{V}(T) > V_0) > 0$.*

A well known related theorem is

Theorem 1. *If there exists a measure $\mathbb{Q} \sim \mathbb{P}$ such that under $\mathbb{Q}$, $\hat{S}$ is a martingale then there is no arbitrage for self-financing portfolios π satisfying*

$$\int_0^T (\pi(s)V(s))^2 ds < \infty \quad a.s..$$

Exercise 10. Prove that the assumption of the previous theorem is satisfied with

$$\frac{d\mathbb{Q}}{d\mathbb{P}} = \exp\left(-\frac{1}{2}\int_0^t \left(\frac{\mu(s)-r}{\sigma(s)}\right)^2 - \int_0^t \frac{\mu(s)-r}{\sigma(s)} dW(s)\right).$$

Now we start to consider the same utility maximization problem but for the insider agent.

3 The Portfolio Problem for the Insider. A Toy Example

To simplify our discussion suppose that the insider has as additional information the value of a random variable $I = S(T) \in \mathcal{F}_T$. This is equivalent to knowing W_T. This example is a "toy" example because it is hard to think of a real example where an insider knows exactly the value of $S(T)$. Nevertheless this example will give us basic information that will be important in what follows.

Exercise 11. (Lévy's Theorem) Let M be a continuous local martingale in a filtration $\mathcal{F}$ such that $\langle M \rangle_t = t$. Then M is an $\mathcal{F}$-Wiener process. Hint: Use Itô's formula to prove that
$$\mathbb{E}\left[\exp\left(i\theta(M_t - M_s)\right)|\mathcal{F}_s\right] = \exp\left(-\frac{\theta^2(t-s)}{2}\right).$$
Therefore the increment $M_t - M_s$ is conditionally independent of $\mathcal{F}_s$ and is a $N(0, t-s)$ random variable.

The natural filtration for the insider is $\mathcal{G}_t = \mathcal{F}_t \vee \sigma(I)$ (the smallest filtration satisfying the usual conditions which contains $\mathcal{F}$ and $\sigma(I)$).

Exercise 12. Prove that $\mathcal{G}_t = \cap_{\varepsilon \geq 0} \sigma\left(\mathcal{F}_{t+\varepsilon} \cup \sigma(I)\right)$.

In general, it is necessary to take the above intersection as the following simple example shows.

Exercise 13. (Barlow-Perkins, also proposed in Williams [46], page 48) Let Y_i be iid Bernoulli random variable's with values $\{-1, 1\}$. Define $X_n = \prod_{i=0}^{n} Y_i$ define $\mathcal{A} = \sigma(Y_i; i \geq 1)$, $\mathcal{B}_n = \sigma(X_r; r > n)$. Prove that $Y_0 \in \mathcal{B}_n \vee \sigma(\mathcal{A})$ for all n but Y_0 is independent of $(\cap \mathcal{B}_n) \vee \sigma(\mathcal{A})$.

Under the enlarged filtration $\mathcal{G}$ the process W is no longer a Wiener process but still a continuous semimartingale. Therefore we are interested in computing the semimartingale decomposition of $W = M + A$ where M is a $\mathcal{G}$-local martingale and A is a $\mathcal{G}$-adapted process of bounded variation. As the quadratic variation is still $\langle W \rangle_t^{\mathcal{G}} = t$ then, by Lévy's theorem M will be a $\mathcal{G}$-Wiener process.

We denote by P_t the regular conditional probability measure of $X = W(T)$ with respect to $\mathcal{F}_t$. That is, $P_t(dx) = P(W(T) \in dx|\mathcal{F}_t)$. Explicitly in our case
$$dP_t(x) = p_{T-t}(W_t, x)dx = \frac{1}{\sqrt{2\pi(T-t)}}\exp\left(-\frac{(x-W(t))^2}{2(T-t)}\right)dx$$

Theorem 2. *The semi-martingale decomposition of W in $\mathcal{G}$ is given by*
$$W(t) = \widehat{W}(t) + \int_0^t \frac{W(T) - W(u)}{T - u} du \tag{10}$$
where $\widehat{W}$ is a Wiener process in the initially enlarged filtration $\mathcal{G}$ in the interval $[0, T]$.

Proof:

We compute $\mathbb{E}([W(t) - W(s)|\mathcal{G}_s]$ for $t \in [0,T)$ and $t > s$. The objective is to obtain an expression of bounded variation which will become the compensator of the process W in the enlarged filtration.

Alternatively we will compute for a measurable bounded function f and a $\mathcal{F}_s$ measurable random variable h_s,

$$\mathbb{E}\left[(W(t) - W(s))f(W(T))h_s\right] = \mathbb{E}\left[(W(t) - W(s))\int f(x)dP_t(x)h_s\right]$$
$$= \mathbb{E}\left[(W(t) - W(s))\int f(x)p_{T-t}(W_t,x)dxh_s\right].$$

Applying Itô's formula to $(W(u) - W(s))p_{T-u}(W_u, x)$, $u \in [s,t]$ and using that p_{T-t} solves the heat equation $\partial_t p_{T-t}(y,x) + \frac{1}{2}\partial_y^2 p_{T-t}(y,x) = 0$, we have

$$\mathbb{E}\left[(W(t) - W(s))f(W(T))h_s\right]$$
$$= \mathbb{E}\left[\int f(x)\left(\int_s^t p_{T-u}(W_u,x) + (W(u) - W(s))\partial_y p_{T-u}(W(u),x)dW(u)\right.\right.$$
$$\left.\left.+ \int_s^t \partial_y p_{T-u}(W(u),x)du\right)dxh_s\right]$$
$$= \mathbb{E}\left[\int f(x)\int_s^t \partial_y p_{T-u}(W(u),x)dudxh_s\right]$$
$$= \mathbb{E}\left[\int_s^t \int f(x)\partial_y \log(p_{T-u}(W(u),x))p_{T-u}(W(u),x)dxduh_s\right]$$
$$= \mathbb{E}\left[f(W(T))\int_s^t \partial_y \log(p_{T-u}(W(u),W(T)))duh_s\right].$$

Therefore by a density argument, one has

$$\mathbb{E}\left[W(t) - W(s) - \int_s^t \partial_y \log(p_{T-u}(W(u),W(T)))du\bigg|\mathcal{G}_s\right] = 0.$$

As $\partial_y \log(p_{T-u}(W(u),W(T))) = \frac{W(T)-W(u)}{T-u} \in \mathcal{G}_u$ then $\hat{W}(t) = W(t) - \int_0^t \frac{W(T)-W(u)}{T-u}du$ is a $\mathcal{G}$-continuous martingale with $\langle \hat{W}\rangle_t = \langle W\rangle_t = t$ and therefore by Lévy's theorem one has that $\hat{W}$ is a $\mathcal{G}$-Wiener process in $[0,T)$. We then define $\hat{W}(T) = \lim_{t\to T} \hat{W}(t)$ and all above properties follow for the closed interval $[0,T]$ as $\mathbb{E}\left[\int_0^T \left|\frac{W(T)-W(u)}{T-u}\right|du\right] < \infty$.

Exercise 14. Prove that $\mathbb{E}\left[\int_0^T \left|\frac{W(T)-W(u)}{T-u}\right|^r du\right] < \infty$ if and only if $r \in [0,2)$ while $\mathbb{E}\left[\left|\int_0^T \frac{W(T)-W(u)}{T-u}du\right|^r\right] < \infty$ for all $r \geq 0$.

The methodology shown in Theorem 2 can be generalized much further in various directions. Some of them are shown in the following exercises.

Exercise 15. Prove that $\{\hat{W}(t); t \in [0,T]\}$ and $W(T)$ are independent random variables.

Exercise 16. (Föllmer-Imkeller, [15]) Prove that there exists a measure $\mathbb{Q} \sim \mathbb{P}$ such that, under $\mathbb{Q}$, $W(t)$ and $W(T)$ are independent.

Exercise 17. (Harnesses) For $s \leq a < b \leq T$, prove that
$$\mathbb{E}\left[\frac{W_a - W_b}{a - b}\bigg|\mathcal{G}_s\right] = \frac{W_T - W_s}{T - s}.$$

Exercise 18. Find the semimartingale decomposition of the Wiener process if $I = \int_0^T h(r) W_r dr$ for a deterministic function $h \in L^2[0,T]$. Impose conditions on h so that the semimartingale decomposition in the enlarged filtration is integrable.

Exercise 19. Suppose that $I = X(T)$ where $X(\cdot)$ is the solution of the stochastic differential equation
$$X(t) = x_0 + \int_0^t b(X(s))ds + \int_0^t \sigma(X(s))dW(s).$$
Here b and $\sigma : \mathbb{R} \to \mathbb{R}$ are smooth functions. Then X is a Markov process. Furthermore suppose that the transition density $p_t(y,x)$ exists, is smooth and satisfies the parabolic partial differential equation
$$\partial_t p_t(y,x) = b(y)\partial_y p_t(y,x) + \frac{1}{2}\sigma(y)^2 \partial_x^2 p_t(y,x) \tag{11}$$
and the Gaussian bounds
$$ct^{-a}\exp\left(-\frac{|x-y|^2}{ct}\right) \leq p_t(x,y) \leq Ct^{-a}\exp\left(-\frac{|x-y|^2}{Ct}\right)$$
$$|\partial_x p_t(x,y)| \leq Ct^{-a}\exp\left(-\frac{|x-y|^2}{Ct}\right) \tag{12}$$
for $t \in (0,T]$, $a, c, C > 0$ and $x, y \in \mathbb{R}$. Prove that
$$\hat{W}(t) = W(t) - \int_0^t \partial_x \log(p_{T-u}(X(u), X(T)))du$$
is a $(\mathcal{G}_t)_{t \in [0,T)}$-Wiener process. Give conditions to obtain a Wiener process in $[0,T]$. In the terminology used in Malliavin calculus, $\partial_x \log(p(u,x,y))$ is called the logarithmic derivative of the density.

Exercise 20. (Ankirchner) Let $I = |W(T)|$. Prove that the compensator in this case is given by
$$\int_0^t W(s)\left(-\frac{1}{T-s} + \frac{|W(T)|}{|W(s)|(T-s)}\tanh\left(\frac{|W(s)W(T)|}{T-s}\right)\right)ds.$$

Exercise 21. Solve equation (10) in $W(t)$ to obtain that for $t < T$

$$\frac{W_T - W_t}{T - t} = \frac{W_T}{T} - \int_0^t \frac{1}{T - r} d\widehat{W}_r.$$

Following the above theorem we have that the evolution of the stock price for the insider is

$$S(t) = S(0) \exp\left(\left(\int_0^t \mu(s) - \frac{1}{2}\sigma^2(s)\right) ds + \int_0^t \sigma(s)\widehat{W}(s) \right.$$
$$\left. + \int_0^t \sigma(u) \frac{W_T - W_u}{T - u} du\right).$$

Note that the values of S observed by both the small investor and the insider are the same. Given that the insider has some extra information about the driving process, this information should give him/her an advantage that can be expressed through the amount of money that he can make using this extra information. To quantify this mathematically, let $\pi = \{\pi(s); 0 \le s \le T\}$ be an $\mathcal{G}$-adapted process that denotes, as before, the proportion of the total wealth that the insider invests in stocks. Then the discounted wealth process $\widehat{V}$ satisfies the equation (7). The solution of this linear equation is

$$\widehat{V}(t) = V_0 \exp\left(\int_0^t \left((\mu(s) - r + \sigma(s)\frac{W_T - W_s}{T - s})\pi(s) - \frac{1}{2}\sigma^2(s)\pi(s)^2\right) ds\right.$$
$$\left. + \int_0^t \sigma(s)\pi(s) d\widehat{W}(s)\right).$$

Using the fact that

$$\mathbb{E}\left[\int_0^t \sigma(s)\pi(s) d\widehat{W}(s)\right] = 0,$$

we can compute the average utility and obtain

$$\mathbb{E}\left[\log(\widehat{V}(t))\right]$$
$$= \log(V_0) + \int_0^t \mathbb{E}\left[\left(\mu(s) - r + \sigma(s)\frac{W_T - W_s}{T - s}\right)\pi(s) - \frac{1}{2}\sigma^2(s)\pi(s)^2\right] ds.$$

where π is $\mathcal{G}$-adapted. As before, it is enough to find the maximum of the function within the expectation. In the case of the insider agent we have to maximize

$$f_s(\pi) = \left(\mu(s) - r + \sigma(s)\frac{W_T - W_s}{T - s}\right)\pi - \frac{1}{2}\sigma^2(s)\pi^2.$$

Note that this function is concave and the optimum is given for

$$\hat{\pi}(s) = \frac{1}{\sigma^2(s)}\left(\mu(s) - r + \sigma(s)\frac{W_T - W_s}{T - s}\right).$$

The maximum utility is

$$\mathbb{E}\left[\log(\hat{V}^*(t))\right] = \log(V_0) + \int_0^t \mathbb{E}\left[\frac{1}{2\sigma^2(s)}\left(\mu(s) - r + \sigma(s)\frac{W_T - W_s}{T - s}\right)^2\right] ds$$

$$= \log(V_0) + \int_0^t \mathbb{E}\left[\frac{(\mu(s) - r)^2}{2\sigma^2(s)}\right] ds + \frac{1}{2}\log\left(\frac{T}{T - t}\right).$$

Exercise 22. Formalize the above argument defining explicitly the class $\mathcal{I}$ of admissible strategies and prove that the optimal portfolio is the one found in the previous informal calculation.

The difference in utility between the insider and the small agent is $\frac{1}{2}\log\left(\frac{T-t}{T}\right)$ which does not depend on the parameters μ or σ. This is easily explained by the fact that the information that the insider possesses does not depend on these parameters. In more complex models this is not expected although a clear example is not available yet (see Exercise 34). This result also shows that the additional information of the insider provides him/her with an infinite wealth as $t \to T$. This is due to the fact that the insider can make use of his information in all the oscillations around the value of $S(T)$. Therefore this market with a small agent and an insider allows for arbitrage in the interval $[0, T]$. This is natural, given the good quality of the extra information.

For any time interval $[0, t]$, $t < T$, it is also clear that the underlying model for the insider becomes a geometric Wiener process with random drift. Therefore all the classical mathematical financial theory applies. For example, option prices (which are the same as Black Scholes prices), hedging strategies, equilibrium theory, portfolio choice theory, etc.

Exercise 23. Find an explicit arbitrage strategy for the portfolio of the insider in $[0, T]$.

This model in the time interval $[0, T]$ corresponds to insider trading in the sense of an agent holding information which is prohibited by law. This point of view may be somewhat interesting if one is interested in detecting this type of insiders.

Instead, we are interested in obtaining models where insiders have finite utility. In fact most insiders are lawful actors in financial markets. So in these notes we will rather discuss how to modify these models so as to obtain finite utilities for insiders. This line will be developed in future sections. First, we will study the mathematical theory around these developments which are based on enlargement of filtrations. Before that, let us propose some exercises.

Exercise 24. Find the optimal strategy for the insider for the utility function $U(x) = x^\theta$ for $\theta \in (0, 1)$.

Note that if one knows the value of W_T then W becomes a Brownian bridge.

Exercise 25. Use an alternative expression for the Brownian bridge (e.g. see page 300 in Protter's book) to prove that the process X defined as

$$X(t) = a\left(1 - \frac{t}{T}\right) + b\frac{t}{T} + (T-t)\int_0^t \frac{1}{T-s}dB(s)$$

is a Brownian bridge starting from $X_0 = a$ and ending at $X(T) = b$ for $t = T$. Compare this exercise with Exercise 21 and point out what is the difference with the present approach?

Exercise 26. Compute the optimal logarithmic utility for the insider in the interval $[0, t]$ conditioned on the value of $S(T) = x$ for $t < T$.

Exercise 27. Define $u^{\mathcal{F}}(t, V_0)$ and $u^{\mathcal{G}}(t, V_0)$ the optimal logarithmic utility (in $[0, t]$) of the small investor and the insider respectively. Obviously as $\mathcal{G}$ is a filtration bigger than $\mathcal{F}$ we have that $u^{t,\mathcal{F}}(x) \leq u^{t,\mathcal{G}}(x)$. Define the fair price of the information $I = W(T)$ as the real number $\rho_t(V_0, I)$ such that $u^{\mathcal{F}}(t, V_0) = u^{\mathcal{G}}(t, V_0 - \rho_t(V_0, I))$. Prove that $\rho_t(V_0, I) = V_0\left(1 - \sqrt{1 - \frac{t}{T}}\right)$. The fair price of information is a fraction that corresponds to how much less money we need in order to make the same amount of wealth as the small trader. Note that $\rho_T(V_0, I) = V_0$ which reflects the fact that there is arbitrage at time T.

Generalizing the previous example leads us to the theory of initial enlargement of filtration and in particular to Jacod's theorem.

4 Jacod's Theorem

As before, let I be a $\mathcal{F}_T = \sigma(W(s); s \leq T)$ adapted random variable and let $P_t(dx)$ denote the regular conditional probability of I given $\mathcal{F}_t$. Define $\mathcal{G}_t$ as the smallest filtration that includes $\mathcal{F}_t$ and $\sigma(I)$ i.e. $\mathcal{G} = \mathcal{F} \vee \sigma(I)$.

Theorem 3. *If there exists a deterministic measure $\eta(dx)$ such that $P_t(dx) \ll \eta(dx)$, then W is a $\mathcal{G}$-semimartingale.*

Exercise 28. Use the ideas given in Theorem 2 to sketch the proof of the above theorem if we further assume that $\mathbb{E}\left[\int_0^t |\alpha(u)|\, du\right] < \infty$. Show that

$$W(t) = \widehat{W}(t) + \int_0^t \alpha(u)du$$

is a $\mathcal{G}$-Wiener process, where

$$\alpha(u) = \frac{\frac{d}{du}\left\langle W, \frac{dP.(I)}{d\eta}\right\rangle_u}{\frac{dP_u(I)}{d\eta}}.$$

In fact, it is better to try to generalize the above result without using the reference measure η. This will become clear later. Before doing the generalization we want to introduce a further example:

Exercise 29. Given a filtration $\mathcal{G} \supseteq \mathcal{F}$. Assume that W is a semimartingale in $\mathcal{G}$ and that its decomposition is given by

$$W_t = \widehat{W}_t + \int_0^t \alpha_u du$$

with $\mathbb{E}\left[\int_0^t |\alpha_u|^2 du\right] < +\infty$ a.s. for all $t < T$. Prove that the optimal portfolio is

$$\hat{\pi}(s) = \frac{\mu(s) - r}{\sigma^2(s)} + \frac{\alpha_s}{\sigma(s)}$$

and the additional utility of the insider in the interval $[0, t]$ is given by $\frac{1}{2} \int_0^t \mathbb{E}\left[\alpha_u^2\right] du$.

We also remark that Jacod's Theorem as stated in [26] is in the semimartingale framework and with slightly less stringent conditions.

5 An Approach Using the Integration by Parts Formula

Jacod's Theorem (see Jacod, [26]) is the basic tool to characterize when a semimartingale keeps this property in a enlarged filtration and what its new decomposition is. Another way to approach this problem is using the integration by parts formula of Malliavin calculus. We show this in the next theorem as a way of illustration. For this reason we do not give full details in this section.

We denote by D the stochastic derivative and by δ_a^b the Skorohod integral from a to b, the dual operator of D in $L^2([a, b] \times \Omega)$. For a general introduction to Malliavin calculus and the notation used here, see Nualart [37]. Now, we define a general integration by parts formula.

Definition 3. *Let (I, Y) be a random vector, measurable with respect to $\mathcal{F}_T$, such that there exists a random variable $H_{u,T}(I, Y) \in L^2(\Omega)$ with the property that for any $f \in C_b^1$ and $A \in \mathcal{F}_u$ it satisfies that*

$$\mathbb{E}\left[f'(I) Y 1_A\right] = \mathbb{E}\left[f(I) H_{u,T}(I, Y) 1_A\right].$$

Then we say that there is an integration by parts formula for (I, Y) in $[u, T]$ and $H_{u,T}$ is the weight associated with the integration by parts formula for (I, Y) in $[u, T]$.

Theorem 4. *Let I be a random variable such that there is an integration by parts formula for $(I, D_u I)$ in $[u, T]$. Then W is a semi-martingale in the enlarged filtration $\{\mathcal{F}_s \vee \sigma(I); s \leq T\}$ and*

$$W_t = \widehat{W}_t + \int_0^t \mathbb{E}\left[H_{u,T}(I, D_u I) | \mathcal{F}_u \vee \sigma(I)\right] du$$

where $\widehat{W}$ is a Wiener process in the enlarged filtration.

Proof:

First, we compute $\mathbb{E}[W_t | \mathcal{F}_s \vee \sigma(I)]$. For this consider for a smooth bounded function f and $A \in \mathcal{F}_s$

$$\mathbb{E}[(W_t - W_s)f(I)1_A] = \mathbb{E}\left[\int_s^t D_u f(I) du 1_A\right]$$

$$= \int_s^t \mathbb{E}[f'(I) D_u I 1_A] du$$

$$= \int_s^t \mathbb{E}[f(I) H_{u,T}(I, D_u I) 1_A] du$$

$$= \mathbb{E}\left[f(I) \int_s^t \mathbb{E}[H_{u,T}(I, D_u I) | \mathcal{F}_u \vee \sigma(I)] du 1_A\right].$$

In conclusion we have that

$$\mathbb{E}([W_t | \mathcal{F}_s \vee \sigma(I)]) = W_s + \int_s^t \mathbb{E}[H_{u,T}(I, D_u I) | \mathcal{F}_u \vee \sigma(I)] du.$$

This means that W is a semi-martingale in the filtration $\{\mathcal{F}_s \vee \sigma(I); s \leq T\}$ and one concludes that its martingale part has to be a Wiener process due to the Lévy characterization theorem.

Under certain conditions we have that the standard integration by parts formula is satisfied if there exists a process h such that

$$\frac{D_u I h(\cdot)}{\int_u^T D_v I h(v) dv} \in Dom(\delta_u^T)$$

in which case

$$H_{u,T}(I, D_u I) = \int_u^T \frac{D_u I h(s)}{\int_u^T D_v I h(v) dv} dW_s.$$

This is the formula of integration by parts using h as a localization function. The usual integration by parts formula is obtained using $h = DI$ which leads to the Malliavin covariance matrix, $\int_u^T (D_u I)^2 du$, in the denominator of the above Skorohod integral. In such a case, if $I \in \mathbb{D}^{2,8}$ and $\mathbb{E}\left[\left(\int_u^T (D_u I)^2 du\right)^{-8}\right] < \infty$ then the integration by parts formula is satisfied for $(X, D_u X)$.

Exercise 30. Obtain Theorem 10 as a consequence of the previous theorem.

6 Weak Information Approach

It is clear that we have the following probability decomposition that represents the Brownian bridge

$$\mathbb{P}(W \in A) = \int \mathbb{P}(W \in A | W(T) = x) P^{W(T)}(dx).$$

That is, under the law $\mathbb{P}(W \in \cdot | W(T) = x)$, W is described by a Brownian bridge. Fabrice Baudoin (see [5]) used this property to construct a "Brownian bridge" where the law of the final random variable is any measure $\nu \sim P^{W(T)}$. That is we define

$$\mathbb{P}^\nu(W \in A) = \int \mathbb{P}(W \in A | W(T) = x) \nu(dx).$$

We denote by $\mathbb{E}^\nu$ the expectation with respect to $\mathbb{P}^\nu$. To obtain the analogue of Theorem 2 in this setting, we compute the Radon Nikodym derivative $\frac{d\mathbb{P}^\nu}{d\mathbb{P}}\big|_{\mathcal{F}_t}$. Let h_t be an $\mathcal{F}_t$-measurable random variable, then

$$\mathbb{E}^\nu[h_t] = \mathbb{E}\left[\frac{d\nu}{dP^{W(T)}}(W(T))h_t\right] = \mathbb{E}\left[\mathbb{E}\left[\frac{d\nu}{dP^{W(T)}}(W(T))\bigg|\mathcal{F}_t\right]h_t\right].$$

Therefore

$$\frac{d\mathbb{P}^\nu}{d\mathbb{P}}\bigg|_{\mathcal{F}_t} = \mathbb{E}\left[\frac{d\nu}{dP^{W(T)}}(W(T))\bigg|\mathcal{F}_t\right].$$

Furthermore using the same technique as in Theorem 2, we have

Theorem 5. *We have* $W(t) = \hat{W}(t) + \int_0^t \alpha(u)du$ *for some Wiener process* $\hat{W}$ *on the stochastic basis* $(\Omega, (\mathcal{F}_t)_{t\in[0,T]}, \mathbb{P}^\nu)$ *and*

$$\alpha(u) = \mathbb{E}\left[\frac{W(T) - W_u}{T - u}\frac{d\nu}{dP^{W(T)}}(W(T))\bigg|\mathcal{F}_u\right]\mathbb{E}\left[\frac{d\nu}{dP^{W(T)}}(W(T))\bigg|\mathcal{F}_u\right]^{-1}$$

Proof:

As before we need to compute $\mathbb{E}^\nu[W(t) - W(s) | \mathcal{F}_s]$. Instead we compute for h_s a $\mathcal{F}_s$ measurable random variable

$$\mathbb{E}^\nu[(W(t) - W(s))h_s] = \mathbb{E}\left[\frac{d\nu}{dP^{W(T)}}(W(T))(W(t) - W(s))h_s\right]$$

$$= \mathbb{E}\left[\frac{d\nu}{dP^{W(T)}}(W(T))\int_s^t \frac{W(T) - W(u)}{T - u}du\, h_s\right]$$

$$= \mathbb{E}^\nu\left[\int_s^t \frac{W(T) - W(u)}{T - u}du\, h_s\right]$$

$$= \mathbb{E}^\nu \left[\int_s^t \mathbb{E}^\nu \left[\frac{W(T) - W(u)}{T - u} \bigg| \mathcal{F}_u \right] duh_s \right]$$

$$= \mathbb{E}^\nu \left[\int_s^t \mathbb{E} \left[\frac{W(T) - W(u)}{T - u} \frac{d\nu}{dP^{W(T)}}(W(T)) \bigg| \mathcal{F}_u \right] \right.$$
$$\left. \mathbb{E} \left[\frac{d\nu}{dP^{W(T)}}(W(T)) \bigg| \mathcal{F}_u \right]^{-1} duh_s \right].$$

where the last equality follows directly if we consider

$$\mathbb{E}^\nu \left[\mathbb{E} \left[\frac{W(T) - W(u)}{T - u} \frac{d\nu}{dP^{W(T)}}(W(T)) \bigg| \mathcal{F}_u \right] \mathbb{E} \left[\frac{d\nu}{dP^{W(T)}}(W(T)) \bigg| \mathcal{F}_u \right]^{-1} h_u \right]$$

$$= \mathbb{E} \left[\frac{d\nu}{dP^{W(T)}}(W(T)) \mathbb{E} \left[\frac{W(T) - W(u)}{T - u} \frac{d\nu}{dP^{W(T)}}(W(T)) \bigg| \mathcal{F}_u \right] \right.$$
$$\left. \mathbb{E} \left[\frac{d\nu}{dP^{W(T)}}(W(T)) \bigg| \mathcal{F}_u \right]^{-1} h_u \right]$$

$$= \mathbb{E} \left[\mathbb{E} \left[\frac{W(T) - W(u)}{T - u} \frac{d\nu}{dP^{W(T)}}(W(T)) \bigg| \mathcal{F}_u \right] h_u \right]$$

$$= \mathbb{E} \left[\frac{W(T) - W(u)}{T - u} \frac{d\nu}{dP^{W(T)}}(W(T)) h_u \right]$$

$$= \mathbb{E}^\nu \left[\frac{W(T) - W(u)}{T - u} h_u \right].$$

The rest of the argument follows as in Theorem 2.

Other cases where ν is not equivalent to the Lebesgue measure can also be treated on a case by case basis. The idea remains the same. The application of this result is also clear. The insider has the information that the law of the final random variable $W(T)$ is ν. Then this restriction means that the insider "knows" the final law of the process $W(T)$ while the small investor thinks it is a Wiener process.

Exercise 31. Find the optimal portfolio and the maximal logarithmic utility of the insider if the law ν is given by a $N(\mu, \sigma^2)$. Find out if the utility is finite in such a case.

Exercise 32. Suppose that

$$\mathbb{E}^\nu \left[\left(\frac{\partial}{\partial x} \log \left(\frac{d\nu}{dP^{W(T)}} \right) \right)^2 (W(T)) \right] < \infty.$$

Prove that the optimal logarithmic utility for the insider possessing the information that the law of $W(T)$ is ν, is finite. Prove that the measure ν given by a $N(\mu, \sigma^2)$ satisfies the above condition.

7 Entropy Characterization of Additional Information

Here we introduce the characterization of additional utility of the insider in the interval $[0, t]$ as the expectation of the entropy of the conditional measure of $W(T)$ with respect to $\mathcal{F}_t$ obtained in Amendiger et. al. [2]. To make things simpler suppose that $P_t \sim P^I$ and that $p_t(x) = \frac{dP_t}{dP^I}(x) > 0$ a.s. (for more on this, see Jacod [26], Lemma 1.8 and Corollary 1.11).

By the Itô representation theorem we further have that there exists an adapted process $\{\alpha_t(x); t \in [0, T)\}$ such that

$$p_t(x) = 1 + \int_0^t \alpha_s(x) dW_s.$$

Theorem 6. *Assume that* $\mathbb{E} \int_0^t \left(\frac{\alpha_s}{p_s}(I)\right)^2 ds < \infty$ *for all* $t \in [0, T)$. *Then the additional logarithmic utility of the insider in the interval* $[0, t]$ *is* $\mathbb{E}\mathbb{E}[\log p_t(I)]$.

Proof:

We compute the compensator of W in $\mathcal{G}$. As before, we have for a measurable bounded function f and an $\mathcal{F}_s$ measurable random variable h_s that

$$\mathbb{E}\left[(W(t) - W(s))f(I)h_s\right] = \mathbb{E}\left[(W(t) - W(s))\int f(x)p_t(x)dP^I(x)h_s\right]$$

$$= \mathbb{E}\left[\int_s^t \int \alpha_\theta(x) f(x) dP^I(x) d\theta h_s\right]$$

$$= \mathbb{E}\left[\int_s^t \int \frac{\alpha_\theta}{p_\theta}(x) f(x) dP_\theta(x) d\theta h_s\right]$$

$$= \mathbb{E}\left[\int_s^t \frac{\alpha_\theta}{p_\theta}(I) d\theta f(I) h_s\right]$$

Therefore

$$W(t) = \hat{W}(t) + \int_0^t \frac{\alpha_s}{p_s}(I) ds$$

where $\hat{W}$ is a $\mathcal{G}$-Wiener process. Then by exercise 29 we have that the additional logarithmic utility of the insider in the interval $[0, t]$ is $\frac{1}{2}\mathbb{E}\left[\int_0^t \left(\frac{\alpha_s}{p_s}(I)\right)^2 ds\right]$. Furthermore, applying Itô's formula to $\log(p_t(x))$, we have

$$\mathbb{E}\left[\log p_t(I)\right] = \frac{1}{2}\mathbb{E}\left[\int_0^t \left(\frac{\alpha_s}{p_s}(I)\right)^2 ds\right].$$

The quantity $\mathbb{E}\left[\log p_t(I)\right] = \mathbb{E}\left[\int p_t(x) \log p_t(x) dP^I(x)\right]$. That is, the gain of the insider is the expectation of the conditional entropy of the random variable I.

Exercise 33. In the case $\mu = r$, apply the same reasoning as above to prove that the optimal logarithmic utility of a weak insider in the sense of Baudoin is given by

$$\mathbb{E}\left[\left(\frac{d\nu}{dP^{W(T)}} \log \frac{d\nu}{dP^{W(T)}}\right)(W(T))\right].$$

Prove that this is not the case if $\mu \neq r$.

8 Finite Utilities for Insiders

8.1 Finite Number of Trades

In stock markets there are heterogenous agents with different types of information which coexist in equilibrium. In this sense the example in Section 3 does not correspond too well to the reality of the coexistence of insiders and small investors. Nevertheless, if one is interested in detecting unlawful insiders then the toy example in Section 3 describes this situation.

Our goal in the sections to follow is to describe ways of modifying the toy example in Section 3 in order that the insider attains a finite optimal utility. The most important reason why the insider does not achieve infinite utility in reality is that the information about $S(T)$ is better for the insider than for the small agent but not an almost sure type of information.

That is, we will consider a model where there is a deformation of information of the insider. We will treat this in the next section (also see the weak information approach in [6]).

In this section, we discuss what happens when the insider is allowed to trade only a fixed number of times up to the date T. Suppose without loss of generality that the times where transactions are allowed are $0 = t_0 < t_1 < ... < t_{n-1}$. Then, using the formulae for wealth from section 2, we have that

$$\mathbb{E}[\log(V_T)] = $$
$$\log(V_0) + \sum_{j=0}^{n-1} \int_{t_j}^{t_{j+1}} \mathbb{E}\left[\left(\mu - r + \sigma \frac{W_T - W_s}{T-s}\right)\pi(t_j) - \frac{1}{2}\sigma^2 \pi(t_j)^2\right] ds.$$

Now note that the decisions of the insider can only be based on his/her information at that time. That is, $\pi(t_j) \in \mathcal{G}_{t_j}$. Therefore, we have to maximize

$$\sum_{j=0}^{n-1} \int_{t_j}^{t_{j+1}} \mathbb{E}\left[\left(\mu - r + \sigma \mathbb{E}\left(\left.\frac{W_T - W_s}{T-s}\right|\mathcal{G}_{t_j}\right)\right)\pi(t_j) - \frac{1}{2}\sigma^2 \pi(t_j)^2\right] ds.$$

As (see exercise 17)

$$\mathbb{E}\left[\left.\frac{W_T - W_s}{T-s}\right|\mathcal{G}_{t_j}\right] = \frac{W_T - W_{t_j}}{T - t_j}.$$

Therefore the function to maximize is

$$f_s(\pi) = \left(\mu(s) - r + \sigma(s)\frac{W_T - W_{t_j}}{T - t_j}\right)\pi - \frac{1}{2}\sigma^2(s)\pi^2.$$

As before the optimal portfolio value is $\hat{\pi}(s) = \frac{\mu(s)-r}{\sigma^2(s)} + \frac{W_T - W_{t_j}}{\sigma(s)(T-t_j)}$ and the optimal logarithmic utility is

$$\log(V_0) + \mathbb{E}\left[\int_0^t \frac{\mu(s) - r}{2\sigma^2(s)}ds\right] + \sum_{j=0}^{n-1} \frac{(t_{j+1} - t_j)}{2(T - t_j)}.$$

This quantity is finite as long as $t_{n-1} < T$. Moreover, this quantity tends to infinity as $t_{n-1} \to T$, therefore this proposed solution is only partial but it also reflects the fact that one possible way to model insiders with finite utility is to allow them to trade only at a finite number of times.

If we want to insist on continuous trades then a possibility is to model the information of the insider in a different fashion. This will be done in the next section. Before this, we give the following interesting exercise.

Exercise 34. Consider the bi-dimensional model

$$S^i(t) = S^i(0)\exp\left(\left(\mu_i - \frac{\sigma_i^2}{2}\right)t + \sigma_i W^i(t)\right)$$

where (W^1, W^2) is a bi-dimensional Wiener process with correlation

$$\mathbb{E}[EW^1(t)W^2(t)] = \rho t.$$

Suppose that the insider has as information $\mathcal{G}_t = \mathcal{F}_t \vee \sigma(W^1(T))$, but that there is a restriction on trading only in the second asset. Find the optimal portfolio with logarithmic utility for the insider. Do the same if the information in the market is only given by $\mathcal{F}_t^2 = \sigma(W^2(s); s \leq t)$.

There are various other simple options to try to limit the behavior of the insider in order to make its utility finite.

Exercise 35. Find the optimal portfolio and the maximal logarithmic utility of the insider if we introduce the restriction $\pi \in [0, 1]$ (no borrowing from the bank or stocks allowed) and prove that the utility is finite in this case.

Exercise 36. Use Exercise 18 to prove that in the case $I = \int_0^T h(s)W(s)ds$, then the optimal logarithmic utility of the insider in the interval $[0, T]$ is ∞. Although this example does not correspond exactly to the average of the stock price, it does show that the utility will also be infinite if the insider has an information on the form of an average.

8.2 Towards a Dynamic Model for Insider Information

Another possibility to obtain finite utilities for the insider is to model the additional information of the insider as $I = f(S(T))$ where f is not a bijection. This alternative modelling has also its weak points. The most important being that the information of the insider does not improve as t approaches T.

Exercise 37. Use Theorem 6 to prove that the optimal logarithmic utility for the insider in the following cases is finite.
1. $I = 1(W(T) \geq a)$ for $a > 0$.
2. $I = W(T) + \varepsilon$ where ε is a $N(0, 1)$ random variable independent of W.

This exercise shows a model which is closer to reality. The information held by the insider is blurred by an additional noise. Nevertheless, this noise does not disappear even when $t = T$. This problem also appears in the weak information approach of F. Baudoin. This is a problem related with the fact that we are doing an initial enlargement of filtration. That is, the filtration is enlarged only once at time $t = 0$.

In relation with this problem we have recently proposed a model of additional information of the type $I(t) = W_T + W'((T-t)^\theta)$ where W' is a Wiener process independent of W (see [10]). This model contains a deformation of information through time. This is equivalent to saying that the information of the insider is $S(T) \exp\left(\sigma W'((T-t)^\theta)\right)$. That is, the blurring is done at logarithmic scale. The filtration, denoted by $\mathcal{G}$, is the smallest filtration that satisfies the usual conditions and that includes $\mathcal{F}_t \vee \sigma(I(s); s \leq t)$.

Theorem 7. Let $I(t) = W_T + W'((T-t)^\theta)$ and $\mathcal{G}_t = \mathcal{F}_t \vee \sigma(I(s); s \leq t)$. Then $\{W(t); t \in [0, T]\}$ is a semimartingale in the enlarged filtration $\mathcal{G}$ and its decomposition is given by

$$W_t = \hat{W}_t + \int_0^t \frac{I(s) - W(s)}{T - s + (T-s)^\theta} ds$$

where $\{\hat{W}(t); t \in [0, T]\}$ is a $\mathcal{G}$ - Wiener process.

Proof:
We try a slight variant of the proofs given before. Consider for $s \leq u \leq t$

$\mathbb{E}\left[W_t - W_u | \mathcal{G}_s\right]$
$= \mathbb{E}\left[W_t - W_u | \mathcal{F}_s \vee \sigma(I(s) - W_s, W'((T-s)^\theta) - W'((T-v)^\theta); v \leq s\right]$
$= \mathbb{E}\left[W_t - W_u | I(s) - W_s\right]$
$= \dfrac{t - u}{T - s + (T-s)^\theta} (I(s) - W_s)$

where we have used the independence of $W_t - W_u$, $\mathcal{F}_s$ and $\sigma(W'((T-s)^\theta) - W'((T-v)^\theta); v \leq s)$ and the formula $\mathbb{E}(X|Y) = \frac{cov(X,Y)}{Var(Y)} Y$ for the conditional expectation of a mean zero Gaussian random vector (X, Y). Similarly,

$$\mathbb{E}\left[W'((T-u)^\theta) \,\middle|\, I(s) - W_s\right] = \frac{(T-u)^\theta}{T-s+(T-s)^\theta} (I(s) - W_s),$$

hence

$$\mathbb{E}\left[W_t - W_u - \int_u^t \frac{I(r) - W(r)}{T-r+(T-r)^\theta} dr \,\middle|\, \mathcal{G}_s\right] = 0.$$

Exercise 38. Prove that if $I(s) = W_T + W'((T-s)^\theta)$, then

$$\mathbb{E}\left[W_t - W_u \,\middle|\, I(s) - W_s\right] = \frac{t-u}{T-s+(T-s)^\theta} (I(s) - W_s)$$

using the time homogeneity property of the Wiener process. Prove that this result is also valid for Lévy processes with finite mean.

As before we can also compute the insider's optimal utility which gives

$$\mathbb{E}\left[\log(\hat{V}^*(t))\right] = \log(x)$$

$$+ \int_0^t \mathbb{E}\left[\frac{1}{2\sigma^2(s)} \left(\mu(s) - r + \sigma(s) \frac{I(s) - W_s}{T-s+(T-s)^\theta}\right)^2\right] ds$$

$$= \log(x) + \int_0^t \mathbb{E}\left[\frac{(\mu(s) - r)^2}{2\sigma^2(s)}\right] ds + \frac{1}{2} \int_0^t \frac{1}{T-s+(T-s)^\theta} ds.$$

This shows that for $\theta < 1$, the utility for this model is finite. In fact, one can even prove that there is absence of arbitrage therefore answering our previous question regarding the coexistence between the insider and the small investor in the same model for the time interval $[0, T]$.

Exercise 39. Prove that the fair price of the flow of information characterized by $\{I(t); t \in [0, T]\}$ (see Exercise 27) is given by

$$p(t, T, I) = V_0 \left(1 - \exp\left(-\frac{1}{2} \int_0^t \frac{1}{T-s+(T-s)^\theta} ds\right)\right).$$

Note that this quantity is not V_0 for $t = T$ and $\theta \in (0, 1)$. This result can be interpreted as the fact that the fair price for the insider's information is less than the wealth of all the other market agents which is the case in Exercise 27.

The semimartingale decomposition for W in the enlarged filtration obtained here is a result of a projection formula. In fact, we have the following result from Corcuera et. al [10]:

Theorem 8. *Let I be an $\mathcal{F}_T$-measurable random variable and let us assume that there exists an $\mathcal{F} \vee \sigma(I)$-progressively measurable process $\alpha = \{\alpha_t, t \in [0,T)\}$ locally in L^1, such that*

$$W_t - \int_0^t \alpha_s ds, \qquad t \in [0, T)$$

is an $\mathcal{F} \vee \sigma(I)$-Wiener process with $\int_0^T |\alpha_u|\, du < +\infty$ a.s.. Then

$$W_t - \int_0^t \mathbb{E}[\alpha_s|\mathcal{G}_s]\, ds, \qquad t \in [0, T)$$

is an $\mathcal{G}$-Wiener process and the additional utility of the insider in the interval $[0,t]$ is given by $\frac{1}{2}\int_0^t \mathbb{E}\left[\mathbb{E}[\alpha_s|\mathcal{G}_s]^2\right] du$.

Proof:
Since W' is independent of $\mathcal{F}_T$, then $\hat{W}_t = W_t - \int_0^t \alpha_s ds$ is a $\mathcal{J}$-Wiener process, with $\mathcal{J} = (\mathcal{F}_t \vee \sigma(I) \vee \sigma(W'_s, s \leq t))_{t \in [0,T)}$. We have that $\mathbb{E}\left[\hat{W}_t|\mathcal{G}_t\right] = W_t - \int_0^t \mathbb{E}[\alpha_s|\mathcal{G}_s]\, ds$, where we can consider an $\mathcal{G}$-progressively measurable version of $\mathbb{E}(\alpha_s|\mathcal{G}_s)$, $s \in [0,T)$ (see Dellacherie and Meyer (1980), page 113), and this will be an $\mathcal{G}$-martingale. In fact, for $0 \leq s < t < T$

$$\mathbb{E}\left[\mathbb{E}\left[\hat{W}_t|\mathcal{G}_t\right]|\mathcal{G}_s\right] = \mathbb{E}\left[\hat{W}_t|\mathcal{G}_s\right] = \mathbb{E}\left[\mathbb{E}\left[\hat{W}_t|\mathcal{J}_s\right]|\mathcal{G}_s\right] = \mathbb{E}\left[\hat{W}_s|\mathcal{G}_s\right].$$

Finally, one concludes using Lévy's characterization theorem.

This idea of adding a vanishing independent Wiener process is useful not only in the example treated in Section 2, but in general in any situation where the semimartingale decomposition of W in the enlarged filtration has an information drift which degenerates at some point.

Nevertheless one awkward point still remains: The optimal portfolios of the insider are highly oscillating. That is

$$\pi(s) = \mu - r + \sigma \frac{I(s) - W_s}{T - s + (T-s)^\theta}$$

is the optimal portfolio of the insider and $\limsup_{s \to T} \pi(s) = +\infty$ and $\liminf_{s \to T} \pi(s) = -\infty$. We will see in the next section that one way to solve this problem is to consider riskier markets. That is, markets with jumps.

9 Insiders in Markets with Jumps

So far all the examples of insider trading have considered $Z \equiv 0$. This assumption is crucial because, when satisfied, the Wiener process has has an explicit density.

However, even if this is not the case some explicit compensator can be calculated. For example, if Z is a Lévy process with finite expectation we have the following theorem. A Lévy process is a càdlàg stochastically continuous stochastic process with independent stationary increments. Basic examples of Lévy process are the Wiener process and the compounded Poisson process introduced in Section 2. Furthermore for any Lévy process we have through the Lévy-Khintchine representation that

$$\mathbb{E}\left[e^{i\theta Z_T}\right] = e^{T\psi(\theta)}$$

$$\psi(\theta) = i\theta b - \frac{\sigma^2\theta^2}{2} + \int_{-\infty}^{+\infty} \left(e^{i\theta x} - 1 - 1\left(|x| \leq 1\right) i\theta x\right) \nu(dx)$$

where $\int_{-\infty}^{+\infty} \left(|x|^2 \wedge 1\right) \nu(dx) < \infty$.

Theorem 9. *Let Z be a Lévy process with $\mathbb{E}|Z_T| < \infty$. If $\mathcal{F}$ denotes the filtration generated by Z and $\mathcal{G}_t = \mathcal{F}_t \vee \sigma(Z_T)$ then*

$$Z_t = \hat{Z}_t + \int_0^t \frac{Z_T - Z_s}{T - s} ds$$

for $t \leq T$ where $\hat{Z}$ is a $\mathcal{G}$-integrable martingale.

Proof:
First note that $\mathbb{E}|Z_T| < \infty$ is equivalent to $\int_{|x|\geq 1} |x| \nu(dx) < \infty$ (see Proposition 25.4 page 159 in [44]). Consider for a $\mathcal{F}_s$-measurable bounded random variable h_s the quantity

$$\mathbb{E}\left[(Z_t - Z_s)e^{i\theta Z_T} h_s\right] = \frac{\partial}{\partial \mu_2} \mathbb{E}\left[e^{i\mu_1(Z_T - Z_t) + i\mu_2(Z_t - Z_s)} e^{i\theta Z_T} h_s\right]\bigg|_{\mu_1 = \mu_2 = 0}$$

$$= \frac{\partial}{\partial \mu_2} e^{(T-t)\psi(\mu_1 + \theta) + (t-s)\psi(\mu_2 + \theta)}\bigg|_{\mu_1 = \mu_2 = 0} \mathbb{E}\left[e^{i\theta Z_s} h_s\right]$$

$$= (t-s)\psi'(\theta)\mathbb{E}\left[e^{i\theta Z_s} h_s\right].$$

From here it follows that

$$\mathbb{E}\left[(Z_t - Z_s)f(Z_T)h_s\right] = \frac{t-s}{T-s}\mathbb{E}\left((Z_T - Z_s)f(Z_T)h_s\right).$$

From this equality one obtains using Fubini's theorem that

$$\mathbb{E}\left[(Z_t - Z_s - \int_s^t \frac{Z_T - Z_u}{T - u} du)f(Z_T)h_s\right] = 0$$

The integrability property follows directly from the definition of $\hat{Z}$.

Exercise 40. (Mansuy-Yor) Define $\mathcal{F}_{s,t} = \mathcal{F}_s \vee \sigma(Z_u; u \geq t)$. Under the same conditions of Theorem 9, prove that for $v \leq s \leq t \leq u$, one has

$$\mathbb{E}\left[\frac{Z_u - Z_v}{u - v}\bigg|\mathcal{F}_{s,t}\right] = \frac{Z_t - Z_s}{t - s}.$$

Exercise 41. (P. Tankov) Prove that in the case that Z is a simple Poisson process, then $\langle Z \rangle_t^{\mathcal{G}} = \int_0^t \frac{Z_T - Z_u}{T - u} du$. Guess the extension of this result for square integrable Lévy process

We consider the pure jump case in order to simplify calculations. Let us suppose that we are given two independent simple Poisson processes N^+ and N^- which count two types of jumps. One of size $a^+ = a \in (0, \ln 2)$ and the other of size $a^- = \ln(2 - e^a) < 0$. The size of the jumps is not very important except that one has to be positive and the other negative. This particular choice simplifies the calculations. Furthermore suppose that the rates of jumps for each type is λ^+ and λ^- respectively. Then, we take the model $S(t) = S_0 \exp(\mu t + N_t)$ where $N_t = a^+ N_t^+ + a^- N_t^-$. As explained in Section 2, the approximative wealth process for transaction at time t_j, $j = 0, ..., i - 1$ is

$$V(t_i) = V_0 \prod_{j=0}^{i-1} \left(1 + (1 - \pi(t_j))\left(e^{r(t_{j+1} - t_j)} - 1\right) + \frac{\pi(t_j)}{S(t_j)}(S(t_{j+1}) - S(t_j))\right).$$

Then

$$\log(V(t_i)) = \log(V_0) +$$

$$\sum_{j=0}^{i-1} \log\left(1 + (1 - \pi(t_j))\left(e^{r(t_{j+1} - t_j)} - 1\right) + \pi(t_j)\left(e^{\mu(t_{j+1} - t_j) + (N_{t_{j+1}} - N_{t_j})} - 1\right)\right).$$

Given that $\left(e^{r(t_{j+1} - t_j)} - 1\right) \approx r(t_{j+1} - t_j)$, $e^{\mu(t_{j+1} - t_j)} - 1 \approx \mu(t_{j+1} - t_j)$ and

$$\log\left(1 + \frac{\pi(t_j)\left(e^{(N_{t_{j+1}} - N_{t_j})} - 1\right)}{1 + (1 - \pi(t_j))\left(e^{r(t_{j+1} - t_j)} - 1\right) + \pi(t_j)\left(e^{\mu(t_{j+1} - t_j)} - 1\right)e^{(N_{t_{j+1}} - N_{t_j})}}\right)$$

$$\approx \log\left(1 + \pi(t_j)\left(e^{(N_{t_{j+1}} - N_{t_j})} - 1\right)\right),$$

we have that

$$\log(V(t_i)) \approx \log(V_0)$$

$$+ \sum_{j=0}^{i-1} \left\{(1 - \pi(t_j))r(t_{j+1} - t_j) + e^{(N_{t_{j+1}} - N_{t_j})}\pi(t_j)\mu(t_{j+1} - t_j)\right\}$$

$$+ \sum_{j=0}^{i-1} \log\left(1 + \pi(t_j)\left(e^{(N_{t_{j+1}} - N_{t_j})} - 1\right)\right).$$

For a *càdlàg* process π such that $\pi(s) \in (-(e^a - 1), e^a - 1)$, this converges to

$$\log(V_t) = \log(V_0) + \int_0^t ((1-\pi(s))r + \pi(s)\mu)\,ds + \sum_{s \le t} \log(1+\pi(s-)(e^{\Delta N(s)}-1)).$$

This is a simple form of Itô's formula for jump processes (for a general formulation, see [21] Theorem 5.1, page 66). As $\Delta N(s) \in \{a^+, a^-\}$ we have that the logarithmic utility becomes

$$\mathbb{E}(\log(\hat{V}_t))$$
$$= \log(V_0) + \mathbb{E}\int_0^t (\mu - r)\pi(s)\,ds + \mathbb{E}\sum_{s \le t} \log(1 + \pi(s-)(e^{\Delta N(s)} - 1))$$
$$= \log(V_0) + \mathbb{E}\int_0^t (\mu - r)\pi(s)\,ds + \sum_{i=-}^{+} \lambda_i \mathbb{E}\int_0^t \log(1 + \pi(s)(e^{a_i} - 1))\,ds.$$

Exercise 42. Use an approximation argument to prove that for any *càdlàg* process π such that the expectations are finite we have that

$$\mathbb{E}\left[\sum_{s \le t} \log(1 + \pi(s-)(e^{\Delta N(s)} - 1))\right]$$
$$= \lambda^+ \mathbb{E}\left[\int_0^t \log(1 + \pi(s)(e^a - 1))\,ds\right] + \lambda^- \mathbb{E}\left[\int_0^t \log(1 + \pi(s)(1 - e^a))\,ds\right].$$

Use Itô's formula for jump processes to derive this result.

As before the solution to the portfolio optimization for the small agent (i.e. non-insider) is obtained by analyzing the function

$$f(\pi) = (\mu - r)\pi + \int_{\mathbb{R}} \log(1 + (e^x - 1)\pi) F(dx)$$

where $F(dx) = \delta_{a^+}(dx)\lambda^+ + \delta_{a^-}(dx)\lambda^-$ and $\pi \in (-\frac{1}{e^a-1}, \frac{1}{e^a-1})$. The function f is a strictly concave function with respect to π and the first order condition $f'(\pi) = 0$ gives

$$\mu - r + \lambda^+ \frac{(e^a - 1)}{1 + (e^a - 1)\pi} + \lambda^- \frac{(1 - e^a)}{1 + (1 - e^a)\pi} = 0.$$

This equation reduces to a quadratic equation with two solutions. Let

$$\pi_+ = -\frac{\lambda^+ + \lambda^-}{2(\mu - r)} + \sqrt{\left(\frac{\lambda^+ + \lambda^-}{2(\mu - r)}\right)^2 + \frac{\lambda^+ - \lambda^-}{(\mu - r)(e^a - 1)} + \frac{1}{(e^a - 1)^2}}.$$

and π_- be the other solution. Then the restriction, $\frac{1}{e^a-1} > \pi > -\frac{1}{e^a-1}$, determines the optimal portfolio, denoted by π^*, as

$$\pi^* = \begin{cases} \pi_+ & \text{if } \mu > r \\ \frac{\lambda^+ - \lambda^-}{\lambda^+ + \lambda^-}(e^a - 1)^{-1} & \text{if } \mu = r \\ \pi_- & \text{if } \mu < r \end{cases}$$

The optimal logarithmic utility is finite (as the portfolio values are bounded) and given by

$$\log(V_0) + \left((\mu - r)\pi^* + \lambda^+ \log(1 + (e^a - 1)\pi^*) + \lambda^- \log(1 + (1 - e^a)\pi^*)\right) T.$$

Note that this result is valid as long as $\lambda^+ > 0$ and $\lambda^- > 0$.

Exercise 43. If $\lambda^- = 0$ prove that the optimal logarithmic utility is infinite if $\mu \geq r$. What happens if $\mu < r$?

The comparison with the Merton problem (see Section 2 and Figure 1) is as follows: While in the continuous model the ratio of investment on the stock grows linearly with the difference between the appreciation rate and the interest rate, in the jump model such growth is limited by the possible jump size in the opposite direction which may make our portfolio not admissible (that is, we may lose all our investment) with positive probability.

An interesting consequence of this analysis is that in models with bounded jump sizes the borrowing/loaning of shares/money is limited according to how big the jumps are. Therefore models where the Lévy measure has unbounded support restrict $\pi \in [0, 1]$.

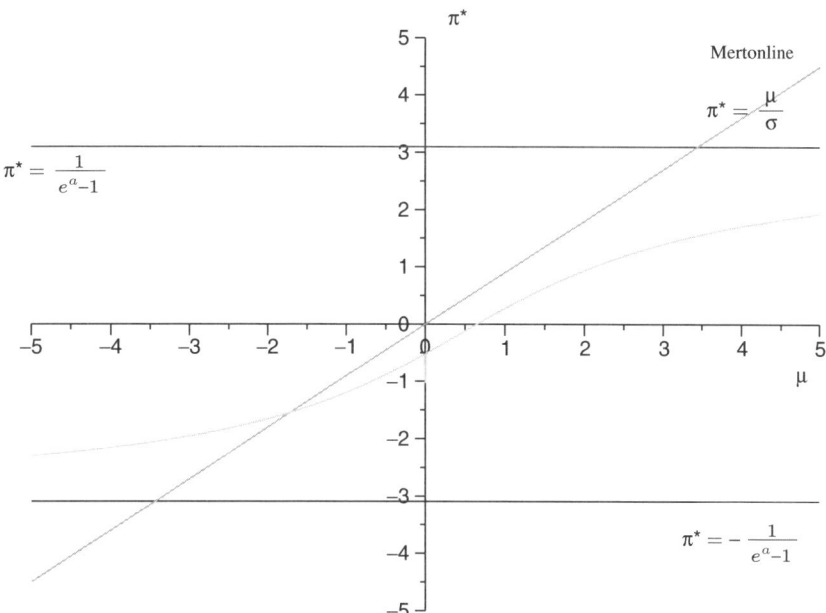

Fig. 1. Merton's line and the optimal portfolio for markets with jumps in the case $\lambda_- > \lambda_+$

This characteristic implies that models with jumps are high risk models. In fact, they lead to the insider to behave cautiously even if he/she has as information the final price of the asset.

That is, define the insider problem as one where the additional information is of the form $I = N(T)$ and $\mathcal{G}_t = \mathcal{F}_t \vee \sigma(I)$. Using Theorem 9, we have that

$$N_t - \int_0^t \frac{N_T - N_s}{T - s} ds = \hat{N}_t$$

is a $\mathcal{G}$-martingale. Nevertheless this will not be enough to compute the logarithmic utility. In fact, we will further enlarge the filtration to $\mathcal{H}_t = \mathcal{F}_t \vee \sigma(N_T^+, N_T^-)$.

Exercise 44. Prove that $\mathcal{H}_t = \mathcal{G}_t$ if it does not exist $a \in (0, \ln 2)$ such that $k_1 a + k_2 \ln(2 - e^a) = 0$ for a pair of integers k_1 and k_2.

Working in the filtration $\mathcal{H}$ we have that

$$N_t^i - \int_0^t \frac{N_T^i - N_s^i}{T - s} ds = \hat{N}_t^i$$

where $\hat{N}^i$ is a $\mathcal{H}$-martingale for $i = +, -$. Then following our previous calculations, we have that the logarithmic utility for the insider can be computed similarly. In fact we have the following result.

Exercise 45. Prove that the utility for the insider is given by

$$\mathbb{E}\left[\log(\hat{V}_t)\right] = \log(V_0) + \mathbb{E}\left[\int_0^t (\mu - r)\pi(s) ds\right] \tag{13}$$

$$+ \sum_{i=-}^{+} \mathbb{E}\left[\int_0^t \log(1 + \pi(s)(e^{a_i} - 1)) \frac{N_T^i - N_s^i}{T - s} ds\right]$$

$$= \log(V_0) + \mathbb{E}\left[\int_0^t (\mu - r)\pi(s) ds\right]$$

$$+ \sum_{i=-}^{+} \mathbb{E}\left[\int_0^t \log(1 + \pi(s)(e^{a_i} - 1)) B_i(s) ds\right],$$

where $B_i(s) = \mathbb{E}\left[\frac{N_T^i - N_s^i}{T - s} \big| \mathcal{G}_s\right]$. Note that the rates λ_i are replaced by the random rates $B_i(s)$.

In general, $B_i(s)$ is positive unless we are in the case described in Exercise 44. In that case, the insider can "count" the jumps and $B_i(s)$ becomes zero after the last jump before T.

As before our objective function is

$$f_s(\pi) = (\mu - r)\pi + \sum_{i=-}^{+} \log(1 + \pi(e^{a_i} - 1)) B_i(s)$$

for
$$\pi \in \begin{cases} (-(e^a-1)^{-1},(e^a-1)^{-1}) & \text{if } B_i(s) > 0 \text{ for } i = -,+ \\ (-(e^a-1)^{-1},+\infty) & \text{if } B_+(s) > 0 \text{ and } B_-(s) = 0 \\ (-\infty,(e^a-1)^{-1}) & \text{if } B_-(s) > 0 \text{ and } B_+(s) = 0 \\ (-\infty,+\infty) & \text{if } B_+(s) = 0 \text{ and } B_-(s) = 0. \end{cases}$$

The function f_s is strictly concave in the first three cases and linear in the last.

CASE I: $B_i(s) > 0$ for $i = -,+$ for all $s \in [0,T]$. In this case the optimal solution is obtained as the solution of $f'_s(\pi) = 0$ which gives for $\mu \neq r$

$$\hat{\pi}_\pm = -\frac{B_+ + B_-}{2(\mu-r)} \pm \sqrt{\left(\frac{B_+ + B_-}{2(\mu-r)}\right)^2 + \frac{B_+ - B_-}{(\mu-r)(e^a-1)} + \frac{1}{(e^a-1)^2}}. \quad (14)$$

As before, the right solutions for the optimal problem are determined with the restriction $\frac{1}{e^a-1} > \pi^*(s) > -\frac{1}{e^a-1}$

$$\hat{\pi}^*(s) = \begin{cases} \hat{\pi}_+(s) & \text{if } \mu > r \\ \frac{B_+ - B_-}{B_+ + B_-}(s)(e^a-1)^{-1} & \text{if } \mu = r \\ \hat{\pi}_-(s) & \text{if } \mu < r \end{cases}$$

In particular, we have the following result:

Theorem 10. *If π is a bounded $\mathcal{G}$-adapted portfolio with $\frac{1}{e^a-1} > \pi(s) > -\frac{1}{e^a-1}$ and $\mathbb{P}\{\omega; B_+(s) > 0 \text{ and } B_-(s) > 0 \text{ for all } s \in [0,T]\} = 1$. Then the utility associated with π is finite.*

Proof:
In fact, from (13) using that $\log(1 + \pi(s)(e^{a_i} - 1)) \leq \pi(s)(e^{a_i} - 1)$ we obtain that

$$\mathbb{E}\left[\int_0^t \log(1+\pi(s)(e^{a_i}-1))B_i(s)ds\right] \leq \mathbb{E}\left[\int_0^t \pi(s)(e^{a_i}-1)B_i(s)ds\right]$$
$$\leq C(a_i)\mathbb{E}\left[\int_0^t \left|\frac{N_T^i - N_s^i}{T-s}\right|ds\right]$$
$$\leq C(a_i)\lambda_i t < +\infty.$$

CASE II(a): In the case that $\lambda \times \mathbb{P}\{(s,\omega); B_+(s) > 0 \text{ and } B_-(s) = 0\} > 0$ and $\mu \geq r$ we have that as $\lim_{\pi \downarrow +\infty} f_s(\pi) = +\infty$ then the maximal utility will be infinite.

CASE II(b): If $\mu < r$ and $\lambda \times \mathbb{P}\{(s,\omega); B_+(s) = 0\} = 0$ then $\pi^*(s) = -(e^a-1)^{-1} - (\mu-r)^{-1}B_-(s)$ and utility is finite.

Exercise 46. Prove something similar for the case $\lambda \times \mathbb{P}\{(s,\omega); B_+(s) = 0$ and $B_-(s) > 0\} > 0$ with $\mu < r$. Also for the case $\lambda \times \mathbb{P}\{(s,\omega); B_+(s) = 0$ and $B_-(s) = 0\} > 0$ prove that the optimal value for $f(\pi)$ is infinite. Do an analysis as in Case II, for the case that $\lambda \times \mathbb{P}\{(s,\omega); B_i(s) = 0\} > 0$ for $i = +, -$.

The situation most typical in financial markets should be that portfolios are bounded a.s. and the logarithmic utility of the $\mathcal{G}$-investor will be finite.

In summary we have proven that the fact that utility is finite or not depends on whether B_+ or B_- are zero or not. To study this problem as we have already seen in Exercise 44 is a matter of obtaining an algebraic characterization of the jump structure. We divide the study in two cases:

Case 1: Assume that there exists $k_1, k_2 \in \mathbb{N}$ such that $k_1 a + k_2 \ln(2 - e^a) = 0$ then for any $x \in \mathbb{N}a + \mathbb{N}\ln(2 - e^a)$, $B_+(s) > 0$ and $B_-(s) > 0$ conditioned on $N_T = x$. Therefore portfolios are bounded and by a similar reasoning as in Theorem 10 maximal logarithmic utility is finite. The existence of a such that there exists $k_1, k_2 \in \mathbb{N}$ with $k_1 a + k_2 \ln(2 - e^a) = 0$ is assured by the continuity of the function $h(a) = -a^{-1} \ln(2 - e^a)$ for $a \in (0, \ln(2))$.

Case 2: On the contrary, if there are no $k_1, k_2 \in \mathbb{N}$ such that $k_1 a + k_2 \ln(2 - e^a) = 0$ then $\mathbb{P}(B_+(s) = 0$ for some subinterval of $[0,T]) > 0$ and $\mathbb{P}(B_-(s) = 0$ for some subinterval of $[0,T]) > 0$. In fact, there is always a time when given the value of $N(T)$ there can be no more both positive and negative jumps but only jumps of one type. Both probabilities being positive assures that the logarithmic utility of the $\mathcal{G}$-investor will be infinite.

Exercise 47. Assume that there exists $k_1, k_2 \in \mathbb{N}$ such that $k_1 a + k_2 \ln(2 - e^a) = 0$, prove that the optimal conditional logarithmic utility of the insider in $[0,T]$ is finite.

Exercise 48. Prove that if the insider knows $N_+(T)$ and $N_-(T)$ then the insider achieves infinite utility.

Exercise 49. Prove that if we add a Wiener process to a Poisson process with positive jumps then there is a case where the utility is infinite.

10 Enlargement of Filtrations for Random Times

10.1 Jacod's Theorem for Random Times

Now we start to describe a general set-up that will become useful later when we treat another case of insider information: That is, the case of positive random variables which we will also called random times. In particular note that, for a simple Poisson process with parameter $\lambda_1 > 0$, and letting T_n denote the time of the n-th jump, we have that

$$\mathbb{P}(T_n \geq x | \mathcal{F}_t) = 1\{x \leq T_n \leq t\} + 1\{T_n > t\} \int_{(x-t)\vee 0}^{\infty} \frac{\lambda_1 (\lambda_1 u)^{n-1-N_t} e^{-\lambda_1 u}}{(n-1-N_t)!} du \quad (15)$$

$$\mathbb{P}(T_n \geq x) = \int_x^{\infty} \frac{\lambda_1 (\lambda_1 u)^{n-1} e^{-\lambda_1 u}}{(n-1)!} du.$$

Here the filtration $\mathcal{F}_t$ is the one generated by the Poisson process N. Therefore the conditional law of the random variable T_n ($\mathbb{P}(T_n \geq x | \mathcal{F}_t)$) is not absolutely continuous with respect to a fixed measure ($\mathbb{P}(T_n \geq x)$) but to a random one. This case can not be handled by Jacod's Theorem as stated. Note that, in this case, N_t does not have a density. In the financial application, this corresponds to the insider that knows the time of the n-th jump of the stock price of size bigger than a certain number (see example 10.2), which is a rough example of market timers. To achieve some generality we work in a semimartingale environment.

Let $Z = \{Z_t, 0 \leq t \leq T\}$ be a d dimensional semimartingale defined on a complete probability space $(\Omega, \mathcal{F}, P)$. Here, $(\mathcal{F}_t)_{t \in [0,T]} \equiv (\mathcal{F}_t^Z)_{t \in [0,T]}$ is the filtration generated by the process Z. We will assume through this Section unless stated otherwise that Z satisfies

$$\sup_{t \in [0,T]} \mathbb{E}[|Z_t|] < \infty. \quad (16)$$

From now on, $\mathcal{G}$ denotes the smallest filtration including $\mathcal{F}$ and $\sigma(I)$.

For each $t \in [0, T]$, we denote by $P_t(\omega, dx)$ a regular version of the conditional law of a random variable τ given the σ-field $\mathcal{F}_t$, abbreviating it by $P_t(dx)$ if its nature as a measure is emphasized. We can choose this version in such a way that the following conditions are satisfied:

1. For every Borel set B on $\mathbb{R}^d$, $\{P_t(B), t \in [0,T]\}$ is an $(\mathcal{F}_t)_{t \in [0,T]}$-progressively measurable process.
2. For every $(t, \omega) \in [0,T] \times \Omega$, $P_t(\omega, dx)$ is a probability measure on $\mathbb{R}^d$.
3. For any bounded and $(\mathcal{F}_t)_{t \in [0,T]}$-adapted process $h : \Omega \times [0,T] \to \mathbb{R}$ and for any bounded and measurable function $f : \mathbb{R}^d \to \mathbb{R}$, we have

$$\mathbb{E}\left[f(\tau) \int_0^T h_t dt\right] = \mathbb{E}\left[\int_0^T \int_{\mathbb{R}^d} f(x) P_t(dx) h_t dt\right].$$

First, we consider the setup $\mathcal{G}_t = \mathcal{F}_t \vee \sigma(\tau)$ for initial enlargement of filtrations. In most situations when a random time is considered, one does not have $P_t^{(1)} \ll P_t$. Nevertheless, this is mostly due to the possible point measure at time τ. Therefore we consider a version of Jacod's theorem which excludes this point. In this section we denote by $P_t(dx)$ the regular conditional probability of τ given $\mathcal{F}_t$.

Definition 4. *We say that a random time τ belongs to the class $\mathcal{L}^*$, fact which we denote by $\tau \in \mathcal{L}^*$, if there exist random kernels $P_t^{(i)}(\omega, dx)$, $i = 1, 2$ such that*

1. *For every Borel set B in the positive real line, $\{P_t^{(i)}(B), t \in [0, T]\}$ is an $(\mathcal{F}_t)_{t \in [0,T)}$-progressively measurable process.*
2. *For every $(t, \omega) \in [0, T) \times \Omega$, $P_t^{(i)}(\omega, dx)$ is a signed measure on the real line.*
3. *For every $t \in [0, T)$, $\mathbb{E}\left[\int_0^t \left|P_u^{(i)}\right| du\right] < \infty$.*
4. *For any bounded and $(\mathcal{F}_t)_{t \in [0,T]}$-adapted process $h : \Omega \times [0, T] \to \mathbb{R}$, for any bounded and measurable function $f : [0, \infty] \to \mathbb{R}$, and for every $t \in [0, T)$, we have*

$$\mathbb{E}\left[f(\tau)1(\tau < s)(Z_t - Z_s)h_s\right] = \mathbb{E}\left[\int_s^t \int_0^s f(x)P_u^{(1)}(dx)du\, h_s\right],$$

$$\mathbb{E}\left[f(\tau)1(t < \tau)(Z_t - Z_s)h_s\right] = \mathbb{E}\left[\int_s^t \int_t^T f(x)P_u^{(2)}(dx)du\, h_s\right].$$

Theorem 11. *Suppose that τ is a random time in the class $\mathcal{L}^*$ and Z is a semimartingale satisfying (16) such that $\mathbb{E}|\Delta Z(\tau)| < \infty$. Assume that for almost all (t, ω), the signed measures $P_t^{(i)}(dx)$, $i = 1, 2$ are absolutely continuous with respect to $P_t(dx)$, and set*

$$\alpha_t^{(i)}(x) = \frac{dP_t^{(i)}}{dP_t}(x).$$

We can choose a version of $\alpha_t^{(i)}(x)$ which is $\mathcal{P} \otimes \mathcal{B}(\mathbb{R})$-measurable, where $\mathcal{P}$ denotes the $\mathcal{F}_t$-progressive σ-field. Define

$$\beta(u) = \alpha_u^{(1)}(\tau)1(u > \tau) + \alpha_u^{(2)}(\tau)1(u < \tau).$$

Then $Z_t - \int_0^t \beta(u)du - \Delta Z(\tau)1(t \geq \tau)$ is a martingale with respect to the filtration $(\mathcal{G}_t)_{t \in [0,T)}$.

Proof:

We choose versions of $\alpha_t^{(i)}(x)$ which is $\mathcal{P} \otimes \mathcal{B}(\mathbb{R})$-measurable for $i = 1, 2$. Let h be a measurable adapted bounded process and f a bounded measurable function on $\mathbb{R}$. Set $F = f(\tau)$. Then we have

$$\mathbb{E}\left[(Z_t - Z_s)F1(t < \tau)h_s\right] = \mathbb{E}\left[(Z_t - Z_s)f(\tau)1(t < \tau)h_s\right]$$

$$= \mathbb{E}\left[\int_s^t \int_t^T f(x)P_u^{(2)}(dx)du\, h_s\right]$$

$$= \mathbb{E}\left[\int_s^t \int_t^T f(x)\alpha_u^{(2)}(x)P_u(dx)du\, h_s\right]$$

$$= \mathbb{E}\left[\int_s^t f(\tau)\alpha_u^{(2)}(\tau)du \mathbf{1}(t<\tau)h_s\right]$$

$$= \mathbb{E}\left[\int_s^t \alpha_u^{(2)}(\tau)duF\mathbf{1}(t<\tau)h_s\right].$$

Similarly, one obtains that

$$\mathbb{E}\left[(Z_t - Z_s)F\mathbf{1}(\tau<s)h_s\right] = \mathbb{E}\left[\int_s^t \alpha_u^{(1)}(\tau)duF\mathbf{1}(\tau<s)h_s\right].$$

To finish the proof we consider the general case. Let

$$\pi = \{t_0 < s = t_1 < \ldots < t_{n-1} = t < t_n\}$$

be a partition with $|\pi| = \max\{t_k - t_{k-1}; 1 \le k \le n\}$.

$$\mathbb{E}\left[(Z_t - Z_s)Fh_s\right]$$
$$= \mathbb{E}\left[\left(\mathbf{1}(\tau \le t_0)\int_s^t \alpha_u^{(1)}(\tau)du + \mathbf{1}(t_n < \tau)\int_s^t \alpha_u^{(2)}(\tau)du\right)Fh_s\right]$$
$$+ \sum_{j=1}^{n-2}\sum_{k=0}^{n-1} \mathbb{E}\left[(Z_{t_{j+1}} - Z_{t_j})F\mathbf{1}(t_k < \tau \le t_{k+1})h_s\right].$$

Let's consider the last term

$$\sum_{j=1}^{n-2}\sum_{k=0}^{n-1} \mathbb{E}\left[(Z_{t_{j+1}} - Z_{t_j})F\mathbf{1}(t_k < \tau \le t_{k+1})h_s\right]$$
$$= \mathbb{E}\sum_{j<k}\left[(Z_{t_{j+1}} - Z_{t_j})F\mathbf{1}(t_k < \tau \le t_{k+1})h_s\right]$$
$$+ \mathbb{E}\sum_{k=1}^{n-2}\left[(Z_{t_{k+1}} - Z_{t_k})F\mathbf{1}(t_k < \tau \le t_{k+1})h_s\right]$$
$$+ \sum_{j>k}\mathbb{E}\left[(Z_{t_{j+1}} - Z_{t_j})F\mathbf{1}(t_k < \tau \le t_{k+1})h_s\right].$$

Now each term can be rewritten as follows:

$$\mathbb{E}\left[\sum_{j<k}(Z_{t_{j+1}} - Z_{t_j})F\mathbf{1}(t_k < \tau \le t_{k+1})h_s\right]$$
$$= \mathbb{E}\left(\sum_{j<k}\int_{t_j}^{t_{j+1}} \alpha_u^{(2)}(\tau)duF\mathbf{1}(t_k < \tau \le t_{k+1})h_s\right)$$
$$= \mathbb{E}\left(\sum_{j=1}^{n-2}\int_{t_j}^{t_{j+1}} \alpha_u^{(2)}(\tau)duF\mathbf{1}(t_{j+1} < \tau \le t_n)h_s\right)$$
$$\to \mathbb{E}\left(\int_s^t \mathbf{1}(u<\tau\le t)\alpha_u^{(2)}(\tau)duFh_s\right),$$

Hence
$$\sum_{k=1}^{n-2} \mathbb{E}\left[(Z_{t_{k+1}} - Z_{t_k})F1(t_k < \tau \leq t_{k+1})h_s\right] \to \mathbb{E}\left[\Delta Z(\tau)F1(s < \tau \leq t)h_s\right]$$

and

$$\mathbb{E}\left[\sum_{j>k}(Z_{t_{j+1}} - Z_{t_j})F1(t_k < \tau \leq t_{k+1})h_s\right]$$

$$= \mathbb{E}\left[\sum_{j>k} F1(t_k < \tau \leq t_{k+1})\int_{t_j}^{t_{j+1}} \alpha_u^{(1)}(\tau) du h_s\right]$$

$$= \mathbb{E}\left[\sum_{j=1}^{n-1} F1(t_0 < \tau \leq t_j)\int_{t_j}^{t_{j+1}} \alpha_u^{(1)}(\tau) du h_s\right]$$

$$\to \mathbb{E}\left[\int_s^t F1(s < \tau \leq u) \alpha_u^{(1)}(\tau) du h_s\right]$$

as $|\pi| \downarrow 0$. Therefore $Z_t - B(t)$ is a martingale in the filtration $(\mathcal{G}_t)_{t \in [0,T)}$ where $B(t) = \int_0^t \beta(u) du + \Delta Z(\tau) 1(t \geq \tau)$.

10.2 Example of Market Timers: n-th Price Jump

Now we consider some simple examples of the above situation. This example treats the situation where the filtration is enlarged by the time of the n-th jump of positive size. To simplify consider the case of the model introduced in Section 9 let N be a compound Poisson process with two types of jumps: $N_t = a^+ N_t^+ + a^- N_t^-$.

Let T_n be the random time associated with the n-th jump associated with the process N_t^+. That is, $T_n = \inf\{s; N_s^+ = n\}$. The random variable T_n is a $\mathcal{F}$ stopping time and in this case, we define $\mathcal{F}_t = \sigma(N_u^i; u \leq t, i = -, +)$ and $\mathcal{G}_t = \mathcal{F}_t \vee \sigma(T_n)$. That is, the insider knows in advance the time at which the stock will jump positively for the n-th time.

Theorem 12. *We have the following decomposition*

$$N_t = \widehat{N}_t + a^- \lambda^- t + a^+ \int_0^{t \wedge T_n} \frac{N_{T_n-}^+ - N_u^+}{T_n - u} du + a^+ \left(\lambda^+(t - T_n) + 1\right) 1(t \geq T_n)$$

where $\widehat{N}$ is a $\mathcal{G}$-martingale. Note that $N_{T_n-}^+ = n - 1$ so that

$$\int_0^{t \wedge T_n} \frac{N_{T_n-}^+ - N_u^+}{T_n - u} du = \int_0^{t \wedge T_{n-1}} \frac{N_{T_n-}^+ - N_u^+}{T_n - u} du.$$

Proof:

We apply Theorem 11 and we start calculating $P_u^{(1)}$:

$$\mathbb{E}\left[f(T_n)1(T_n < s)(N_t - N_s)h_s\right] = \mathbb{E}\left[f(T_n)1(T_n < s)\mathbb{E}\left[N_t - N_s|\mathcal{F}_s\right]h_s\right]$$
$$= \lambda(t-s)\mathbb{E}\left[f(T_n)1(T_n < s)h_s\right].$$

Therefore $dP_u^{(1)} = \lambda dP_u$. Next we compute

$$\mathbb{E}\left[f(T_n)1(T_n > t)(N_t - N_s)h_s\right]$$
$$= \mathbb{E}\left[\mathbb{E}\left[f(T_n)1(T_n > t)|\mathcal{F}_t\right](N_t - N_s)h_s\right]$$
$$= \mathbb{E}\left[\int_0^{+\infty} f(u+t)\frac{\lambda^+\left(\lambda^+ u\right)^{n-1-N_t^+} e^{-\lambda^+ u}}{(n-1-N_t^+)!}du\,(N_t - N_s)h_s\right]$$
$$= a^-\lambda^-(t-s)\mathbb{E}\left[f(T_n)1(T_n > t)h_s\right]$$
$$+ a^+\mathbb{E}\left[\int_0^{+\infty} f(u+t)\frac{\lambda^+\left(\lambda^+ u\right)^{n-1-N_t^+} e^{-\lambda^+ u}}{(n-1-N_t^+)!}du\,\left(N_t^+ - N_s^+\right)h_s\right].$$

Here we have used the conditional distribution of τ given $\mathcal{F}_t$ (see equation (15)) and the independence of N^+ and N^-. Now we compute the second expectation on the right hand side above. This gives using the probability distribution of $N_t^+ - N_s^+$

$$\mathbb{E}\left[\int_0^{+\infty} f(u+t)\frac{\lambda^+\left(\lambda^+ u\right)^{n-1-N_t^+} e^{-\lambda^+ u}}{(n-1-N_t^+)!}du\,\left(N_t^+ - N_s^+\right)h_s\right]$$
$$= \sum_{j=1}^{\infty} \frac{e^{-\lambda^+(t-s)}(\lambda^+(t-s))^j}{(j-1)!}\mathbb{E}\left[\int_0^{+\infty} f(u+t)\frac{\lambda^+\left(\lambda^+ u\right)^{n-1-j-N_s^+} e^{-\lambda^+ u}}{(n-1-j-N_s^+)!}duh_s\right]$$
$$= \left(\lambda^+\right)^2(t-s)\mathbb{E}\left[\int_0^{+\infty} f(u+t)\frac{\left(\lambda^+ u\right)^{n-2-N_t^+} e^{-\lambda^+ u}}{(n-2-N_t^+)!}duh_s\right]$$
$$= (t-s)\mathbb{E}\left[\int_0^{+\infty} f(u+t)\frac{N_{T_n-}^+ - N_t^+}{u}\frac{\lambda^+\left(\lambda^+ u\right)^{n-1-N_t^+} e^{-\lambda^+ u}}{(n-1-N_t^+)!}duh_s\right]$$
$$= (t-s)\mathbb{E}\left[\frac{N_{T_n-}^+ - N_t^+}{T_n - t}f(T_n)1(T_n > t)h_s\right].$$

Then, one can verify that

$$\mathbb{E}\left[f(T_n)1(T_n > t)\left(N_t - N_s - a^-\lambda^-(t-s) - a^+\int_s^t \frac{N_{T_n-}^+ - N_u^+}{T_n - u}du\right)h_s\right] = 0. \tag{17}$$

Therefore

$$dP_u^{(2)} = a^+\frac{N_{T_n-}^+ - N_u^+}{T_n - u}dP_u,$$

and the conclusion follows from Theorem 11.

Exercise 50. Verify the previous equality (17).

Note that the above Theorem is also valid for $\lambda^- = 0$ or $\lambda^+ = 0$ To adapt to the fact that T_n takes values in $[0, +\infty)$ we maximize the following utility

$$\max_{\pi \in \mathcal{G}} \int_0^{+\infty} e^{-rs} \mathbb{E}\left[\log(\hat{V}_s)\right] ds.$$

Following a similar discussion as in the previous example one finds that the optimal portfolio for $s \leq T_{n-1}$ is given as in the previous example by

$$\hat{\pi}^*(s) = \begin{cases} \hat{\pi}_+(s) & \text{if } \mu > r \\ \frac{n-1-N_s^+ - \lambda^-(T_n-s)}{n-1-N_s^+ + \lambda^-(T_n-s)}(e^a - 1)^{-1} & \text{if } \mu = r \\ \hat{\pi}_-(s) & \text{if } \mu < r \end{cases}$$

where $\hat{\pi}_\pm(s)$ is also the solution of the corresponding quadratic equation (14) with $B_+(u) = \frac{N_{T_n}^+ - N_u^+}{T_n - u}$ and $B_-(u) = \lambda^-$.

Exercise 51. Prove that the optimal utility of the insider in the interval $[0, T_{n-1}]$ is finite for $n \geq 2$.

Next for $t \in (T_{n-1}, T_n)$ we have that if $\mu > r$ then $\hat{\pi}^*(s) = (e^a - 1)^{-1} - \lambda^-(\mu - r)^{-1}$. In contrast, if $\mu \leq r$ then there is no optimal value and the maximal logarithmic utility is infinite. The optimal portfolio after T_n of the $\mathcal{G}$-investor and the $\mathcal{F}$-investor coincide. In conclusion if $\mu > r$ then the optimal logarithmic utility of the insider in the interval $[0, T_n)$ is finite if and only if $\mu > r$. Clearly the analysis of the optimal portfolio and the utility after T_n is like in the beginning of Section 9. In the interval $[0, T_n]$ there is no optimal portfolio in the case $\mu > r$.

Exercise 52. Provide a careful analysis to prove that the logarithmic utility of the insider is finite in the interval $[0, T_n)$ if and only if $\mu > r$.

11 The Insider as a Large Trader

In this chapter we initiate a first glance of the interactions between three distinguished elements which where somewhat independent in the previous chapters: the stock price, the insider strategy and the small trader.

First we will start with a study of the effect of the insider's strategy on the stock price dynamics. That is, suppose that the insider information is modelled using a filtration $\mathcal{G}$ and that the insider's strategy, π, is adapted to $\mathcal{G}$. Then we first consider the following discrete model

$$S_n(t_{i+1}) = S_n(t_i)\left(1 + (\mu + b\pi(t_i))(t_{i+1} - t_i) + \sigma\left(W(t_{i+1}) - W(t_i)\right)\right)$$

then we have that

$$S_n(t_n) = S_0 \prod_{i=0}^{n-1} \left(1 + (\mu + b\pi(t_i))(t_{i+1} - t_i) + \sigma\left(W(t_{i+1}) - W(t_i)\right)\right)$$

$$= S_0 \exp\left(\sum_{i=0}^{n-1} \log\left(1 + (\mu + b\pi(t_i))(t_{i+1} - t_i) + \sigma\left(W(t_{i+1}) - W(t_i)\right)\right)\right)$$

$$\approx S_0 \exp\left(\sum_{i=0}^{n-1} \left(\left(\mu + b\pi(t_i) - \frac{\sigma^2}{2}\right)(t_{i+1} - t_i) + \sigma\left(W(t_{i+1}) - W(t_i)\right)\right)\right).$$

Therefore taking limits we have that S_n converges a.s. to

$$S(t) = S_0 \exp\left(\int_0^t \left(\mu + b\pi(s) - \frac{\sigma^2}{2}\right) ds + \sigma W(t)\right).$$

In order to formalize the idea that S is the solution of a linear stochastic differential equation we need the notion of the forward integral.

Definition 5. *Let $\phi : [0,T] \times \Omega \to \mathbb{R}$ be a measurable (non necessarily adapted) continuous process. The forward integral of ϕ with respect $W(.)$ is defined by*

$$\int_0^T \phi(t) d^-W(t) = \lim_{n \to +\infty} \sum_{i=0}^{n-1} \phi(t_i)(W(t_{i+1}) - W(t_i)), \qquad (18)$$

if the limit exists in $L^1(\Omega)$ and is independent of the partition sequence taken.

This definition does not coincide exactly with the definition of Russo-Vallois. Under some more assumptions one can prove that this definition coincides with the original definition of Russo-Vallois.

Now, we can say that the class of admissible portfolios for the large trader are the ones such that S is the unique solution to the following model for the stock price

$$S(t) = S(0) + \int_0^t (\mu + b\pi(s)) S(s) ds + \int_0^t \sigma S(s) d^-W(s). \qquad (19)$$

Note that since π is not adapted to the filtration generated by W, $\mathcal{F}$, the usual rules of stochastic calculus do not apply. In particular, one has the following result

Exercise 53. For $f \in C_b^1([0,T] \times \mathbb{R})$ prove that

$$\mathbb{E}\left[\int_0^T f(t, W(T)) d^-W(t)\right] = \mathbb{E}\left[\int_0^T \frac{\partial}{\partial x} f(t, W(T)) dt\right].$$

That is, forward integrals do not have expectation zero.

The situation described above models the fact that the insiders policies have an effect on the price dynamics. From now on we assume that $0 < b < \sigma^2/2$. The condition $b > 0$ expresses that as the insider increases his/her portfolio holdings then the price of the stock increases. The condition $b < \sigma^2/2$ expresses that the volatility has to be big enough to "hide" the insider's behavior.

Now in order to express the wealth process we need to further assume that W is a semimartingale in $\mathcal{G}$ with the decomposition $W(t) = \hat{W}(t) + \int_0^t \alpha(s)ds$ where α is a $\mathcal{G}$-adapted process and $\hat{W}$ is a $\mathcal{G}$ Wiener process. Then we have in the above expression that

$$\int_0^t \sigma S(s)dW(s) = \int_0^t \sigma S(s)d\widehat{W}(s) + \int_0^t \sigma S(s)\alpha(s)ds.$$

As in Section 2, one defines the optimal logarithmic utility problem for the insider. In such a case, we obtain that the logarithmic utility is

$$\mathbb{E}\left[\log(\hat{V}(t))\right] = \log(V_0) + \int_0^t \mathbb{E}\left[\pi(s)(\mu - r + \sigma\alpha(s)) + \left(b - \frac{1}{2}\sigma^2\right)\pi(s)^2\right]ds.$$

As before, one considers the strictly concave function $f_s(\pi) = \pi(\mu - r + \sigma\alpha(s)) + \left(b - \frac{1}{2}\sigma^2\right)\pi^2$. Note that here we use that $b < \sigma^2/2$. Then the optimal portfolio for the insider is

$$\hat{\pi}(s) = \frac{\mu - r + \sigma\alpha(s)}{\sigma^2 - 2b}$$

and the optimal logarithmic utility is

$$\log(V_0) + \frac{(\mu - r)^2 t}{2(\sigma^2 - 2b)} + \int_0^t \mathbb{E}\left[\frac{\sigma^2 \alpha(s)^2}{2(\sigma^2 - 2b)}\right]ds.$$

Exercise 54. Consider the case that the large investor is not an insider. That is, the admissible portfolios are $(\mathcal{F}_t)_{t \in [0,T]}$–adapted portfolios. Prove that the optimal portfolio in this case is

$$\hat{\pi} = \frac{\mu - r}{\sigma^2 - 2b}$$

and the optimal logarithmic utility is

$$\log(V_0) + \frac{(\mu - r)^2 t}{2(\sigma^2 - 2b)}.$$

In this case the model for the underlying increases its instantaneous return if and only if $\mu \geq r$.

Note that in the case that $b \geq \sigma^2/2$ then the system explodes as the function f_s becomes convex. One first example of application may be the classical case $\mathcal{G}_t = \mathcal{F}_t \vee \sigma(W(T))$. Nevertheless this example loses some of its interest because it is clear that $\sigma(W(T)) \neq \sigma(S(T))$. Therefore the information held by the insider is not clearly interpretable from a financial point of view.

Exercise 55. Prove that $\sigma(S(T)) = \sigma(\int_0^T \frac{b\alpha(r)}{\sigma^2-2b}dr + W(T))$.

Clearly the problem we are proposing here is a fixed point problem and related with some type of equilibrium concept. If the information of the insider is $\mathcal{F}_t \vee \sigma(S(T))$, but the price is also influenced by the information itself through the portfolio π. In order to solve this situation we will use Example 18 to obtain the following result:

Theorem 13. *Suppose that $0 < b < \sigma^2/2$ and define the following portfolio for the insider*

$$\hat{\pi}(t) = \frac{\mu - r}{\sigma^2 - 2b} + \frac{\sigma}{\sigma^2 - 2b} a(t) A(t)^{-1} \int_t^T a(r) dW(r)$$

$$a(t) = (T-t)^\theta$$

$$A(t) = \int_t^T a(r)^2 dr$$

$$\theta = -b\sigma^{-2} \in (-0.5, 0).$$

Then the portfolio $\hat{\pi}$ is the optimal portfolio for the large investor under model (19) with information $\mathcal{F}_t \vee \sigma(S(T))$.

Proof:
First suppose that we are given the portfolio $\hat{\pi}$. Then, we have that

$$\log(S(T)/S(0)) = \int_0^T \left(\mu + b\hat{\pi}(s) - \frac{\sigma^2}{2}\right) ds + \sigma W(T).$$

Therefore the sigma field generated by $S(T)$ is the same as the sigma field generated by the random variable

$$Y = \frac{b\sigma}{\sigma^2 - 2b} \int_0^T a(t) A(t)^{-1} \int_t^T a(r) dW(r) dt + \sigma W(T)$$

$$= \frac{b\sigma}{\sigma^2 - 2b} \int_0^T a(r) \int_0^r a(t) A(t)^{-1} dt dW(r) + \sigma W(T).$$

After some calculations, we obtain that

$$A(t) = \int_t^T (T-u)^{2\theta} du = (2\theta + 1)^{-1} (T-t)^{2\theta+1}$$

$$\int_0^r a(t) A(t)^{-1} dt = (2\theta + 1)\theta^{-1} \left((T-r)^{-\theta} - T^{-\theta}\right)$$

$$Y = \int_0^T \left(\frac{b\sigma}{\sigma^2 - 2b}(2\theta + 1)\theta^{-1}\left(1 - T^{-\theta}(T-r)^\theta\right) + \sigma\right) dW(r)$$

$$= \sigma T^{-\theta} \int_0^T (T-r)^\theta dW(r)$$

Therefore the filtration $\mathcal{F}_t \vee \sigma(Y)$ is of the type of Exercise 18 which gives that the compensator is

$$\alpha(t) = a(t)A(t)^{-1} \int_t^T a(r)dW(r).$$

Note that $\hat{\pi}$ is admissible as it is generated through an enlargement of filtrations procedure.

Exercise 56. Prove that $\sigma(S(s); s \leq t) \vee \sigma(S(T)) = \mathcal{F}_t \vee \sigma(Y)$.

One natural question after this calculation is what the small investor can do in this situation? One should then note that the small investor does not have access to $\hat{\pi}$ or to the filtration $\mathcal{G}$. That is, the small investor may try to do his "best" possible model with the data he/she possesses which is $\mathcal{H}_t = \sigma(S(s); s \leq t) = \sigma\left(\int_0^s \frac{b\alpha(r)}{\sigma^2 - 2b}dr + W(s); s \leq t\right) \subseteq \mathcal{G}$. Then the model that the small investor will use is

$$S(t) = S_0 + \int_0^t \mathbb{E}\left[\mu + b\hat{\pi}(s)|\mathcal{H}_s\right]S(s)ds + \int_0^t \sigma S(s)d\widetilde{W}(s) \quad (20)$$

where $\widetilde{W}$ is a Wiener process on $\mathcal{H}$ (here we are assuming that $\mathcal{H}$ can support a Wiener process). In this situation the small investor will use as optimal portfolio

$$\widetilde{\pi}^*(s) = \frac{\mu - r}{\sigma^2} + \frac{b}{\sigma^2}\mathbb{E}\left[\hat{\pi}(s)|\mathcal{H}_s\right]$$
$$= \frac{(\mu - r)(\sigma^2 - b)}{\sigma^2(\sigma^2 - 2b)} + \frac{b}{\sigma(\sigma^2 - 2b)}\mathbb{E}\left[\alpha(s)|\mathcal{H}_s\right],$$

and the optimal logarithmic utility in $[0, t]$ the small trader expects to gain with model (20) is

$$A = \log(V_0) + \frac{(\mu - r)^2(\sigma^2 - b)^2}{2\sigma^2(\sigma^2 - 2b)^2}t + \frac{b^2}{2(\sigma^2 - 2b)^2}\int_0^t \mathbb{E}\left[\mathbb{E}\left[\alpha(s)|\mathcal{H}_s\right]^2\right]ds.$$

Nevertheless as the actual model for S is the one in (19) we have that the actual utility of the small investor is

$$A + \frac{b}{\sigma^2 - 2b}\int_0^t \mathbb{E}\left[\mathbb{E}\left[\alpha(s)|\mathcal{H}_s\right]^2\right]ds.$$

The interpretation of this result is the following: If you make more money than what you expect with your adapted model, it may be because there is a large investor exerting an influence on the price of the market. This may look at bit odd but in fact the risk of the small trader is also bigger than he/she thinks it is.

Obviously, if the small trader is able to guess the model of the large trader his expected logarithmic utility will further increase (but always finite) and the optimal portfolio will be

$$\pi^*(t) = \widetilde{\pi}^*(s) + \sigma^{-1}\mathbb{E}\left[\alpha(s)|\mathcal{H}_s\right]$$

One of the remaining problems with this model is that it still explodes when the logarithmic utility for the insider/large trader is considered in the interval $[0, T]$. Nevertheless, the problem can also be solved using the techniques explained in Section 8. For this reason, we study some simpler models to try to understand better the structure of the enlargement of filtrations approach within this model. Somewhat this is the goal of the next section. Before that we make a remark.

Remark 1. We have assumed that the filtration $\mathcal{H}$ is rich enough to support a Wiener process $\widetilde{W}$. In fact, one can prove that if there exists an optimal portfolio $\widetilde{\pi}^*$ for the small trader leading to a finite utility, then this is the case and furthermore the drift of the model is

$$\mathbb{E}\left[\mu + b\widehat{\pi}(s)|\mathcal{H}_s\right] = \sigma^2\widetilde{\pi}^*(s) + r.$$

That is, the projection of the anticipating large trader insider model into the filtration given by the price gives as result that the optimization problem for the small trader is the Merton problem.

12 Continuous Stream of Information

In the particular case where $\mathcal{G}_t = \mathcal{F}_t \vee \sigma(W(T))$, the optimal portfolio for the insider is a function of $[W(T) - W(t)]/(\sigma(T-t))$ (similarly for Theorem 13, the optimal portfolio is anticipating). Roughly speaking, the above model allows to introduce the anticipation through the drift of the stochastic differential equation defining the stock price model. Therefore one way to introduce a continuous information in the market is by taking a drift that depends on the variables representing his/her additional information. To look at a concrete "toy" example (think of various reasons why this is a toy example) consider for $\delta > T$ fixed

$$S(t) = S(0) + \int_0^t (\mu + bW(s+\delta))\, S(s)ds + \int_0^t \sigma S(s)d^-W(s). \qquad (21)$$

In this model, the insider has an effect on the drift of the diffusion through information that is δ units of time in the future. This continuous deformation of information may be used to model streams of information rather than one single piece of information. In this case, it is difficult to see what is the information held by the insider but his/her effect on the market is known.

Exercise 57. Prove that W is not a semi-martingale in the filtration $(\mathcal{F}_{t+\delta})_{t\in[0,T]}$.

In such a situation we are interested in looking at the optimal policy of the small investor. That is, the small investor filtration is $\mathcal{H}_t = \sigma(S_s; s \leq t)$. The above stochastic integral can be treated as in the previous section and this gives as solution

$$S(t) = S(0) \exp\left(\left(\mu - \frac{1}{2}\sigma^2\right)t + b\int_\delta^{t+\delta} W(s)ds + \sigma W(t)\right).$$

Therefore $\mathcal{H}_t = \sigma\left(b\int_\delta^{s+\delta} W(r)dr + \sigma W(s); s \leq t\right)$.

We now describe the wealth process for a strategy π such that $E\int_0^T \pi^2(s)ds < \infty$ through the discrete time argument, as in (4) (recall also a similar argument in Section 9), to obtain that

$$\log(V(t_i)) \approx \log(V_0) + \sum_{j=0}^{i-1}\left(\left((1-\pi(t_j))r + \pi(t_j)\left(\mu - \frac{1}{2}\sigma^2\right)\right)(t_{j+1} - t_j)\right.$$

$$\left. + \pi(t_j)b\int_{t_j+\delta}^{t_{j+1}+\delta} W(s)ds\right)$$

$$+ \sum_{j=0}^{i-1}\log\left(1 + \pi(t_j)\left(\exp\left(\sigma\left(W(t_{j+1}) - W(t_j)\right)\right) - 1\right)\right)$$

As before the first integral will converge to the Lebesgue integral. For the second, we will have to make assumptions on π to obtain the convergence as this sum will tend to an anticipating stochastic integral (that is, there may be correlations between $\pi(t_j)$ and the increments $(W(t_{j+1}) - W(t_j))$). To see this consider the Taylor expansion approximation of the last term above

$$\sum_{j=0}^{i-1}\left(\pi(t_j)\left(\sigma\left(W(t_{j+1}) - W(t_j)\right)\right)\right.$$

$$\left. + \frac{1}{2}\pi(t_j)\left(1 - \pi(t_j)\right)\sigma^2\left(W(t_{j+1}) - W(t_j)\right)^2\right).$$

Now we suppose that

$$\sum_{j=0}^{i-1}\left(\pi(t_j)\left(W(t_{j+1}) - W(t_j)\right)\right)$$

converges in L^1 to an integrable random variable denoted by $\int_0^t \pi(s)dW(s)$. Note in particular that the expectation of this random variable is not necessarily zero as there may be covariances between $\pi(t_j)$ and $(W(t_{j+1}) - W(t_j))$. Also suppose that

$$\sum_{j=0}^{i-1}\left(\frac{1}{2}\pi(t_j)\left(1 - \pi(t_j)\right)\sigma^2\left(\left(W(t_{j+1}) - W(t_j)\right)^2 - (t_{j+1} - t_j)\right)\right)$$

converges in L^1 to zero. We will show later that the optimal portfolios proposed satisfy this condition. With these assumptions, we have that the limit of the logarithmic wealth process can be written as

$$J(t,\pi) := \mathbb{E}\left[\log(\hat{V}(t))\right] \tag{22}$$

$$= \log(V_0) + \mathbb{E}\left[\int_0^t \left(\pi(s)(\mu - r + bW(s+\delta)) - \frac{1}{2}\sigma^2\pi(s)^2\right)ds\right]$$

$$+ \sigma\mathbb{E}\left[\int_0^t \pi(s)d^-W(s)\right].$$

Here $d^-W(s)$ denotes the forward stochastic integral of Russo-Vallois.

The previous discussion can be carried out in full generality without going through the above approximative argument (see Kohatsu-Sulem) but we have preferred the above approach as to convince the reader that the concept of forward integral is natural for the above financial problem and that it is not an artificial mathematical construct.

For the rest of the discussion suppose that the optimization problem $\max_\pi J(t,\pi)$ has a solution. Then we apply a variational argument to J to obtain for any $\mathcal{H}$ adapted process v so that the forward integral of this process exists. Then the first order condition is

$$\left.\frac{\partial J(t, \pi^* + \epsilon v)}{\partial \epsilon}\right|_{\epsilon=0} = \mathbb{E}\left[\int_0^t \left(v(s)(\mu - r + bW(s+\delta)) - \sigma^2\pi^*(s)v(s)\right)ds\right]$$

$$+ \sigma\mathbb{E}\left[\int_0^t v(s)d^-W(s)\right]$$

$$= 0 \tag{23}$$

Note that the second order condition is satisfied (J is a concave functional). Now we consider $v(s) = X1(s \geq \theta)$ for $\theta \leq t$ fixed and X an $\mathcal{H}_\theta$ measurable random variable which is forward integrable. Then we have that

$$\mathbb{E}\left[X\left(\int_\theta^t \left((\mu - r + bW(s+\delta)) - \sigma^2\pi^*(s)\right)ds + \sigma\left(W(t) - W(\theta)\right)\right)\right] = 0.$$

This gives that

$$\mathbb{E}\left[\int_\theta^t \left((\mu - r + bW(s+\delta)) - \sigma^2\pi^*(s)\right)ds + \sigma\left(W(t) - W(\theta)\right)\bigg|\mathcal{H}_\theta\right] = 0. \tag{24}$$

Therefore this also proves that $\mathbb{E}\left(W(t) - W(\theta)|\mathcal{H}_\theta\right)$ is differentiable in t and furthermore that

$$\pi^*(\theta) = \frac{\mu - r}{\sigma^2} + \frac{b}{\sigma^2}\mathbb{E}\left(W(\theta+\delta)|\mathcal{H}_\theta\right) + \frac{1}{\sigma}\lim_{t\to\theta}\mathbb{E}\left[\frac{W(t) - W(\theta)}{t - \theta}\bigg|\mathcal{H}_\theta\right].$$

We are then reduced to the computation of $\mathbb{E}\left[W(s)|\mathcal{H}_t\right]$ for $s \geq t$.

Lemma 2. *Define* $Y(t) = b \int_\delta^{t+\delta} W(r)dr + \sigma W(t)$. *Then for* $\delta \geq T$

$$\lim_{s \downarrow t} \mathbb{E}\left[\left.\frac{W(s) - W(t)}{s - t}\right| \mathcal{H}_t\right] = bM \int_0^t g(t, u) dY(u).$$

$$\mathbb{E}\left[W(t + \delta) | \mathcal{H}_t\right] = (b(t + \delta) + \delta) M \int_0^t g(t, u) dY(u)$$

where $M \equiv M_t = \sigma^{-1} b \left((b\delta + 2\sigma)\left(e^{\frac{2bt}{\sigma}} - 1\right) + \sigma \left(e^{\frac{2bt}{\sigma}} + 1\right)\right)^{-1}$ *and* $g(t, u) = e^{\frac{b}{\sigma}(2t-u)} + e^{\frac{b}{\sigma}u}$.

Proof:

First note that Y is a Gaussian process. Consequently

$$\mathbb{E}\left[W(s) | \mathcal{H}_t\right] = \int_0^t h(s, t, u) dY(u)$$

for a deterministic function h. To compute h we compute the covariances between $W(s)$ and the stochastic integral and $Y(v)$ for some $v \leq t \leq s$. First

$$\mathbb{E}\left[W(s) Y(v)\right] = bsv + \sigma(s \wedge v). \tag{25}$$

Also

$$\mathbb{E}\left[\int_0^t h(s, t, u) dY(u) Y(v)\right] = b^2 \int_0^t \int_0^u h(s, t, \theta_1)(\theta_1 \wedge \theta_2 + \delta) d\theta_2 d\theta_1$$
$$+ 2b\sigma v \int_0^t h(s, t, \theta) d\theta + \sigma^2 \int_0^u h(s, t, \theta) d\theta. \tag{26}$$

Therefore the above two expressions have to be equal. After differentiation of the equality wrt $v \leq t$ three times, we obtain

$$-b^2 h(s, t, u) + \sigma^2 \frac{\partial^2 h}{\partial u^2}(s, t, u) = 0.$$

Solving this differential equation gives

$$h(s, t, u) = C_1(s, t) e^{-\frac{b}{\sigma} u} + C_2(s, t) e^{\frac{b}{\sigma} u}. \tag{27}$$

Next one verifies that for the following constants, equations (25) and (26) coincide.

$$C_2(s, t) = \sigma^{-1} b(bs + \delta) \left((b\delta + 2\sigma)\left(e^{\frac{2bt}{\sigma}} - 1\right) + \sigma \left(\sigma e^{\frac{2bt}{\sigma}} + 1\right)\right)^{-1}$$
$$C_1(s, t) = e^{\frac{2bt}{\sigma}} C_2(s, t).$$

Therefore, we have that

$$\mathbb{E}\left(\left.\frac{W(s) - W(t)}{s - t}\right| \mathcal{H}_t\right) = \int_0^t \frac{h(s, t, u) - h(t, t, u)}{s - t} dY(u).$$

Then the result follows.

Now that we have a proposed solution one can prove that the pertinent hypotheses are all satisfied.

Lemma 3. *The portfolio π^* defined by*

$$\pi^*(t) = \frac{\mu - r}{\sigma^2} + bM\left(\sigma + \frac{b(t+\delta)+\delta}{\sigma^2}\right)\int_0^t g(t,u)dY(u).$$

satisfies that $\mathbb{E}\int_0^T \pi^*(s)^2 ds < \infty$ *and the vector*

$$\left(\sum_{j=0}^{i-1}\left(\pi(t_j)\left(W(t_{j+1}) - W(t_j)\right)\right),\right.$$

$$\left.\sum_{j=0}^{i-1}\left(\frac{1}{2}\pi(t_j)\left(1 - \pi(t_j)\right)\sigma^2\left(\left(W(t_{j+1}) - W(t_j)\right)^2 - (t_{j+1} - t_j)\right)\right)\right)$$

converges in L^1 to a random variable $(X, 0)$ with

$$\mathbb{E}[X] = b\int_0^t (t-u)g(t,u)du.$$

Proof:

Proving that $\mathbb{E}\int_0^T \pi^*(s)^2 ds < \infty$ is easy. We only give the sketch of the proof of the L^1-convergence. It is just a matter of separating conveniently the covariance structure between π^* and the Wiener increments. That is,

$$\int_0^{t_j} g(t_j, u)dY(u)\left(W(t_{j+1}) - W(t_j)\right)$$

$$= b\int_0^{t_j} g(t_j, u)\left((W(u+\delta) - W(t_{j+1}))\right.$$

$$\left. + (W(t_{j+1}) - W(t_j)) + W(t_j)\right)du\left(W(t_{j+1}) - W(t_j)\right)$$

$$+ \sigma\int_0^{t_j} g(t_j, u)dW(u)\left(W(t_{j+1}) - W(t_j)\right).$$

The sum (for $j = 0, ..., n-1$) of each of the four terms in the above sum converge in L^2. The first to a backward integral, the second to the quadratic variation and the last two to an adapted integrand. In fact, the L^2-limit is

$$b\int_0^T\int_0^t g(t,u)(W(u+\delta) - W(t))dud^-W(t) + b\int_0^T\int_0^t g(t,u)dudt$$

$$+ b\int_0^T\int_0^t g(t,u)duW(t)dW(t) + \sigma\int_0^T\int_0^t g(t,u)dW(u)dW(t).$$

Except for the second term all the integrals above have expectation zero. The terms in the second component of the vector are similarly treated (although long to write!). This second term shows that in general expectations of forward integrals are not zero and that in fact their expectations are a result of "trace"terms.

Theorem 14. *Define the class of admissible portfolios as*

$$\mathcal{A} = \left\{ \pi; \mathcal{H}\text{-adapted}, \mathbb{E} \int_0^T \pi(s)^2 ds < \infty \text{ and } \mathbb{E} \left| \int_0^T \pi(s) d^- W(s) \right| < \infty \right\}.$$

Then the optimal portfolio for the logarithmic utility is given by

$$\pi^*(t) = \frac{\mu - r}{\sigma^2} + bM \left(\sigma + \frac{b(t+\delta) + \delta}{\sigma^2} \right) \int_0^t g(t, u) dY(u)$$

and the optimal utility is finite and given by

$$J(t, \pi^*) = \log(V_0) + \frac{\sigma^2}{2} \mathbb{E} \left[\int_0^t \pi^*(s)^2 ds \right]$$

Proof:
First one has that the functional

$$\mathbb{E} \left(\int_0^t \left(\pi(s)(\mu - r + bW(s+\delta)) - \frac{1}{2} \sigma^2 \pi(s)^2 \right) ds + \sigma \int_0^t \pi(s) d^- W(s) \right)$$

is strictly concave. As π^* satisfies the first order condition then a standard argument leads to the optimality of π^*. The utility associated with π^* is finite due to the previous Lemma. To evaluate the utility we use again (23) with $v = \pi^*$ which gives

$$\mathbb{E} \left[\int_0^t (\pi^*(s)(\mu - r + bW(s+\delta)) - \sigma^2 \pi^*(s)^2) ds \right] + \sigma \mathbb{E} \left[\int_0^t \pi^*(s) d^- W(s) \right] = 0.$$

This replaced in the expression for the logarithmic utility (22) gives the result.

Exercise 58. Prove the convergence part of Lemma 3 using Malliavin Calculus techniques (in particular the duality principle). Hint: Use formula (1.12) in page 130 in Nualart [38].

Exercise 59. Prove as in the end of Section 11 that if the small trader makes an inference of his best model in the $\mathcal{H}$ filtration then his expected utility with this model will be smaller than the utility obtained through the "actual" market driving model (21). That is, the model that the small trader proposes is

$$d\widetilde{S}(t) = \mathbb{E}\left[\mu + bW(t+\delta)|\mathcal{H}_t\right] \widetilde{S}(t) dt + \sigma \widetilde{S}(t) dW_{\mathcal{H}}(t)$$

where $W_{\mathcal{H}}$ is a Wiener process in $\mathcal{H}$ (supposing this exists). Find the portfolio $\widetilde{\pi}^*$ that optimizes the logarithmic utility $\widetilde{J}(t, \pi) = \mathbb{E}\left[\log\left(\widetilde{V}^\pi(T)\right)\right]$ where $\widetilde{V}^\pi$ denotes the discounted wealth process using the price process $\widetilde{S}$. Prove that $\widetilde{J}(t, \widetilde{\pi}^*) \le J(t, \widetilde{\pi}^*)$.

A more general situation with other examples is studied in [32]. The case $\delta \leq T$ can also be studied although explicit expressions are difficult to write. There is an important issue that we have not addressed so far: the existence of arbitrage. In fact, given that in principle we are not in a standard set-up one does not know if Girsanov's theorem can be applied and therefore the non-existence of arbitrage is an interesting issue. In fact, W is not adapted to $\mathcal{H}$ and therefore this setup can not be considered as an enlargement of filtration approach.

Theorem 15. *If the logarithmic utility is finite then there is no arbitrage.*

Proof:

We have that the filtration $\mathcal{H}$ is generated by the process Y. Now we compute the semimartingale decomposition of the process Y. That is,

$$\mathbb{E}\left[Y(t) - Y(s) | \mathcal{H}_s\right] = b \left(\int_t^{t+\delta} \int_0^s h(r,s,u) dY(u) dr \right.$$

$$\left. - \int_s^{s+\delta} \int_0^s h(r,s,u) dY(u) dr \right)$$

$$+ \sigma \int_0^s \left(h(t,s,u) - h(s,s,u)\right) dY(u) =: A(t)$$

where h is given in (27). It is not difficult to prove that A is differentiable and therefore this gives the semimartingale decomposition of the process Y. For this, define

$$B(t) = \int_0^t \int_0^r h(r+\delta, r, u) dY(u) dr - \int_0^t \int_0^r h(r, r, u) dY(u) dr$$

$$+ \sigma \int_0^t \int_0^r D_1 h(r, r, u) dY(u) dr.$$

Here D_1 denotes the derivative with respect to the first variable in h. In fact, $Y - B$ is a continuous $\mathcal{H}$-martingale. To prove this is enough to prove that given a sequence of partitions $s = t_0 < ... < t_n = t$ whose norm is tending to zero, we have that

$$\lim_{n \to \infty} \mathbb{E}\left[\sum_{i=0}^{n-1} \mathbb{E}\left[Y(t_{i+1}) - Y(t_i) - B(t_{i+1}) + B(t_i) | \mathcal{H}_{t_i}\right] \mathcal{H}_s\right] = 0.$$

Furthermore $\langle Y - B \rangle_t = t\sigma^2$. Therefore by Lévy's characterization theorem (see Example 11) we have that there exists a $\mathcal{H}$- Wiener process $W_\mathcal{H}$ such that $Y - A = \sigma W_\mathcal{H}$. Therefore the price process becomes

$$S(t) = S(0) \exp\left(\left(\mu - \frac{1}{2}\sigma^2\right) t + A(t) + \sigma W_\mathcal{H}(t)\right).$$

Therefore the classical theory of no-arbitrage applies.

This proof may lead to the misconception that the above explicit calculations are not necessary because everything becomes a consequence of the previous theorem. In general, this is not so when the calculations are not so explicit. In fact, we have the following exercise.

Exercise 60. Prove, without using the explicit optimal portfolio, that if there is an optimal portfolio leading to a finite logarithmic utility that satisfies (24) then there is no arbitrage. Hint: follow the same structure of proof as above without the explicit calculations.

Exercise 61. For $\mu \geq r$ prove that

$$\mathbb{E}\int_0^T \pi^*(s)^2 ds \leq \mathbb{E}\left(\int_0^T \pi^*(s) d^-W(s)\right)^2.$$

Interpret this result as a risk issue of the small trader in an insider influenced model. Link this risk with the existence of the "trace" terms as explained at the end of the proof of Lemma 3.

Exercise 62. Set $\delta = T/2$. Using the ideas of Girsanov's theorem in an anticipating setting to prove that the model

$$dS(t) = (\mu + bW(t+\delta))S(t) + \sigma S(t) d^-W(t)$$

does not allow for arbitrage strategies for the small trader in the interval $[0, T]$, who uses the filtration $\mathcal{H}_t = \sigma(S(s); s \leq t)$, inside a certain class of portfolio strategies.

References

1. Amendinger, J., Imkeller, P., Schweizer, M., 1998. Additional logarithmic utility of an insider. Stochastic Proc. Appl., 75, 263-286.
2. Amendinger, J., 2000. Martingale representations theorems for initially enlarged filtrations. Stochastic Proc. Appl., 89, 101-116.
3. Amendinger, J., Becherer D., Schweizer, M., 2003. A monetary value for initial information in portfolio optimization. Finance and Stochastics, 7, 29-46.
4. Ankirchner, S., Dereich S., Imkeller P., 2006. The Shannon information of filtrations and the additional logarithmic utility of insiders. Ann. Probab., 34, 743-778.
5. Baudoin, F., 2003. Modelling anticipations in financial markets. In Paris-Princeton Lectures on Mathematical Finance 2002. Lecture Notes in Mathematics 1814, Springer-Verlag. Berlin.
6. Baudoin, F., 2002. Conditioned Stochastic Differential Equations and Application to Finance, Stochastic Proc. Appl., 100, 109-145.
7. Baudoin, F. Nguyen-Noc, L., 2004. The financial value of a weak information on a financial market. Finance and Stochastics 8, 415-435.

8. Back, K., 1992. Insider Trading in Continuous Time. Review of Financial Studies 5, 387-409.
9. Biagini, F. and Øksendal B., 2005. A general stochastic calculus approach to insider trading. Applied Mathematics and Optimization, 52, 167-181.
10. Corcuera J. M., Imkeller P., Kohatsu-Higa A., Nualart D., 2004. Additional utility of insiders with imperfect dynamical information. Finance and Stochastics, 8, 437-450, 2004.
11. Corcuera J.M., Guerra J., Nualart D., Schoutens W., 2006. Optimal investment in a Lévy market. Applied Mathematics and Optimization, 53, 279-309.
12. Chaumont L., Yor M., 2004. Exercises in Probability, Cambridge University Press.
13. Elliot R.J., Geman H., Korkie B.M., 1997. Portfolio optimization and contingent claim pricing with differential information. Stochastics and Stochastics Reports 60, 185-203.
14. Elliot R.J., Jeanblanc M., 1999. Incomplete markets with jumps and informed agents. Math. Meth. Oper. Res. 50, 475-492.
15. Föllmer, H. Imkeller P., 1993. Anticipation cancelled by a Girsanov transformation : a paradox on Wiener space", Ann. Inst. Henri Poincaré, 29-4, 569-586.
16. Framstad N.C., Oksendal B., Sulem A., 1998. Optimal consumption and portfolio in a jump diffusion model. In Shiryaev, A., Sulem, A. (eds.), Proceedings of the Workshop on Mathematical Finance, INRIA, Paris.
17. Goll T., Kallsen J. 2000. Optimal Portfolios for Logarithmic Utility. Stochastic Processes and their Applications 89 (1), 31-48.
18. Goll T., Kallsen J. 2003. A Complete Explicit Solution to the Log-Optimal Portfolio Problem. The Annals of Applied Probability 13, 774-799.
19. Grorud, A., 2000. Asymmetric information in a financial market with jumps. International Journal of Theoretical and Applied Finance, 3, 641-659.
20. Grorud, A., Pontier, M., 1998. Insider Trading in a continuous Time Market Model. International Journal of Theoretical and Applied Finance 1, 331-347.
21. Ikeda, N., Watanabe, S., 1989. Stochastic Differential Equations and Diffusion Processes. North-Holland-Kodansha.
22. Imkeller, P., 1996. Enlargement of the Wiener filtration by an absolutely continuous random variable via Malliavin's calculus. Probab. Th. Rel . Fields 106, 105-135.
23. Imkeller, P., 1997. Enlargement of the Wiener filtration by a manifold valued random element via Malliavin's calculus. In Statistics and Control of Stochastic Processes. The Lipster Festschrift, Y.M. Kabanov, B.L. Rosovskii, A.N. Shiryaev (eds.), World Scientific, Singapore.
24. Imkeller, P., 2002. Random times at which insiders can have free lunches. Stochastics and Stochastics Reports, 74, 465-487.
25. Imkeller, P., Pontier, M., Weisz, F., 2001. Free lunch and arbitrage possibilities in a financial market with an insider. Stochastic Proc. Appl. 92, 103-130.
26. Jacod, J., 1985. Grossissement initial, hypothèse (H'), et théorème de Girsanov. In : Grossissements de filtrations: exemples et applications. T Jeulin, M. Yor (eds.) Lect. Notes in Maths. 1118. Springer-Verlag. Berlin.
27. Jacod J.,Protter P. 1988. Time reversal of Lévy processes. Ann. Probab. 16, 620-641.
28. Jeulin, T., 1980. Semi-martingales et groissessement de filtration. Lect. Notes in Maths. 833. Springer-Verlag, Berlin

29. Karatzas, I., Pikovsky, I., 1996. Anticipative portfolio optimization. Adv. Appl. Prob. 28, 1095-1122.
30. Kohatsu-Higa A., Yamazato M., 2007. Enlargement of filtrations with random times for processes with jumps. Preprint.
31. Kohatsu-Higa A., Yamazato M., 2007. Insider modelling and logarithmic utility in markets with jumps, Preprint.
32. Kohatsu-Higa A., Sulem, A., 2006. Utility maximization in an insider influenced market. Mathematical Finance, 16, 153-179.
33. Kyle, A., 1985. Continuous Auctions and Insider Trading. Econometrica 53, 1315-1335.
34. Kunita H., 2003: Mathematical finance for price processes with jumps. Proccedings of the Kusatsu Congress. World Scientific 2003.
35. Mansuy, R., Yor, M., 2005. Harnesses, Lévy processes and Monsieur Jourdain. Stochastic Process. Appl. 115, 329-338.
36. Meyer, P. A., 1996. Probability and Potentials. Blaisdell Publishing Company, Waltham, Mass.
37. Nualart, D. , 1995. The Malliavin Calculus and Related Topics. Springer-Verlag. Berlin.
38. Nualart, D., 1995. Analysis on Wiener space and anticipating calculus. In Lectures on Probability Theory and Statistics. Ecole d'eté de Probabilités de Saint-Flour XXV. Lecture Notes in Mathematics 1690. Springer-Verlag.
39. Øksendal B., Sulem, A., 2005. Partial observation in an anticipative environment. To appear in Proceedings of the Kolmogorov conference 2003.
40. Protter, P., 2005. Stochastic Integration and Differential Equations. Springer-Verlag. New York.
41. Russo F., Vallois P., 1993. Forward, backward and symmetric stochastic integration. Probab. Th. Rel. Fields, 97, 403-421.
42. Russo F., Vallois P., 2000. Stochastic calculus with respect to continuous finite quadratic variation processes. Stochastics and Stochastics Reports, 70, 1-40.
43. Russo F., Vallois P., 1995. The generalized covariation process and Itô formula. Stochastic Process. Appl., 59, 81-104.
44. Sato K. 1999: Lévy processes and infinite divisible distributions. Cambridge University Press.
45. Séminaire de Calcul Stochastique 1982/83, Université Paris VI, 1985. Grossissements de fitrations: exemples et applications. T Jeulin, M. Yor (eds.). Lect. Notes in Maths. 1118. Springer-Velag. Berlin
46. Williams, D. 1991. Probability with martingales. Cambridge University Press.
47. Yor, M. Grossissement de filtrations et absolue continuite de noyaux. In : Grossissements de filtrations: exemples et applications. T Jeulin, M. Yor (eds.) Lect. Notes in Maths. 1118. Springer-Verlag. Berlin.

13 Solutions and Hints for the Proposed Exercises

Solution 1. Suppose by contradiction that $p_1(t_i) < 0$ for some $i = 0, ..., n-1$. Then $\mathbb{P}(V(t_i) + p_1(t_i)S(t) + p_0(t_i) < 0 \ for \ some \ t \in [t_i, t_{i+1}]) > 0$.

Solution 2. Define the measure

$$\frac{dR}{d\mathbb{P}} = \exp\left(-\int_0^t \theta(s)dW(s) - \frac{1}{2}\int_0^t \theta(s)^2 ds\right),$$

where $\theta(s) = \sigma(s)^{-1}\left(\mu(s) - r + \lambda(\mathbb{E}(e^X) - 1)\right)$. Under R the dynamics of $\hat{S}$ are given by

$$d\log(\hat{S})(t) = -\frac{1}{2}\sigma^2(t)dt + \sigma(t)d\tilde{W}(t) - \lambda(\mathbb{E}(e^X) - 1)dt + dZ(t),$$

where $\tilde{W}$ denotes a Wiener process on the space $(\Omega, \mathcal{F}, R)$. As before, consider for $s < t$

$$\mathbb{E}_R\left[\hat{S}(t)\big|\mathcal{F}_s\right] = \hat{S}(s)\mathbb{E}_R\left[\exp\left(-\frac{1}{2}\int_s^t \sigma^2(u)du + \int_s^t \sigma(u)d\tilde{W}(u)\right.\right.$$

$$\left.\left.-\lambda(\mathbb{E}(e^X) - 1)(t-s) + Z(t) - Z(s)\big|\mathcal{F}_s\right)\right]$$

$$= \hat{S}(s).$$

Here we have used Itô's formula for jump type processes.

Solution 3. For $n \in \mathbb{N}$, define the stopping time $T_n = \inf\{t_i; |p_0(t_i)| \wedge |p_1(t_i)| > n\}$. Then $\mathbb{E}[|p_0(t_i \wedge T_n)| + |p_1(t_i \wedge T_n)|] \leq n$. Therefore the argument in the proof can applied to obtain that $\hat{V}$ is a discrete time martingale.

$$\mathbb{E}_{\mathbb{Q}}\left[\hat{V}(t_{i+1} \wedge T_n)\big|\mathcal{F}_{t_i}\right] = \hat{V}(t_i \wedge T_n)).$$

As before one obtains that $\mathbb{E}_{\mathbb{Q}}[\hat{V}(T_n)] = V_0$. Then the contradiction follows after an application of Fatou's lemma. In fact, $\hat{V}$ is a supermartingale.

Solution 4. In fact, one can also perform a change of measure on the compound Poisson process. Suppose that X has a density given by f and let g be another density such that f/g is well defined. Then the Girsanov's theorem in this setting can be applied to obtain the following equation in θ, λ_1 and g:

$$\mu - r - \sigma(s)\theta(s) + \lambda_1\left(\int e^x g(x)dx - 1\right) = 0$$

Obviously this equation has an infinite number of solutions except for trivial cases (such as $\lambda = 0$). Then the change of measure is given by

$$\frac{dR}{d\mathbb{P}} = \exp\left(-\int_0^t \theta(s)^2 ds - \int_0^t \theta(s)dW(s)\right)$$

$$\exp\left(-(\lambda - \lambda_1)t + \log(\lambda/\lambda_1)N(t)\right)\prod_{i=1}^{N(t)}\frac{f}{g}(X_i)$$

Solution 5. We give the idea of the solution. First, we define $\eta(t) = \sup\{t_i; t_i \leq t\}$ and write equation (5) in differential form as

$$\hat{V}^n(t) = V_0 + \int_0^t \frac{\pi(\eta(s))\hat{V}^n(\eta(s))}{\hat{S}(\eta(s))} d\hat{S}(s). \tag{28}$$

$\hat{V}^n$ is a continuous time extension of $\hat{V}$ defined in equation (5). The idea of the proof is to take the difference between equations (6) and (28) taking into account that

$$\int_0^t \frac{u(s-)}{\hat{S}(s-)} d\hat{S}(s) = \int_0^t u(s-)\left(\mu(s-) - r - \frac{1}{2}\sigma^2(s-)\right) ds$$
$$+ \int_0^t u(s-)\sigma(s-)dW(s) + \sum_{s \leq t} u(s-)\left(e^{\Delta Z(s)} - 1\right).$$

Here $\Delta Z(s) = Z(s) - Z(s-)$. It has to be proven that the last sum above is well defined. The final estimates are carried through $L^2(\Omega, \mathcal{F}, \mathbb{P})$ estimates of the differences assuming that π is integrable enough. The general case is carried out through a classical stopping time argument.

Solution 6. Evaluating π^* in the logarithmic utility we have

$$\mathbb{E}\left[\log(\hat{V}^*(t))\right] = \log(V_0) + \mathbb{E}\left[\int_0^t \frac{(\mu(s) - r)^2}{\sigma^2(s)} ds\right].$$

Here $\hat{V}^*$ denotes the wealth associated with the optimal portfolio π^*.

Solution 7. Let $\pi \in \mathcal{A}(t)$ then the stochastic integral $\int_0^{\cdot} \sigma(s)\pi(s)dW(s)$ is well defined and using Itô's formula we have due to the strict concavity of f_s that

$$\mathbb{E}\left[\log(\hat{V}(t))\right] = \log(V_0) + \mathbb{E}\left(\int_0^t \left((\mu(s) - r)\pi(s) - \frac{1}{2}\sigma^2(s)\pi(s)^2\right) ds\right)$$
$$\leq \mathbb{E}\left[\log(\hat{V}^*(t))\right].$$

Solution 8. Consider the stopping time $\tau_n = \inf\{t \geq 0; \int_0^t \pi(s)^2 ds \leq n\}$ then as before $\int_0^{\cdot \wedge \tau_n} \pi(s)dW(s)$ is well defined and is a martingale. Therefore as before

$$\mathbb{E}\left[\log(\hat{V}(t \wedge \tau_n))\right] \leq \mathbb{E}\left[\log(\hat{V}^*(t \wedge \tau_n))\right] \leq \mathbb{E}\left[\log(\hat{V}^*(t))\right] < \infty.$$

By taking limits the result follows.

Solution 9. Define the class

$$\mathcal{A}_\theta(t) = \{\pi : \pi \text{ is } \mathcal{F} - \text{adapted}, \int_0^t \pi(s)^2 ds < \infty \text{ a.s. and } \mathbb{E}\hat{V}(t)^\theta < \infty\}.$$

As before consider $\tau_n = \inf\{r \geq 0; \int_0^r \pi(s)^2 ds \leq n\}$. First note that for any portfolio π we have that

$$\mathbb{E}\left[\hat{V}(t \wedge \tau_n)^\theta\right]$$
$$= V_0^\theta \mathbb{E}\left[\exp\left(\int_0^{t\wedge\tau_n} \theta\left((\mu(s) - r)\pi(s) - \frac{1}{2}\sigma^2(s)\pi(s)^2\right) ds\right.\right.$$
$$\left.\left. + \theta \int_0^{t\wedge\tau_n} \sigma(s)\pi(s)dW(s)\right)\right].$$
$$= V_0^\theta \mathbb{E}^{\mathbb{Q}}\left[\exp\left(\theta \int_0^{t\wedge\tau_n} \left((\mu(s) - r)\pi(s) - \frac{1}{2}(1-\theta)\sigma(s)^2\pi(s)^2\right) ds\right)\right]$$

where

$$\frac{d\mathbb{Q}^n}{d\mathbb{P}} = \exp\left(-\frac{1}{2}\int_0^{t\wedge\tau_n}\sigma^2(s)\theta^2\pi(s)^2 ds + \theta\int_0^{t\wedge\tau_n}\sigma(s)\pi(s)dW(s)\right).$$

As before the function $f_s(\pi) = \theta(\mu(s)-r)\pi - \frac{1}{2}\theta(1-\theta)\sigma^2(s)\pi^2$ is a strictly concave function for $\theta \in (0,1)$ and its maximal value is attained by $\pi^* = \frac{(\mu(s)-r)}{(1-\theta)\sigma^2(s)}$. Then $\mathbb{E}\hat{V}(t \wedge \tau_n)^\theta \leq \mathbb{E}\hat{V}^*(t \wedge \tau_n)^\theta$ where

$$\mathbb{E}\left[\hat{V}^*(t\wedge\tau_n)^\theta\right] = V_0^\theta \mathbb{E}^{\mathbb{Q}^n}\left[\exp\left(\int_0^{t\wedge\tau_n} \frac{(\theta - 1/2)(\mu(s) - r)^2}{2(1-\theta)\sigma^2(s)}\right)\right]$$
$$\leq V_0^\theta \mathbb{E}^{\mathbb{Q}^n}\left[\exp\left(\int_0^t \frac{|\theta - 1/2|(\mu(s) - r)^2}{2(1-\theta)\sigma^2(s)} ds\right)\right].$$

We can therefore take limits to obtain that π^* is the optimal portfolio. The optimal wealth is then given by $\mathbb{E}\left[\hat{V}^*(t)^\theta\right] = V_0^\theta \exp\left(\frac{\theta(\mu-r)^2}{2(1-\theta)\sigma^2}t\right)$ in the case that μ and σ are constant.

Solution 11. Applying the Itô's formula to $\exp(i\theta(M_t - M_s))$ we obtain

$$e^{(i\theta(M_t - M_s))} = 1 + \int_s^t i\theta e^{(i\theta(M_u - M_s))} dM_u - \frac{\theta^2}{2}\int_s^t e^{(i\theta(M_u - M_s))} du.$$

Taking conditional expectations we obtain

$$\mathbb{E}\left[\exp\left(i\theta(M_t - M_s)\right)|\mathcal{F}_s\right] = 1 - \frac{\theta^2}{2}\int_s^t \mathbb{E}\left[\exp\left(i\theta(M_u - M_s)\right)|\mathcal{F}_s\right] du.$$

Solving this equation we obtain the result.

Solution 12. First we have that $\mathcal{F}_{t+\varepsilon} \cup \sigma(I) \supseteq \mathcal{F}_t \cup \sigma(I)$ for all $\varepsilon > 0$ therefore $\mathcal{F}_t \vee \sigma(I) \subset \cap_{\varepsilon \geq 0}\sigma\left(\mathcal{F}_{t+\varepsilon} \cup \sigma(I)\right)$.

Solution 14. It is enough to note first that $\mathbb{E}\left[|W(T) - W(u)|^r\right] = (T-u)^{r/2}C_r$ where $C_r = \int_0^\infty \sqrt{\frac{2}{\pi}} x^r e^{-\frac{x^2}{2}} dx$. Therefore if $r \in [0,2)$

$$\mathbb{E}\left[\int_0^T \left|\frac{W(T)-W(u)}{T-u}\right|^r du\right] = C_r \int_0^T (T-u)^{-r/2} du = \frac{C_r}{1-r/2}T^{1-r/2}.$$

If $r \geq 2$ then the above integral diverges and if $r < 0$ then the expectation is infinite. For the second part we use Hölder inequality with r a positive integer to obtain that

$$\mathbb{E}\left[\left|\int_0^T \frac{W(T)-W(u)}{T-u} du\right|^r\right] \leq C_r \int_{[0,T]^r} \prod_{i=1}^r (T-u_i)^{-1/2} du_i < \infty.$$

Solution 15. As $\hat{W}$ is a $\mathcal{G}$ Wiener process and $\mathcal{G}_0 = \sigma(W(T))$ then the independence follows.

Solution 16. Define the measure

$$\frac{d\mathbb{Q}}{d\mathbb{P}} = \frac{d\mathbb{P}_0}{d\mathbb{P}_t},$$

then for two measurable bounded functions f and g

$$\mathbb{E}^{\mathbb{Q}}\left[f(W(t))g(W(T))\right] = \mathbb{E}\left[f(W(t)) \int g(x) \frac{dP_0}{dP_t}(x) dP_t(x)\right]$$
$$= \mathbb{E}\left[f(W(t))\right] E(\left[g(W(T))\right].$$

Taking g a constant one obtains that $\mathbb{E}^{\mathbb{Q}}\left[f(W(t))\right] = \mathbb{E}\left[f(W(t))\right]$ and similarly for g. From here the conclusion follows.

Solution 17. The above property called the harness property is closely tied with the enlargement of filtrations for Lévy processes. For more on this see [12] and [35]. To prove this property we consider

$$W_T - W_b = \hat{W}_T - \hat{W}_b + \int_b^T \frac{W_T - W_u}{T-u} du.$$

Taking conditional expectations we obtain the equation

$$\frac{d}{db}\mathbb{E}\left[W_T - W_b | \mathcal{G}_s\right] = -\frac{\mathbb{E}\left[W_T - W_b | \mathcal{G}_s\right]}{T-b},$$

on $[s,T]$ with initial condition $\mathbb{E}\left(W_T - W_s | \mathcal{G}_s\right) = W_T - W_s$. The solution is

$$\mathbb{E}\left[W_T - W_b | \mathcal{G}_s\right] = \frac{W_T - W_s}{T-s}(T-b).$$

To finish, one only needs to note that

$$\mathbb{E}\left[W_a - W_b | \mathcal{G}_s\right] = \int_b^a \frac{\mathbb{E}\left[W_T - W_u | \mathcal{G}_s\right]}{T-u} du.$$

From here the formula follows.

Solution 18. By the integration by parts formula, we have that $I = \int_0^T h(r)W_r dr = \int_0^T a(r)dW_r$, where we denote $a(t) = \int_t^T h(s)ds$. Then we have for $A(t) = \int_t^T a(r)^2 dr$

$$\frac{dP_t(x)}{dx} = \sqrt{\frac{1}{2\pi A(t)}} \exp\left(-\frac{\left(x - \int_0^t a(r)dW_r\right)^2}{2A(t)}\right).$$

Applying the same sequence of ideas as before we have that

$$\mathbb{E}\left[(W(t) - W(s))f(I)h_s\right] = \mathbb{E}\left[\int_s^t \frac{\int_u^T a(r)dW_r}{A(u)} a(u) du f(I) h_s\right].$$

Therefore as the process $\int_\cdot^T a(r)dW_r$ is $\mathcal{G}$ adapted we finally have that

$$W_t = \hat{W}_t + \int_0^t \frac{\int_u^T a(r)dW_r}{A(u)} a(u) du$$

where $\hat{W}$ is a $\mathcal{G}$ Wiener process in $[0,T)$. In order to prove that the definition is valid in the closed interval we have

$$\mathbb{E}\left[\int_0^T \frac{\left|\int_u^T a(r)dW_r\right|}{A(u)} |a(u)| du\right] = \sqrt{\frac{2}{\pi}} \int_0^T \frac{|a(u)|}{\sqrt{A(u)}} du < \infty.$$

Therefore the needed condition is $\int_0^T \frac{|a(u)|}{\sqrt{A(u)}} du < \infty$. For example, if $h(r) = (T-r)^\theta$ with $\theta > -1/2$, this condition is satisfied.

Solution 19. As in the proof of Theorem 2 consider

$$\mathbb{E}\left[(W(t) - W(s))f(X(T))h_s\right] = \mathbb{E}\left[(W(t) - W(s))\int f(x)dP_t(x)h_s\right]$$

$$= \mathbb{E}\left[(W(t) - W(s))\int f(x)p_{T-t}(X_t, x)dx h_s\right].$$

Applying Itô's formula to $(W(t) - W(s))p_{T-t}(W_t, x)$ in the interval $[s,t]$, and using that p_{T-t} solves the parabolic equation (11), we have

$$\mathbb{E}\left[(W(t) - W(s))f(X(T))h_s\right]$$

$$= \mathbb{E}\left[\int f(x) \int_s^t \partial_y p_{T-u}(X(u), x) du dx h_s\right]$$

$$= \mathbb{E}\left[\int_s^t \int f(x) \partial_y \log(p_{T-u}(X(u), x)) p_u(X(u), x) dx du h_s\right]$$

$$= \mathbb{E}\left[f(W(T)) \int_s^t \partial_y \log(p_{T-u}(X(u), X(T))) du h_s\right].$$

Therefore by a density argument, one has

$$\mathbb{E}\left[W(t) - W(s) - \int_s^t \partial_y \log(p_{T-u}(X(u), X(T)))du \bigg| \mathcal{G}_s\right] = 0.$$

As $\partial_y \log(p_{T-u}(X(u), X(T))) \in \mathcal{G}_u$ then

$$\hat{W}(t) = W(t) - \int_0^t \partial_y \log(p_{T-u}(X(u), X(T)))du$$

is a $\mathcal{G}$-continuous martingale with $\langle \hat{W} \rangle_t = \langle W \rangle_t = t$ and therefore by Lévy's theorem one has that $\hat{W}$ is a $\mathcal{G}$-Wiener process in $[0, T)$. We then define $\hat{W}(T) = \lim_{t \to T} \hat{W}(t)$ and all above properties follow for the closed interval $[0, T]$ (sure?). Note that given the estimates on the density function p and its derivative one has that as

$$\mathbb{E}\left[\int_0^T |\partial_y \log(p_{T-u}(X(u), X(T)))| du\right]$$
$$= \int_0^T \int |\partial_y p_{T-u}(x, y)| p_u(X_0, x) dx dy du$$
$$\leq C \int_0^T u^{1/2-a}(T-u)^{1/2-a} du < \infty$$

if $a < 3/2$. Therefore in relation with the estimate (12) there is a need to improve to be able to apply it to the hypoelliptic case.

Solution 22. To simplify the notation define $\alpha_s = \frac{W_T - W_s}{T-s}$. Define $\mathcal{I} = \{\pi : \pi \text{ is } \mathcal{G}\text{-adapted}, \int_0^t \pi(s)^2 (\alpha_s^2 + 1) ds < \infty \text{ a.s. and } \mathbb{E}[\log(\hat{V}(t))] < \infty\}$. Then, as before, we define the $\mathcal{G}$ stopping times $\tau_n = \inf\{t \geq 0; \int_0^t \pi(s)^2 ds \leq n\}$. First note that for any portfolio π we have that

$$\mathbb{E}\left[\log\left(\hat{V}(t \wedge \tau_n)\right)\right]$$
$$= \log V_0 + \mathbb{E}\left[\left(\int_0^{t \wedge \tau_n} \left((\mu(s) - r + \sigma(s)\alpha_s)\pi(s) - \frac{1}{2}\sigma^2(s)\pi(s)^2\right) ds\right.\right.$$
$$\left.\left. + \int_0^{t \wedge \tau_n} \sigma(s)\pi(s) d\widehat{W}(s)\right)\right].$$
$$\leq \log V_0 + \mathbb{E}\left[\int_0^{t \wedge \tau_n} \left((\mu(s) - r + \sigma(s)\alpha_s)\hat{\pi}(s) - \frac{1}{2}\sigma^2(s)\hat{\pi}(s)^2\right) ds\right]$$
$$= \log V_0 + \mathbb{E}\left[\int_0^{t \wedge \tau_n} \frac{(\mu(s) - r + \sigma(s)\alpha_s)^2}{2\sigma^2(s)} ds\right]$$

$$\leq \log V_0 + \mathbb{E}\left[\int_0^t \frac{(\mu(s)-r)^2}{2\sigma^2(s)} ds\right] + \mathbb{E}\left[\int_0^t \frac{\alpha_s^2}{2} ds\right]$$

$$= \log V_0 + \mathbb{E}\left[\int_0^t \frac{(\mu(s)-r)^2}{2\sigma^2(s)} ds\right] + \frac{1}{2}\log\left(\frac{T-t}{T}\right).$$

From here the result follows.

Solution 24. We consider the maximization of $E(V(t)^\theta)$ for $t < T$. As in Exercise 9, we have

$$\mathcal{A}_\theta(t) = \{\pi : \pi \text{ is } \mathcal{G}-\text{adapted}, \int_0^t \pi(s)^2 ds < \infty,$$

$$\int_0^t |\pi(s)\alpha(s)|\, ds < \infty \text{ and } E\hat{V}(t)^\theta < \infty\}.$$

As before, consider $\tau_n = \inf\{r \geq 0; \int_0^r \pi(s)^2 ds + \int_0^r |\pi(s)\alpha(s)|\, ds \leq n\}$ First note that for any portfolio π we have that

$$\mathbb{E}\left[\hat{V}(t\wedge\tau_n)^\theta\right] = V_0^\alpha \mathbb{E}^{\mathbb{Q}^n}\left[\exp\left(\theta \int_0^{t\wedge\tau_n}\left((\mu(s)-r+\sigma(s)\alpha(s))\pi(s) - \frac{1}{2}(1-\theta)\sigma^2(s)\pi(s)^2\right)ds\right)\right]$$

where

$$\frac{d\mathbb{Q}^n}{d\mathbb{P}} = \exp\left(-\frac{1}{2}\int_0^{t\wedge\tau_n}\sigma^2(s)\theta^2\pi(s)^2 ds + \theta\int_0^{t\wedge\tau_n}\sigma(s)\pi(s)d\widehat{W}(s)\right).$$

The function $f_s(\pi) = \theta(\mu(s)-r+\sigma(s)\alpha(s))\pi - \frac{1}{2}\theta(1-\theta)\sigma^2(s)\pi^2$ is a strictly concave function for $\theta \in (0,1)$ and its optimum portfolio process is given by $\pi^*(s) = \frac{\mu(s)-r+\sigma(s)\alpha(s)}{(1-\theta)\sigma^2(s)}$.

Solution 26. Considering $S(T) = x$ is equivalent to

$$W(T) = \sigma^{-1}\left(\log(x/S_0) - \mu T\right).$$

Therefore without loss of generality we consider $W(T) = x$. Repeating the same calculations as before we have that the conditional expectation of the logarithmic wealth is

$$\mathbb{E}\left[\log(\hat{V}(t))\Big| W(T) = x\right] = \log(V_0)$$

$$+ \int_0^t \mathbb{E}\left[\left(\mu(s)-r+\sigma(s)\frac{x-W_s}{T-s}\right)\pi(s) - \frac{1}{2}\sigma^2(s)\pi(s)^2\Big| W(T) = x\right] ds.$$

The optimal portfolio is also $\hat{\pi}(s) = \frac{1}{\sigma^2(s)}\left(\mu(s) - r + \sigma(s)\frac{x-W_s}{T-s}\right)$. The optimal portfolio value is

$$\log(V_0) + \mathbb{E}\left[\int_0^t \frac{(\mu(s)-r)^2}{2\sigma^2(s)}ds\right]$$
$$+ \int_0^t \mathbb{E}\left[(\mu(s)-r)\frac{x-W_s}{\sigma(s)(T-s)} + \frac{1}{2}\left(\frac{x-W_s}{T-s}\right)^2 \bigg| W(T) = x\right]ds.$$

Using conditional expectation of Gaussian random vectors we have that

$$\mathbb{E}\left[W(T) - W_s | W(T)\right] = W(T)(T-s)/T$$

and

$$\mathbb{E}\left[(W(T) - W_s)^2 \big| W(T)\right] = (T-s)s/T + (W(T)(T-s)/T)^2.$$

Therefore the optimal utility is

$$\log(V_0) + \mathbb{E}\left[\int_0^t \frac{(\mu(s)-r)^2}{2\sigma^2(s)} + \frac{x(\mu(s)-r)}{\sigma(s)T}ds\right] + \frac{x^2 t}{2T^2} - \frac{t}{2T} + \frac{1}{2}\log\left(\frac{T}{T-t}\right).$$

Solution 27. The equation $u^{\mathcal{F}}(t, V_0) = u^{\mathcal{G}}(t, V_0 - \rho_t(V_0, I))$ can be rewritten as

$$\log(V_0) = \log(V_0 - \rho_t(V_0, I)) + \frac{1}{2}\log\left(\frac{T}{T-t}\right).$$

Therefore

$$V_0\left(1 - \sqrt{\frac{T-t}{T}}\right) = \rho_t(V_0, I).$$

Note that as $t \to T$ then the value of the information I for the insider is V_0 as he can perform arbitrage. That is, the insider will be willing to exchange his/her information only if offered all the money of the other market players is transferred to him/her.

Solution 28.

$$\mathbb{E}\left[(W(t) - W(s))f(I)h_s\right] = \mathbb{E}\left[(W(t) - W(s))\int f(x)\frac{dP_t}{d\eta}(x)d\eta(x)h_s\right]$$
$$= \mathbb{E}\left[\int_s^t \int f(x)\frac{d}{du}\left\langle W, \frac{dP_\cdot}{d\eta}(x)\right\rangle_u d\eta(x)du h_s\right]$$
$$= \mathbb{E}\left[\int_s^t \int f(x)\beta(u,x)dP_u(x)du h_s\right],$$

where $\beta(u, I) = \alpha(u)$.

Solution 29. First we use the same arguments as before to obtain that

$$\mathbb{E}\left[\log(\hat{V}(t))\right] = \log(V_0) + \mathbb{E}\left[\int_0^t \left((\mu(s) - r + \sigma(s)\alpha_s)\pi(s) - \frac{1}{2}\sigma^2(s)\pi(s)^2\right) ds\right].$$

The optimal value of the strictly concave function $f_s(\pi) = (\mu(s) - r + \sigma(s)\alpha_s)\pi - \frac{1}{2}\sigma^2(s)\pi^2$ is $\hat{\pi}(s) = \frac{\mu(s)-r}{\sigma^2(s)} + \frac{\alpha_s}{\sigma(s)}$. Next, we note that $E(\alpha_s) = 0$ which follows from the semimartingale decomposition. That is, we have that for any $\mathcal{F}_s$ measurable random variable h_s

$$0 = \mathbb{E}[(W_t - W_s)h_s] = \mathbb{E}[(\widehat{W}_t - \widehat{W}_s)h_s] + \int_s^t \mathbb{E}[h_s \alpha_u] du.$$

Therefore $\mathbb{E}\left[\alpha(s)|\mathcal{F}_s\right] = 0$. Now we compute the optimal utility as

$$\mathbb{E}\left[\log(\hat{V}(t))\right] = \log(V_0) + \mathbb{E}\left[\int_0^t \left(\frac{1}{2\sigma^2(s)}(\mu(s) - r + \sigma(s)\alpha_s)^2\right) ds\right],$$

and the result follows.

Solution 31. The compensator is given by

$$\alpha(u) = \frac{W(u)(\sigma^2 - T) + \mu T}{(\sigma^2 - T)(u) + T^2}.$$

The optimal logarithmic utility in the interval $[0,T]$ is finite. Note that in this approach the knowledge of the insider given by μ and σ^2 are always constant throughout the interval $[0,T]$. Compare with Section 8.2.

Solution 32. Note that the numerator of the compensator is given by

$$\int \frac{x}{T-u} \frac{d\nu}{dP^{W(T)}}(x + W(u)) \frac{\exp\left(-\frac{x^2}{2(T-u)}\right)}{\sqrt{2\pi(T-u)}} dx.$$

Integrating by parts with respect to x and taking in consideration that the denominator normalizes the measure, we have that

$$\mathbb{E}^\nu\left[\alpha(u)^2\right] \leq \mathbb{E}^\nu\left[\left(\frac{\partial}{\partial x}\log\left(\frac{d\nu}{dP^{W(T)}}\right)\right)^2 (W(T))\right].$$

Solution 34. First, we compute the compensator of W^2 in the enlarged filtration:

$$\mathbb{E}\left[(W^2(t) - W^2(s))f(W^1(T))h_s\right]$$
$$= \mathbb{E}\left[\int_s^t \mathbb{E}\left[\frac{W^2(T) - W^2(u)}{T-u}\bigg|W^1(T) - W^1(u)\right] du f(W^1(T))h_s\right].$$

Since $\mathbb{E}\left(W^2(T) - W^2(u)\mid W^1(T) - W^1(u)\right) = \rho\left(W^1(T) - W^1(u)\right)$, repeating the same sequel of calculations as in Exercise 22, the optimal portfolio is

$$\widehat{\pi}(s) = \frac{1}{\sigma^2(s)}\left(\mu(s) - r + \sigma(s)\rho\frac{W_T^1 - W_s^1}{T - s}\right).$$

The expected logarithmic utility in $[0, T]$ is infinite for any $\rho > 0$.
In the second case the calculation is similar, except that the compensator will be

$$\mathbb{E}\left[\frac{W^2(T) - W^2(u)}{T - u}\bigg| \mathcal{F}_u^2 \vee \sigma\left(W^1(T)\right)\right] = \rho\frac{W^1(T) - \rho W^2(u)}{T - \rho^2 u}.$$

Therefore

$$\widehat{\pi}(s) = \frac{1}{\sigma^2(s)}\left(\mu(s) - r + \sigma(s)\rho\frac{W^1(T) - \rho W^2(s)}{T - \rho^2 s}\right)$$

and the optimal logarithmic utility is finite and given by

$$\log(V_0) + \mathbb{E}\left[\int_0^t \frac{(\mu(s) - r)^2}{2\sigma^2(s)}ds\right] - \rho^2\log\left(1 - \rho^2\right).$$

The first model corresponds to two correlated assets where the insider has information about the first but he/she is restricted to trade only on the second asset while observing the evolution of the first. In the second model the insider has related information about the first asset which is not traded in the market (for example, the volatility of the asset).

Solution 35. The argument is the same as in Section 3. In fact the optimal solution is just the projection on the interval $[0, 1]$ of the solution without constraints.

Solution 36. Following the result of exercise 18 and 29 we have to compute

$$\mathbb{E}\left[\int_0^T \left(\frac{\int_u^T a(r)dW_r}{A(u)} a(u)\right)^2 du\right] = -\left[\log(A(u))\right]_0^T = +\infty.$$

Solution 37. For case 1 we use Theorem 6. Due to this theorem we have that the extra utility of the insider is given by $\mathbb{E}[\log p_T(I)]$. First we have that

$$p_t(x) = \frac{\mathbb{P}(W(T) \geq a\mid \mathcal{F}_t)}{\mathbb{P}(W(T) \geq a)}1(x = 1) + \frac{\mathbb{P}(W(T) < a\mid \mathcal{F}_t)}{\mathbb{P}(W(T) < a)}1(x = 0).$$

Therefore

$$\mathbb{E}[\log p_t(I)] = \mathbb{E}\left[\mathbb{P}(W(T) \geq a\mid \mathcal{F}_t)\log\left(\mathbb{P}(W(T) \geq a\mid \mathcal{F}_t)\right)\right]$$
$$+ \mathbb{E}\left[\mathbb{P}(W(T) < a\mid \mathcal{F}_t)\log\left(\mathbb{P}(W(T) < a\mid \mathcal{F}_t)\right)\right]$$
$$- \left(\mathbb{P}(W(T) \geq a)\log\left(\mathbb{P}(W(T) \geq a)\right) + \mathbb{P}(W(T) < a)\log\left(\mathbb{P}(W(T) < a)\right)\right).$$

In particular

$$\mathbb{E}[\log p_T(I)]$$
$$= -\left(\mathbb{P}(W(T) \geq a) \log\left(\mathbb{P}(W(T) \geq a)\right) + \mathbb{P}(W(T) < a) \log\left(\mathbb{P}(W(T) < a)\right)\right)$$

which is obviously finite. For the second case, we obtain that

$$P_t^I(dx) = \frac{e^{-\frac{(x-W_t)^2}{2(T-t+1)}}}{\sqrt{2\pi(T-t+1)}} dx.$$

Applying Jacod's theorem (see Theorem 3) with $\eta(dx) = dx$ we have that

$$W_t = \hat{W}_t + \int_0^t \frac{W_T + \varepsilon - W_u}{T - t + 1} du$$

and the optimal logarithmic utility is given by

$$\log(V_0) + \mathbb{E}\left[\int_0^T \frac{(\mu(s)-r)^2}{2\sigma^2(s)} ds\right] + \log\left(\frac{T+1}{T-t+1}\right).$$

Note that this is finite even for $t = T$ but in this model $\mathcal{G}_t$ gives always the same deformed information, $W(T) + \varepsilon$, to the insider even when t is close to T.

Solution 38. We leave the first part to the reader. Suppose that Z and Z' are two Lévy processes with the same characteristics. Then we want to compute for $X(s) = Z(T) + Z'((T-s)^\theta)$ the quantity

$$\mathbb{E}\left[(Z_t - Z_u) f(Z_T - Z_s + Z'((T-s)^\theta))\right].$$

Due to the independence of the increments and the invariance of the law of the increments (which only depend on the size of the interval) we have that if $s < t < u < T$ with $t - u = \frac{T-s}{2}$ then

$$2\mathbb{E}\left[(Z_t - Z_u) f(Z_T - Z_s + Z'((T-s)^\theta))\right]$$
$$= \mathbb{E}\left[(Z_T - Z_s) f(Z_T - Z_s + Z'((T-s)^\theta))\right].$$

Then by continuity of the expectation in the time variables we have that

$$\mathbb{E}\left[(Z_t - Z_u) f(Z_T - Z_s + Z'((T-s)^\theta))\right]$$
$$= \frac{t-u}{T-s} \mathbb{E}\left[(Z_T - Z_s) f(Z_T - Z_s + Z'((T-s)^\theta))\right].$$

A similar argument also gives that

$$\mathbb{E}\left[(Z_t - Z_u) f(Z_T - Z_s + Z'((T-s)^\theta))\right]$$
$$= \frac{t-u}{T-s+(T-s)^\theta} \mathbb{E}\left[(Z_T - Z_s + Z'((T-s)^\theta)) f(Z_T - Z_s + Z'((T-s)^\theta))\right].$$

From here the result follows.

Solution 40. First note that $\mathcal{F}_{s,T} = \mathcal{F}_s \vee \sigma(Z_r; r \geq T) = \mathcal{F}_s \vee \sigma(Z_T) \vee \sigma(Z_T - Z_t; r \geq t)$, where the last sigma algebra is independent of the other two. Therefore we have using Theorem 9 that

$$\mathbb{E}[Z_t - Z_s | \mathcal{F}_s \vee \sigma(Z_T)]$$
$$= \mathbb{E}\left[\widehat{Z}_t - \widehat{Z}_s \middle| \mathcal{F}_s \vee \sigma(Z_T)\right] + \int_s^t \frac{\mathbb{E}[Z_T - Z_u | \mathcal{F}_s \vee \sigma(Z_T)]}{T - u} du.$$

Then setting $\phi(t) = \mathbb{E}[Z_t - Z_s | \mathcal{F}_s \vee \sigma(Z_T)]$, we have that the following ordinary differential equation is satisfied

$$\phi(t) = \int_s^t \frac{Z_T - Z_s}{T - u} du - \int_s^t \frac{\phi(u)}{T - u} du,$$

whose unique solution is

$$\mathbb{E}[Z_t - Z_s | \mathcal{F}_s \vee \sigma(Z_T)] = \frac{Z_T - Z_s}{T - s}(t - s).$$

Therefore

$$\mathbb{E}\left[\frac{Z_t - Z_s}{t - s} \middle| \mathcal{F}_{s,T}\right] = \frac{Z_T - Z_s}{T - s}.$$

Furthermore for $u \in (s, t)$

$$\mathbb{E}\left[\frac{Z_t - Z_u}{t - u} \middle| \mathcal{F}_{s,T}\right] = \mathbb{E}\left[\mathbb{E}\left[\frac{Z_t - Z_u}{t - u} \middle| \mathcal{F}_{u,T}\right] \middle| \mathcal{F}_{s,T}\right]$$
$$= \mathbb{E}\left[\frac{Z_T - Z_u}{T - u} \middle| \mathcal{F}_{s,T}\right]$$
$$= \frac{Z_T - Z_s}{T - u} + \mathbb{E}\left[\frac{Z_s - Z_u}{T - u} \middle| \mathcal{F}_{s,T}\right]$$
$$= \frac{Z_T - Z_s}{T - s}.$$

Solution 41. First prove that for $s < t_1 < t_2 < t$

$$\mathbb{E}\left[(Z_{t_2} - Z_{t_1})^2 \middle| \mathcal{G}_s\right]$$
$$= \mathbb{E}\left[(Z_{t_2} - Z_{t_1})^2 \middle| Z_T - Z_s\right]$$
$$= (Z_T - Z_s)(Z_T - Z_s - 1)\left(\frac{t_2 - t_1}{T - s}\right)^2 + (Z_T - Z_s)\frac{t_2 - t_1}{T - s}.$$

Consider for the partition $s = t_0 < t_1 < ... < t_n = t$ then the quantity

$$\mathbb{E}\left[\sum_{i=0}^{n-1}\left\{(Z_{t_{i+1}} - Z_{t_i})^2 - \frac{Z_T - Z_{t_i}}{T - t_i}(t_{i+1} - t_i)\right\} \middle| \mathcal{G}_s\right]$$
$$= \mathbb{E}\left[\sum_{i=0}^{n-1} (Z_T - Z_{t_i})(Z_T - Z_{t_i} - 1)\left(\frac{t_{i+1} - t_i}{T - t_i}\right)^2 \middle| \mathcal{G}_s\right]$$

goes to zero a.s. and therefore taking limits with respect to the norm of the partition we have that

$$\mathbb{E}\left[[Z]_t - [Z]_s - \int_s^t \frac{Z_T - Z_u}{T-u} du \bigg| \mathcal{G}_s \right] = 0.$$

From here the result follows. Similarly be taking approximations of pure jump Lévy processes one can also prove that

$$\mathbb{E}\left[[Z]_t - [Z]_s - \int_s^t \frac{[Z]_T - [Z]_u}{T-u} du \bigg| \mathcal{G}_s \right] = 0.$$

A similar calculation also leads to the fact that the formula is valid for general square integrable Lévy processes.

Solution 43. If there are only positive jumps and $\mu - r \geq 0$ then there is arbitrage and in fact any investment on the underlying will provide a positive return. On the contrary if $\mu - r < 0$ then the function f becomes

$$f(\pi) = (\mu - r)\pi + \lambda^+ \log\left(1 + (e^a - 1)\pi\right).$$

This strictly concave function has its optimum at $\pi^* = -\frac{1}{e^a-1} - \frac{\lambda^+}{\mu-r}$ and the optimal logarithmic utility is

$$\log(V_0) - \left(\lambda^+ + \frac{\mu-r}{e^a-1} - \lambda^+ \log\left(-\frac{\lambda^+(e^a-1)}{\mu-r}\right)\right)T$$

Solution 44. Let x such that $\mathbb{P}(N(T) = x) > 0$, then there exists a unique pair (l_1, l_2) of natural numbers such that $l_1 a + l_2 \ln(2 - e^a) = x$. Then $N(T) = x$ implies that $N^+(T) = l_1$ and $N^-(T) = l_2$ and therefore the two filtrations coincide.

Solution 47. It is enough to note that the conditional logarithmic utility can be written as

$$\mathbb{E}\left[\log(\hat{V}_t) \bigg| N(T)\right] = \log(V_0) + \mathbb{E}\left[\int_0^t (\mu - r)\pi(s)ds \bigg| N(T)\right]$$
$$+ \sum_{i=-}^{+} \mathbb{E}\left[\int_0^t \log(1 + \pi(s)(e^{a_i} - 1))B_i(s)ds \bigg| N(T)\right].$$

Solution 48. In such a case, we have that $B_i(s) = \frac{N_i(T) - N_i(s)}{T-s}$ then an arbitrage is to wait until $B_i(s) = 0$ and invest all resources in the asset if $i = -$ or borrow the asset if $i = +$.

Solution 49. Try the following portfolio:

$$\pi_s = \theta(a\frac{W(T) - W(s)}{T-s} + b\frac{N_T - N_s}{T-s}) \vee 0.$$

Find a set of constants a and b such that the utility is infinite.

Solution 50. We only give the main idea of the solution. First obtain that on the set $1(T_n > t)$ we have that

$$\mathbb{E}\left[\frac{N_t^+}{T_n - t} - \frac{N_s^+}{T_n - s} \Big/ \mathcal{G}_s\right] = \frac{(n-1)(t-s)}{(T_n - t)(T_n - s)}.$$

Finally prove that $\frac{N_{T_n^-}^+ - N_u^+}{T_n - u}$ is a $\mathcal{G}_u$ martingale on $T_n > u$ to conclude.

Solution 51. This exercise follows as in the proof of Theorem 10 and at the end is necessary to compute the joint law of (T_{n-1}, T_n) to prove that

$$\mathbb{E}\Big(\int_0^{+\infty} \frac{1(T_{n-1} > u)}{T_n - u} du\Big) < \infty.$$

In fact, in the interval $[0, T_{n-1}]$ we will have that the wealth is smaller than

$$\log(V_0) + \lambda^- \log(2)\mathbb{E}\left[\int_0^\infty 1(s < T_{n-1}) ds\right]$$
$$+ \log(2)\mathbb{E}\left[\int_0^\infty 1(s < T_{n-1})\frac{n-1-N_s^+}{T_n - s} ds\right].$$

The last expectation above is bounded by

$$\int_{t_1 < \ldots < t_n} \frac{e^{-\lambda^+ t_n} 1(t_{n-1} > s)}{t_n - s} (\lambda^+)^n dt_1 \ldots dt_n$$
$$= (\lambda^+)^n \int_s^\infty \frac{e^{-\lambda^+ t_n} (t_n^{n-1} - s^{n-1})}{(n-1)!(t_n - s)} dt_n < \infty.$$

Solution 52. The analysis in the interval $[T_{n-1}, T_n]$ follows as in Case II(b) analyzed previously. For $\mu \leq r$ is enough to note that there is a positive probability that there is going to be a negative jump in the interval $[T_{n-1}, T_n]$ which gives an infinite utility. It is also interesting that in the case that the interval goes beyond T_n then there is no optimum for the problem in the case $\mu > r$.

Solution 53. Note that

$$\mathbb{E}\left[f(t, W(T))(W(t_{i+1}) - W(t_i))\right] = (t_{i+1} - t_i)\mathbb{E}\left[\frac{\partial}{\partial x} f(t, W(T))\right].$$

Solution 56. Find the solution of equation (19) as explicitly as possible and prove that it generates the same filtration as the one generated by $Y - \sigma(T-t)^\theta \int_0^s (T-r)^\theta dW(r)$, $s \leq t$ and conclude.

Solution 57. Consider the definition of semimartingale as given in Protter page 52. If W is a $(\mathcal{F}_{t+\delta})$-semimartingale, then for any partition whose norm tends to zero and always smaller than δ, consider the process

$$H(t) = \sum_{i=0}^{n-1}(W(t_{i+1}) - W(t_i))1_{(t_i, t_{i+1}]}(t).$$

This process is then $(\mathcal{F}_{t+\delta})$-adapted and converges uniformly to zero but its stochastic integral converges to the quadratic variation of W leading to a contradiction.

Solution 59. The model proposed by the small trader is an adapted model with random drift. Therefore the analysis to obtain the optimal portfolio for the logarithmic utility follows the same lines as in Section 2. The optimal porfolio is given by

$$\widetilde{\pi}^*(t) = \frac{\mu - r}{\sigma^2} + \frac{b\mathbb{E}\left[W(t+\delta)|\mathcal{H}_t\right]}{\sigma^2}.$$

This gives

$$\widetilde{J}(t, \widetilde{\pi}^*) = \log(V_0) + \frac{1}{2\sigma^2}\mathbb{E}\left[\int_0^t (\mu - r + b\mathbb{E}\left[W(s+\delta)|\mathcal{H}_s\right])^2 ds\right],$$

while

$$J(t, \widetilde{\pi}^*) = \widetilde{J}(t, \widetilde{\pi}^*) + \sigma \mathbb{E}\left[\int_0^t \widetilde{\pi}^*(s) d^- W(s)\right].$$

Finally as in Exercise 53, one proves that

$$\mathbb{E}\left[\int_0^t \widetilde{\pi}^*(s) d^- W(s)\right] = \int_0^t D_{s+}\widetilde{\pi}^*(s) ds,$$

where

$$D_{s+}\widetilde{\pi}^*(s) = bM\left(\sigma + \frac{b(s+\delta) + \delta}{\sigma^2}\right)\int_{t-\delta}^t bg(t, u) du \geq 0.$$

For more details on the notation D_{s+} see Kohatsu-Sulem.

Solution 60. As in the proof of the Theorem 15 $\mathcal{H}$ is generated by Y. Using equation (24) we have that

$$\mathbb{E}\left[Y(t) - Y(s)|\mathcal{H}_s\right] = \sigma^2 \mathbb{E}\left[\int_s^t \pi^*(s) ds \Big| \mathcal{H}_s\right] - (\mu - r)(t-s).$$

Define $B(t) = \sigma^2 \int_0^t \pi^*(s) ds - (\mu - r)t$ and finish the proof as in the proof of the Theorem 15.

Solution 61. The solution of this exercise is long and requires some knowledge of anticipating calculus. The main steps are as follows. First rewrite

$$\int_0^T \pi^*(s) d^- W(s) = \int_0^T \phi_1(t) dW(t) + \int_0^T \phi_2(t) d_- W(t),$$

for some specific stochastic processes ϕ_1 and ϕ_2 where ϕ_1 is adapted and ϕ_2 is adapted to the backward filtration. The notation d_- denotes the backward Itô integral. Then prove that

$$\mathbb{E}\left[\int_0^T \phi_1(t)dW(t) \int_0^T \phi_2(t)d_-W(t)\right]$$
$$= \mathbb{E}\left[\int_0^T \phi_1(t)\phi_2(t)dt + \int_0^T \int_0^s D_s\phi_2(u)D_u\phi_1(s)duds\right]$$

and that the expectation of the last term above is strictly positive for $\mu \geq r$.

Solution 61. We only sketch the solution: In this case note that for the small trader we will have that

$$\mathcal{H}_t = \sigma(S(s); s \leq t) = \sigma(\int_0^s bW(\theta + \delta)d\theta + \sigma W(s); s \leq t)$$

We will now deduce an anticipating Girsanov's theorem that will allow us to apply it to the above model.

For this, set $I(t) = \mu + bW(t + \delta)$. As the portfolios of the small trader have to be adapted to $\mathcal{H}$, let us suppose that

$$\pi(t_i) = \pi(\sum_{k=0}^{j-1} I(t_k)\Delta + \sigma W(t_j), j \leq i - 1)$$
$$I(t_j) = I(t_j, W(t_{i+1}) - W(t_i); i = 0, ..., n - 1)$$

where $t_{j+1} - t_j = \Delta = T/n$. Now consider the following expression

$$\mathbb{E}^{\mathbb{Q}^n}\left[\sum_{j=0}^{n-1}\{\pi(t_j)I(t_j)\Delta + \sigma\pi(t_j)(W(t_{j+1}) - W(t_j))\}\right]$$
$$= \int_{\mathbb{R}^{2n}} \sum_{j=0}^{n-1} \pi(t_j)\{I(t_j)\Delta + \sigma z_j\} \frac{\exp\left(-\frac{|z|^2}{2\Delta}\right)}{(2\pi\Delta)^{n/2}} \frac{d\mathbb{Q}^n}{d\mathbb{P}} dz,$$

where

$$\frac{d\mathbb{Q}^n}{d\mathbb{P}} = \exp\left(-\sum_{i=0}^{n-1} \frac{z_i\sigma^{-1}I(t_i)\Delta + (\sigma^{-1}I(t_i)\Delta)^2}{2\Delta}\right).$$

Now we will perform the following change of variables for $j = 0, ..., n - 1$

$$w_i = \sigma^{-1}I(t_i)\Delta + z_i.$$

To compute the inverse of the jacobian $J_n = \left(\frac{\partial w_i}{\partial z_j}\right)$, we need to compute

$$\frac{\partial w_i}{\partial z_j} = \sigma^{-1}\frac{\partial I(t_i)}{\partial z_j}\Delta + I_{ij}.$$

Now we consider the particular case that $I(t) = \mu + bW(t+\delta)$ with fixed $\delta = T/2$. Then we can rewrite

$$\frac{\partial w_i}{\partial z_j} = \sigma^{-1}bI(t_{j+1} \leq t_i + \delta)\Delta + I_{ij}.$$

After some heavy algebraic manipulation with the jacobian matrix J one finds that

$$\det(J_n) = \left\{1 + \sigma^{-1}b\sum_{j=0}^{j_0}\left(1 + \frac{\sigma^{-1}b\Delta j}{\sigma^{-1}b\Delta + 1}\right)^{-1}\Delta\right\}\left(1 + \sigma^{-1}b\Delta(j_0 + 2)\right),$$

if $b > 0$ then the above quantity is strictly positive and the change of variables is allowed. Therefore we have that

$$\mathbb{E}^{\mathbb{Q}^n}\left[\sum_{j=0}^{n-1}\{\pi(t_j)I(t_j)\Delta + \sigma\pi(t_j)(W(t_{j+1}) - W(t_j))\}\right]$$

$$= \mathbb{E}^{\mathbb{Q}^n}\left[\sum_{j=0}^{n-1}\left\{\pi(t_j)(\widehat{W}(t_{j+1}) - \widehat{W}(t_j))\right\}\right]\det(J_n)^{-1},$$

where $\widehat{W}(t_k) = W(t_k) + \sum_{j=0}^{j=k-1}I(t_j)\sigma^{-1}\Delta$ is a Wiener process in the filtration $\mathcal{H}$. Therefore by taking limits we have that there exists an equivalent martingale measure therefore not allowing for arbitrage.

Large Investor Trading Impacts on Volatility

Pierre-Louis Lions[1] and Jean-Michel Lasry[2]

[1] CEREMADE - UMR C.N.R.S. 7534 and Collège de France
Université Paris 9 - Dauphine   11, place Marcelin Berthelot
Place du Maréchal de Lattre de Tassigny   75005 Paris
75775 Paris Cedex 16

[2] CALYON
9, Quai du Président Paul Doumer
92920 PARIS-LA DÉFENSE CEDEX

Summary. This is the first paper in a series devoted to a tentative model for the influence of hedging on the dynamics of an asset. We study here the case of a "large" investor and solve two problems in the context of such a model: the question of the fair value (or liquidation value) of a "large" position and the question of pricing or hedging an option. In order to do so, we use a utility maximization approach and some new results in stochastic control theory.

Keywords: Stochastic control, Large investor impact, Utility pricing, Liquidation value
Mathematics Subject Classification 91B24

1 Introduction

This paper is the first in a series devoted to attempt to reconcile the classical Black-Scholes theory of option pricing and hedging with various phenomena observed in the markets such as the influence of trading and hedging on the dynamics of an asset. Assuming, for simplicity, that the price of an asset is modified by trading and hedging through a linear "elastic" law, we shall prove that the optimal hedging strategy derived from a utility maximization principle impacts upon the volatility of the asset.

More precisely, we shall consider here the case of a "large" investor whose trading influences the dynamics of an asset price. We study two related problems:

i) what is the "liquidation" value of a position? As is well known, this value that might be called the fair value is not simply derived from the market price (as is natural, since selling a sizable number will immediately make the price of the asset go down $\cdots$). Our approach will, in some sense, allow to reconcile the notion of fair value with market prices (mark to market). And we shall also see the impact on volatility;

ii) how is the fundamental option pricing and hedging theory of F. Black and M. Scholes [2], and R. Merton [9] modified as a consequence of our basic assumptions on the price dynamics?

At this stage, and before describing more precisely our main results, we wish to mention some previous related works concerned with some "phenomenological" models of the influence factor we are investigating and we refer to A.S. Kyle [10], K. Back [2], K. Back, C.H. Cao and G. Willard [3], H. Föllmer [5], R. Frey and A. Stremme [9]. In fact, we discovered these references after introducing and testing closely related phenomenological models for some particular type of financial products. And, when we observed some practical relevance, our goal was to study this type of phenomenon and derive rigorously from first principles such models or rigorous models related to those phenomenological ones.

Let us also point out that our second paper in this series [11] will be devoted to an attempt at a rigorous theory incorporating the price influences of a "large" number of "small" (compared to the total market activity) investors/traders in a self-consistent way and leading to a tentative theory for the formation of volatility.

Finally, we wish to emphasize the fact that, although we modify drastically the basic hypothesis on the asset price dynamics (as an exogenous diffusion process a priori determined), the classical Black-Scholes theory is quite resilient and our results will clearly show that its basic structure remains the same even though the precise details are indeed completely changed.

We conclude this long introduction by introducing our main notation and assumption, and explaining more precisely the problems we solve below. We begin with a simple model due to A.S. Kyle [10] (see also K. Back [2], K. Back, C.H. Cao and G. Willard [3]) for the dynamics of a single underlying asset which we assume to be given by

$$dS_t = \sigma(S_t)dB_t - ku_t dt, \qquad (1)$$

where $(\Omega, \mathcal{F}, \mathcal{F}_t, \mathbb{P}, B_t)$ is a standard Wiener space (B_t is a standard Brownian motion), σ is a given function (assumed, for instance, to be bounded from above and from below away from 0, and Lipschitz on $\mathbb{R}$) that corresponds to some a priori exogenous volatility and k is a positive parameter. The term $(u_t dt)$ corresponds to the rate at which a single "large" investor sells or buys the asset, thus modifying instantaneously the price S_T by a simple linear law (of offer and demand). The quantity of the asset owned by this investor is then denoted by α_t and we thus have

$$d\alpha_t = -u_t dt. \qquad (2)$$

The corresponding wealth in the classical Black-Scholes theory is always taken to be $\alpha_t S_t$ i.e. the amount of cash obtained from liquidating the position. This is of course meaningless if $k \neq 0$ and we thus cannot define a priori the wealth (it is in fact precisely one of the issues we consider and solve below···). We thus simply consider the cash C_t used in the trading of S_t namely

$$dC_t = u_t S_t dt. \qquad (3)$$

Observe that, in order to simplify both the presentation and the notation, we do not incorporate any interest rate (taken to be 0).

We may now formulate explicitly the first problem considered in this paper: let $T > 0$, given a position α at time t, what is the "maximal" amount of cash we may generate if we request that $\alpha_T = 0$. In order to formulate precisely this problem that corresponds to a notion of fair value or liquidation value, we use a utility function maximization approach and we consider the following optimal stochastic control problem

$$V = \sup_{u \in \mathcal{U}} \left\{ \mathbb{E}\left[U(C_T) \,|\, C_t = C,\, S_t = S,\, \alpha_t = \alpha \right] \right\} \tag{4}$$

where

$$\mathcal{U} = \left\{ u \text{ adapted},\, \mathbb{E}\left[\int_t^T |u_s|\, ds \right] < \infty; \alpha_T = 0 \right\}, \quad t < T, \tag{5}$$

and $V = V(C, S, \alpha, t) \in (-\infty, +\infty]$, where U is a given utility function that we assume, in order to simplify the presentation (although we shall consider more general ones below), to be C^2 on $\mathbb{R}$, strictly increasing on $\mathbb{R}$ and $U''(z) < 0$ for all $z \in \mathbb{R}$. A typical example is the exponential utility namely

$$U(z) = U_\lambda(z) = 1 - \exp(-\lambda z) \quad \text{for all } z \in \mathbb{R}, \tag{6}$$

where $\lambda > 0$ is the absolute risk aversion.

In Section 2 below, we completely solve this problem and we consider as well extensions allowing for some drift terms, or relaxing the final target constraint namely $\alpha_T = 0$, or replacing (1) by

$$\frac{dS_t}{S_t} = \sigma(S_t)\, dB_t - k u_t dt \tag{1'}$$

As we shall see, we derive the following explicit "fair" values

$$P = \alpha S - k \frac{\alpha^2}{2} \tag{6}$$

in the case of (1), and

$$P = \frac{1 - e^{-k\alpha}}{\alpha} S \tag{6'}$$

in the case of (1').

Next, in Section 3, we study the issue of option pricing in such a context. Namely, we replace (4) by

$$V = \sup_{u \in \mathcal{U}} \left\{ \mathbb{E}\left[U(C_T - \Phi(S_T)) \,|\, C_t = C,\, S_t = S,\, \alpha_t = \alpha \right] \right\} \tag{7}$$

where Φ is the pay-off of a European option. We shall also consider the case when k is small ($k = \epsilon$) or, more interestingly, when Φ is replaced by $\epsilon \Phi$ and U by $U\left(\frac{\cdot}{\epsilon}\right)$, and then we compute the corrections to the classical Black-Scholes theory. Let us mention that we also consider in that section the more natural situation when we do not require α_T to vanish.

As we shall see, both in Sections 2 and 3, not only do we solve the above stochastic control problems, but we deduce through the optimal strategy, the resulting price dynamics. And, as we shall explain, this class of stochastic control problems is a rather new class of *singular* stochastic control problems which cannot be handled by the classical methods (e.g. Hamilton-Jacobi-Bellman equations) and we have to rely upon the theory we built in [12] in order to accommodate such problems. However, for the "simple" problems above, it is possible to make a direct proof that we shall present, although, this argument uses some tricks whose explanation is a direct consequence of [12]. At this stage, we simply announce that the resulting price dynamics correspond to a *modified* volatility!

Finally, in Section 4, we briefly mention some possible extensions to an arbitrary number of assets, more general "influence" models, and models with interest rates ···

After this series of works was completed, we became aware of some independent works by M. Avellaneda and M.D. Lipkin [1] and by G. Lasserre [13] that are somewhat related to the topics addressed here.

2 Fair Values

We begin with the problem (4), or with a more general version where we relax the final target constraint namely $\alpha_T = 0$:

$$V = \sup_u \left\{ \mathbb{E} \left[U(C_T - \frac{W}{2} \alpha_T^2) \,|\, C_t = C,\, S_t = S,\, \alpha_t = \alpha \right] \right\}. \tag{8}$$

At least formally, we recover the problem (4) letting the positive (weight) parameter W go to $+\infty$. And we first consider the case when $\sigma(S)$ is a constant : $\sigma(S) = \sigma > 0$.

As we mentioned in the introduction, this type of stochastic control problem cannot be handled by classical methods such as, e.g., the corresponding Hamilton-Jacobi-Bellman equation (HJB in short). We refer the reader to the monographs by W.H. Fleming and H.M. Soner [7], M. Bardi and I. Capuzzo-Dolcetta [4] for a presentation of the classical methods. Indeed, at least formally, the dynamic programming principle "leads" to the following (formal) HJB equation

$$\frac{\partial V}{\partial t} + \frac{\sigma^2}{2} \frac{\partial^2 V}{\partial S^2} + \sup_{u \in \mathbb{R}} \left\{ u \left(S \frac{\partial V}{\partial C} - k \frac{\partial V}{\partial S} - \frac{\partial V}{\partial \alpha} \right) \right\} = 0$$

with $V\big|_{t=T} = U\left(C - \frac{W}{2} \alpha^2\right)$. This equation does not contain enough information in order to characterize V since the supremum in u is always infinite unless we have

$$S \frac{\partial V}{\partial C} - k \frac{\partial V}{\partial S} - \frac{\partial V}{\partial \alpha} = 0.$$

This equation means, at least formally, that V should be left invariant by the flow

$$\Psi_\mu(C, S, \alpha) = (C + \mu S - k\frac{\mu^2}{2}, S - k\mu, \alpha - \mu) \quad (\mu \in \mathbb{R}) \tag{9}$$

which is nothing else than the integral flow associated to

$$\dot{S} = -k, \quad \dot{\alpha} = -1, \quad \dot{C} = S$$

(or (1)-(3) with $\sigma \equiv 0$ and $u \equiv 1$). Observe also that V at time T (namely $V = V(C - \frac{W}{2}\alpha^2)$) is not invariant by this flow!

This type of stochastic control problem is completely solved in J.-M. Lasry and P.-L. Lions [12] where it is shown that V is indeed invariant by the flow (9) for $t < T$ (with a terminal boundary layer at $t = T$) and that it is possible to write for V a *reduced* stochastic control problem (in some vague sense, a stochastic control problem on the quotient space induced by the orbits of the flow) which can then be solved by the classical methods of stochastic control theory.

Instead of applying blindly this general theory, we extract from it some crucial facts and present some direct arguments (let us emphasize however that this is possible here only because we carefully selected problems that are simple enough). Indeed, the "expected" invariance of V by the flow means that V should only depend upon two reduced variables and a natural choice, in view of (9), is

$$P = C + \alpha S - k\frac{\alpha^2}{2}, \quad \text{and} \quad X = S - k\alpha. \tag{10}$$

As we shall see later on, P corresponds to the total wealth (or cash) and X is the price of the asset when one takes α instantaneously to 0. Let us mention at this stage that similar choices of "reduced" variables were introduced previously by K. Back, C.H. Cao and G.A. Willard [3].

We thus reformulate (1)-(3), (4) and (8) with this new set of variables: we obviously find

$$dX_t = \sigma \, dB_t \tag{11}$$

(here we use the fact that σ is a constant), and

$$dP_t = dC_t + \alpha_t dS_t + S_t d\alpha_t - k\alpha_t d\alpha_t$$
$$= u_t S_t dt + \alpha_t(\sigma dB_t - ku_t dt) - S_t u_t dt + ku_t \alpha_t dt$$

or

$$dP_t = \alpha_t \sigma dB_t = \alpha_t dX_t, \tag{12}$$

i.e. the classical evolution of wealth in terms of the hedge α_t.

Using the invariance and the constraint $\alpha_T = 0$, we then deduce that (4) is "equivalent" for $t < T$ to

$$V(P, X, t) = \sup \left\{ \mathbb{E}\left[U(P_T)\, \big|\, P_t = P, \ X_t = X\right] \big/ \alpha_t \text{ adapted} \right\}. \tag{13}$$

In the case of (8), we need to introduce

$$\sup_{\mu \in \mathbb{R}} U\left(C_\mu - \frac{W}{2}\alpha_\mu^2\right) = U\left(\sup_{\mu \in \mathbb{R}} \left(C_\mu - \frac{W}{2}\alpha_\mu^2\right)\right)$$

where $C_\mu = C + \mu S - k\frac{\mu^2}{2}$, $\alpha_\mu = \alpha - \mu$. We observe that

$$\begin{aligned}
\sup_{\mu \in \mathbb{R}} \left(C_\mu - \frac{W}{2}\alpha_\mu^2\right) &= C - \frac{W}{2}\alpha^2 + \frac{1}{2}\frac{(S+\alpha W)^2}{k+W} \\
&= P - \alpha S + \frac{k\alpha^2}{2} - \frac{W}{2}\alpha^2 + \frac{1}{2}\frac{(X+\alpha(k+W))^2}{k+W} \\
&= P - \alpha X - \frac{k+W}{2}\alpha^2 + \frac{1}{2}\frac{(X+\alpha(k+W))^2}{k+W} \\
&= P + \frac{1}{2}\frac{X^2}{k+W}.
\end{aligned}$$

And we claim that (8) is "equivalent" for $t < T$ to

$$\begin{cases} V(P,X,t) = \sup\left\{\mathbb{E}\left[U\left(P_T + \frac{1}{2(k+W)}X_T^2\right) \,\Big|\, P_t = P, X_t = X\right]; \\ \alpha_t \text{ adapted}\right\}. \end{cases} \quad (14)$$

From a rigorous mathematical viewpoint, the equivalence is a simple application of our results in [12] and we do not repeat the detailed proof here. However, it is possible to make a convincing argument for the above modification of the terminal reward function $U(C - \frac{W}{2}\alpha^2)$: indeed, as soon as t is strictly smaller than T (and very close to T, $t = T - \epsilon$, we may use as possible controls $u_\epsilon \equiv \frac{\mu}{\epsilon}$ on $[T-\epsilon, T]$ in which case we deduce easily that $\alpha_T = \alpha - \mu$, $S_T \approx S - k\mu$ and $C_T \approx C + \mu S - k\frac{\mu^2}{2}$. Therefore, we may "jump instantanously" (or almost) along the orbit of the flow and thus we should only consider the supremum along the orbit of the flow of the final reward function which is precisely the quantity we computed above.

We are led to the following result:

Theorem 1. *Assume that σ is constant.*

1) The value function V given by (13) satisfies

$$V = U\left(C + \alpha S - k\frac{\alpha^2}{2}\right) \quad \text{for all } C, S \in \mathbb{R} \quad (15)$$

2) The value function V given by (14) satisfies

$$V = U\left(C + \frac{S^2}{2(k+W)} + \frac{W}{k+W}\alpha S - \frac{kW}{k+W} - (T-t)\frac{\sigma^2}{2(k+W)}\right), \quad (16)$$

the associated optimal strategy α_t satisfies

$$d\alpha_t = -\frac{\sigma}{k+W} dB_t, \qquad (17)$$

and the induced price dynamics are given by

$$dS_t = \sigma \frac{W}{k+W} dB_t \qquad (18)$$

Remark II.1: Part 1) means that the fair value or liquidation value is given by $\alpha S - k\frac{\alpha^2}{2}$. Obviously, when $k = 0$, we recover the usual "market" value αS. As we shall see below, this is in fact the case for an arbitrary volatility σ (we may already observe that it is independent of the constant σ).

Remark II.2: All the above explicit formulas are independent of the utility function.

Remark II.3: It is not difficult to extend the above result to the case when we add a drift term to (1) i.e.

$$dS_t = \sigma \, dB_t + b \, dt - ku_t \, dt$$

(where b is a constant). We leave such easy extensions to the reader.

Conclusion of the proof of Theorem 1:
There only remains to show that the value function V given by (14) satisfies

$$V = U\left(P + \frac{X^2}{2(k+W)} - (T-t)\frac{\sigma^2}{2(k+W)}\right) \qquad (19)$$

with an optimal strategy α given by : $\alpha = -\frac{X}{k+W}$, since (16)-(18) then follow from (10), (1) and (2). Finally, (15) may be, for instance, deduced from (16) upon letting W go to $+\infty$ (or directly from (13) which yields easily, since U is strictly concave, that $V = U(P)$).

Finally, in order to prove the above claims, we may simply use HJB equations which, here, take the following form

$$\frac{\partial V}{\partial t} + \frac{1}{2}\sigma^2 \frac{\partial^2 V}{\partial X^2} + \sup_{\alpha \in \mathbb{R}}\left\{\sigma^2 \frac{\alpha^2}{2}\frac{\partial^2 V}{\partial P^2} + \sigma^2 \alpha \frac{\partial^2 V}{\partial X \partial P}\right\} = 0$$

with $V\big|_{t=T} = U\left(P + \frac{X^2}{2(k+W)}\right)$. The above claims then follow from a simple verification that we skip. □

We now check that the formula (15) is in fact valid for an arbitrary $\sigma(S)$.

Theorem 2. *The value function defined by (4) is given by (15).*

Proof: The reduction made before Theorem II.2 remains valid provided we replace σ by $\sigma(S_t) = \sigma(X_t + k\alpha_t)$. Then, V solves the following HJB equation (one may also argue directly using the concavity of the utility function U).

$$\frac{\partial V}{\partial t} + \sup_{\alpha \in \mathbb{R}} \left\{ \frac{1}{2} \sigma^2 \left(X + k\alpha \right) \left(\frac{\partial^2 V}{\partial X^2} + \alpha^2 \frac{\partial^2 V}{\partial P^2} + \alpha \frac{\partial^2 V}{\partial X \partial P} \right) \right\} = 0 \quad (20)$$

and $V\big|_{t=T} = U(P)$. Obviously, we have : $V = U(P)$ and (15) is shown. □

Remark II.4: In the results above (and in those of the next section) one obtains an optimal hedging strategy which is not of bounded variation (in fact, in section III, we shall see some exemple where α_t may even have no differential properties whatsoever ...). In other words, there is no possible interpretation in terms of an optimal u_t (with $d\alpha_t = u_t dt$)! This is a consequence of the singular nature of the optimal stochastic control problems we are solving, which transforms a state variable into a control. A related, eve simpler, example is given by

$$\begin{cases} dx_t = dW_{t_1}, \ x_0 = x \in \mathbb{R} \ ; \ dy_y = u_t dt, \ y_0 = y \in \mathbb{R} \\ \sup\left\{ \mathbb{E}\left[F(x_T, y_T)\right]; u_t \text{ adapted}, u_t \in L^1(0, T; \mathbb{R}) \right\} = V(x, y, T), \end{cases}$$

where F is, say, continuous and bounded from above on $\mathbb{R}^2$. As is easily checked, the problem reduces to the following one for $T > 0$

$$V(x, y, T) = V(x, T) = \sup\left\{ \mathbb{E}\left[F(x_T, y_T)\right]; y_t \text{ adapted}, y_T \in L^1(0, T; \mathbb{R}) \right\}$$
$$= \mathbb{E}\left[\sup_{y \in \mathbb{R}} F(x_T, y) \right].$$

The state variable y has become a control and y_t is optimal if y_t is a maximum point of $F(x_t, y)$ over $\mathbb{R}$. For instance, if $F(x, y) = -|x|\frac{y^2}{2} + xy - \frac{|x|}{2}$, the optimal feedback is given by $y = \text{sign}(x)$ and thus $y_t = \text{sign}(x + W_t)$ which has no differentiability properties...! □

We conclude this section by considering the case when (1) is replaced by (1'). Then, the value function V defined by (4) is invariant by the flow

$$\Psi_\mu(C, S, \alpha) = \left(C + \frac{1 - e^{-k\mu}}{k} S, \ e^{-k\mu} S, \ \alpha - \mu \right), \quad (9')$$

$$P = C + \frac{1 - e^{-k\alpha}}{k} S, \ X = e^{-k\alpha} S. \quad (10')$$

We now find

$$dX_t = e^{-k\alpha_t} dS_t - k e^{-k\alpha_t} S_t \, d\alpha_t$$

or
$$\frac{dX_t}{X_t} = \sigma\left(e^{k\alpha_t} X_t\right) dB_t, \qquad (11')$$

and
$$dP_t = u_t s_t dt + \frac{1 - e^{-k\alpha_t}}{k} dS_t + e^{-k\alpha_t} S_t d\alpha_t$$
$$= \frac{1 - e^{-k\alpha_t}}{k}\left(\sigma(e^{k\alpha_t} X_t) S_t dB_t - k u_t S_t dt\right) + u_t S_t(1 - e^{-k\alpha_t}) dt$$

or
$$dP_t = \frac{e^{k\alpha_t} - 1}{k} X_t \sigma\left(e^{k\alpha_t} X_t\right) dB_t = \frac{e^{k\alpha_t} - 1}{k} dX_t. \qquad (12')$$

Once more, we come back to the "usual" dynamics for wealth except that the "hedge" is now given by $\frac{e^{k\alpha} - 1}{k}$ (instead of α_t...). And, V is still given by (13). We then have the

Theorem 3. *In the case of the dynamics (1'), the value function defined by (4) is given by*
$$V = U\left(C + \frac{1 - e^{-k\alpha}}{k} S\right) \qquad (15')$$

Remark II.5: The analogues of Remarks II.1 - II.3 are valid in the case of dynamics given by (1').

The proof of Theorem 3 is similar to those made before and we skip it (observe that $V = U(P)$).

3 Option Pricing

We now turn to the issue of option pricing and hedging under the dynamics (1) (or (1')). We thus consider the stochastic control problem (7) where Φ is the pay-off of the option. In order to avoid technical considerations, we assume that Φ is uniformly continuous on $\mathbb{R}$ (although discontinuities could be allowed).

We may still follow the reduction performed in Section 2, introducing
$$P_t = C_t + \alpha_t S_t - k \frac{\alpha_t^2}{2}, \quad \text{and} \quad X_t = S_t - k\alpha_t,$$

and exactly as in the preceding section, we find that we have
$$V(P, X, t) = \sup_{\alpha_t \text{ adapted}} \mathbb{E}\left[U(P_T - \Phi(X_T)) \mid P_t = P, X_t = X\right] \qquad (21)$$

with

$$\begin{cases} dX_s = \sigma(X_s + k\alpha_s)\, dB_s \\ dP_s = \alpha_s \sigma(X_s + k\alpha_s)\, dB_s = \alpha_s\, dX_s \end{cases} \quad (22)$$

It is worth observing that P_t is precisely the total *wealth* corresponding to the addition of the cash C_t and the *value* of the quantity α_t of the asset whose price is S_t (namely $\alpha_t S_t - k\frac{\alpha_t^2}{2}$ as shown in Section 2). Then, as in Constantinides and Zariphopoulou [6] (see also Musiela and Zariphopoulou [15]), we define the *indifference price* π of the option by the relationship

$$V = U(P - \pi) = U\left(C + \alpha S - k\frac{\alpha^2}{2} - \pi\right) \quad (23)$$

since $U(P)$ is the value function corresponding to $\Phi \equiv 0$ (see section II). The *hedge* is given by the optimal control in (21) provided it exists.

At this stage it is worth explaining the role of the somewhat unnatural constraint contained in the definition of the class $\mathcal{U}$ of admissible controls namely $\alpha_T = 0$. First of all, if we consider the cash at time T namely C_T (as in (7)), it is obvious that it does not amount to the total wealth unless $\alpha_T = 0$. Therefore, if we wish to suppress the constraint $\alpha_T = 0$, we may do so provided we replace in (7) C_T by the effective total wealth at time T namely $C_T + \alpha_T S_T - k\frac{\alpha_T^2}{2}$ since we now know from Section 2 that $\alpha_T S_T - k\frac{\alpha_T^2}{2}$ is precisely the "fair" or "liquidation" value corresponding to owning α_T of the asset whose price is S_T. In addition, a *fair* contract should be written in terms of $\Phi(X_T) = \Phi(S_T - k\alpha_T)$ and not of $\Phi(S_T)$ (otherwise the trader could "push" instantaneously just before T to a price corresponding to the minimum value of Φ i.e. 0 for a call option, an obviously unfair behavior!). But, if the option is written as $\Phi(S_T - k\alpha_T)$, we obtain a fair contract in which the trader has no interest in trying to manipulate the price since we have for all $\mu \in \mathbb{R}$

$$S_T - k\mu - k(\alpha_T - \mu) = S_T - k\,\alpha_T.$$

We are thus led to the utility maximization problem (21). In other words, relaxing the terminal constraint ($\alpha_T = 0$) is possible provided one defines a <u>fair</u> contract (as discussed above) in which case the two problems are in fact equivalent!
Our first result is the following

Theorem 4. *We assume that the utility function is of exponential type (6). Then*

$$V = U\left(C + \alpha S - k\frac{\alpha^2}{2} - \pi(S - k\alpha, t)\right) \quad (24)$$

where the indifference price π is the unique viscosity solution (uniformly continuous in x, uniformly in $t \leq T$ bounded) of the following HJB equation

$$\frac{\partial \pi}{\partial t} + \frac{1}{2} \inf_{\alpha \in \mathbb{R}} \left\{ \sigma^2(X + k\alpha) \left[\frac{\partial^2 \pi}{\partial X^2} + \lambda(\alpha - \frac{\partial \pi}{\partial X})^2 \right] \right\} = 0 \qquad (25)$$
$$\text{for } X \in \mathbb{R}, \, t < T,$$

such that
$$\pi \Big|_{t=T} = \Phi(X). \qquad (26)$$

Corollary 1. *If σ is independent of S, the indifference price reduces to the Black-Scholes price*

$$\pi(X,t) = E\Big[\Phi(X_T) \,/ X_t = X\Big] = E\Big[\Phi(X + \sigma(W_T - W_t))\Big]$$

and the optimal strategy is unique and given by $\alpha = \dfrac{\partial \pi}{\partial X}(X,t)$; hence, the induced dynamics of the price S_t of the asset become

$$dS_t = \sigma\left(1 + k\frac{\partial^2 \pi}{\partial X^2}(X_t, t)\right)dB_t, \quad dX_t = \sigma dB_t. \qquad (27)$$

In addition, the above facts are true for any utility function.

Corollary 2. *1) As λ goes to 0_+ (if $k > 0$), the indifference price π goes (uniformly on compact sets) to the unique viscosity solution of*

$$\frac{\partial \pi}{\partial t} + \frac{1}{2} \inf_{y \in \mathbb{R}} \left\{ \sigma^2(y) \frac{\partial^2 \pi}{\partial X^2} \right\} = 0 \quad \text{for } X \in \mathbb{R}, \, t < T, \qquad (28)$$

satisfying (26);

2) As λ goes to $+\infty$, the indifference price π goes (uniformly on compact sets) to the unique viscosity solution of

$$\frac{\partial \pi}{\partial t} + \frac{1}{2} \sigma^2\left(x + k\frac{\partial \pi}{\partial X}\right) \frac{\partial^2 \pi}{\partial X^2} = 0 \quad \text{for } X \in \mathbb{R}, \, t < T, \qquad (29)$$

satisfying (26).

Remark III.1: Obviously, when $k = 0$, we recover the usual Black-Scholes theory for the price and the hedge since the optimal strategy is then simply given by the unique infimum in α (in (25)) namely $\alpha = \dfrac{\partial \pi}{\partial X}$. We shall come back below to the expansion (correction) as k goes to 0_+.

Remark III.2: In general, the existence (and meaning) of an optimal hedge may be delicate since the infimum in α (in (25)) does not yield in general a unique minimum. Formally, the infimum in α yields the optimal hedging strategy. When there exists a unique optimal α that is smooth in (X, t), the hedge $\alpha_t = \alpha(X_t, t)$ leads to the following price dynamics

$$dS_t = \sigma(S_t)\, dB_t + k\, d\alpha_t \quad, \quad dX_t = \sigma(S_t)\, dB_t$$

and thus we have

$$\begin{cases} dS_t = \sigma(S_t)\left(1 + k\, \dfrac{\partial \alpha}{\partial X}(X_t, t)\right) dB_t + b_t\, dt \\ dX_t = \sigma(S_t)\, dB_t, \end{cases}$$

for some drift $b_t \left(= \dfrac{\partial \alpha}{\partial t}(X_t, t) + \dfrac{1}{2}\sigma^2(S_t)\dfrac{\partial^2 \alpha}{\partial X^2}(X_t, t)\right)$. Finally, let us remark that if we assume that σ and Φ are of class $C_b^{1,1}$ then, if $\dfrac{k}{\lambda}\|\sigma'\|_\infty$ is small enough, there exists a unique such optimal strategy (which is smooth in X, t)...

Remark III.3 : It is possible to adapt the above results to the case when we incorporate a drift term in (1). Although somewhat more technical, the approach is exactly the same and the computations are straightforward. Let us only mention here that Corollary 1 remains valid in the case when we replace σdB_t by $\sigma dB_t + b dt$ where σ and b are constants. Of course, when we add a drift term, we need to modify the implicit definition (23) of the indifference price : indeed, denoting by V_0 the same value function as V but with $\Phi \equiv 0$, the indifference price is then defined by

$$V(C, S, t) = V_0(C - \pi, S, t). \tag{30}$$

Proof of Theorem 4 It is a straightforward application of stochastic control theory and viscosity solutions once we have obtained (21)-(22). Indeed, V solves the HJB equation (20) with $V\big|_{t=0} = U(P - \Phi))$. Next, as is well-known, one looks for a solution V of the following form : $V = 1 - e^{-\lambda(P - \pi(X, t))}$. And we obtain (25)!

Proof of Corollary 1 One simply observes that (25) reduces to the usual Black-Scholes theory with a unique optimal α given by $\alpha = \dfrac{\partial \pi}{\partial X}(X, t)$. And (27) follows easily.

Proof of Corollary 2 Both statements are formally obvious since the equation (25) goes to the equation (28) as λ goes to 0_+ and to the equation (29) as λ goes to $+\infty$. The justification of these formal asymptotics is a simple application of viscosity solutions theory. □

We now explain some asymptotics as k goes to 0 that are valid for an arbitrary utility function (and for the case when we add a drift term to (1)). Using the HJB equations satisfied by V, one shows easily, at least formally, that the indifference price satisfies

$$\pi = \pi_{BS}(S, t) + O(k) \tag{31}$$

(it is in fact possible to make a more precise expansion but we shall not do so here), where π_{BS} is the Black-Scholes price namely $\pi_{BS}(S,t) = \mathbb{E}[\Phi(S_T)|S_t = S]$ where S_t solves (1) with $k = 0$. This formal argument is once more automatically justified by viscosity solutions arguments - for instance -). If we assume that $\sigma \in C_b^{1,1}$, then, for k small enough, we have a unique optimal strategy α (see also Remark III.2) and we deduce easily that we have as k goes to 0_+

$$\alpha = \frac{\partial \pi_{BS}}{\partial X}(X,t) + O(k) = \frac{\partial \pi_{BS}}{\partial S}(S,t) + O(k). \tag{32}$$

Finally, this leads to the following dynamics (or effective dynamics) for S_t namely

$$dS_t = \sigma(S_t)\left(1 + k\frac{\partial^2 \pi_{BS}}{\partial S^2}(S_t,t) + O(k^2)\right) dB_t + \tilde{b}_t\, dt \tag{33}$$

for some $\tilde{b}_t$ that we do not detail here. In other words, the resulting volatility (that we may call effective volatility) is given by

$$\sigma_{\text{eff}}(S) = \sigma(S)\left\{1 + k\frac{\partial^2 \pi_{BS}}{\partial S^2}(S,t) + O(k^2)\right\}. \tag{34}$$

Let us also mention that it is possible to obtain higher order terms in an asymptotic expansion in k.

Remark III.4 : Let us observe that everything we did above (and in the preceding section) extends mutatis mutandis to situations where σ depends also on t ($\sigma = \sigma(S,t)$), and where the pay-off is of the form $\int_t^T \Phi(s,S_s)ds + \Phi(S_T)$. □

A closely related asymptotic result holds when we consider a "small" pay-off of typical size ϵ. In other words, when we consider $\epsilon\Phi$ instead of Φ. Then, in order to retain some risk aversion, we scale accordingly the utility function replacing $U(z)$ by $U\left(\frac{z}{\epsilon}\right)$. We may now perform a simple scaling analysis that shows that the indifference price may be written as $\epsilon\pi$, the control α_t as $\epsilon\alpha_t$ and thus this amounts to replace k by ϵk and (31)-(34) remain true replacing k by $k\epsilon$. In particular, we have

$$\sigma_{\text{eff}} = \sigma\left\{1 + k\epsilon\frac{\partial^2 \pi_{BS}}{\partial S^2} + O(\epsilon^2)\right\}. \tag{35}$$

We now conclude this section by explaining the modifications to the above results required if we use (1') instead of (1). The reductions made above lead to (21) with

$$\begin{cases} dX_s = \sigma\left(e^{k\alpha_s}X_s\right) X_s\, dB_s \\ dP_s = \dfrac{e^{k\alpha_s}-1}{k} X_s\, \sigma\left(e^{k\alpha_s}X_s\right) dB_s = \dfrac{e^{k\alpha_s}-1}{k} dX_s\, . \end{cases} \tag{36}$$

The first modification we notice is that the quantity that plays the role of the hedge in Black-Scholes theory is now $\beta = \dfrac{e^{k\alpha}-1}{k}$ while α is really the hedge and that, necessarily, we have $\beta \geq -1/k$. Then, Theorem 4 is still true provided we replace (25) by

$$\begin{cases} \dfrac{\partial \pi}{\partial t} + \dfrac{1}{2} \inf_{\alpha \in \mathbb{R}} \left\{ \sigma^2(e^{k\alpha}X)X^2 \left[\dfrac{\partial^2 \pi}{\partial X^2} + \lambda \left(\dfrac{e^{k\alpha}-1}{k} - \dfrac{\partial \pi}{\partial x} \right)^2 \right] \right\} = 0 \\ \text{or} \\ \dfrac{\partial \pi}{\partial t} + \dfrac{1}{2} \inf_{\beta \geq -1/k} \left\{ \sigma^2((1+k\beta)X)X^2 \left[\dfrac{\partial^2 \pi}{\partial X^2} + \lambda \left(\beta - \dfrac{\partial \pi}{\partial x} \right)^2 \right] \right\} = 0. \end{cases} \quad (37)$$

In particular, if σ is independent of S, the equation reduces to

$$\dfrac{\partial \pi}{\partial t} + \dfrac{1}{2} \sigma^2 X^2 \left\{ \dfrac{\partial^2 \pi}{\partial X^2} + \lambda \left(\dfrac{\partial \pi}{\partial X} + \dfrac{1}{k} \right)_+^2 \right\} = 0, \quad (38)$$

an equation that remains nonlinear in general because of the constraint ($\beta \geq -1/k$). However, if we assume in addition that we have

$$\Phi'(X) > -\dfrac{1}{k} \quad \text{for all } X \geq 0, \quad (39)$$

then, because of the strict maximum principle, the indifference price is simply given by the Black-Scholes price π_{BS} that is the solution of

$$\dfrac{\partial \pi_{BS}}{\partial t} + \dfrac{1}{2} \sigma^2 X^2 \dfrac{\partial^2 \pi_{BS}}{\partial X^2} = 0 \quad (40)$$

with $\pi_{BS}\big|_{t=T} = \Phi$. Indeed, we then have

$$\dfrac{\partial \pi_{BS}}{\partial X} > -\dfrac{1}{k} \quad \text{for all } X, \, t < T$$

and thus π_{BS} solves (38) (or (37)). In addition, the optimal feedback is given by $\dfrac{e^{k\alpha}-1}{k} = \dfrac{\partial \pi_{BS}}{\partial X}$ or in other words

$$\alpha = \dfrac{1}{k} \log \left(1 + k \dfrac{\partial \pi_{BS}}{\partial X} \right). \quad (41)$$

And the effective price dynamics become

$$\dfrac{\partial S_t}{S_t} = \sigma \, dB_t + k \, d\alpha_t$$

$$= \sigma \, dB_t + \dfrac{k}{1 + k \dfrac{\partial \pi_{BS}}{\partial X}} \dfrac{\partial^2 \pi_{BS}}{\partial X^2} dX_t + b_t^1 \, dt$$

or

$$\dfrac{\partial S_t}{S_t} = \sigma \left(1 + \dfrac{k X_t}{1 + k \dfrac{\partial \pi_{BS}}{\partial X}(X_t, t)} \dfrac{\partial^2 \pi_{BS}}{\partial X^2}(X_t, t) \right) dB_t + \tilde{b}_t \, dt \quad (42)$$

for some $b_t^1, \tilde{b}_t$ that we do not detail here, and $dX_t = \sigma X_t dB_t$.
◇ As λ goes to 0, the reservation price goes to the solution of

$$\frac{\partial \pi}{\partial t} + \frac{1}{2} \inf_{y \in \mathbb{R}} \left\{ \sigma^2(y) X^2 \frac{\partial^2 \pi}{\partial X^2} \right\} = 0 \qquad (28')$$

with $\pi\big|_{t=0} = \Phi$.

◇ The limit λ goes to $+\infty$ is much more delicate. We need to consider for all $X \geq 0$

$$\widehat{\Phi}(X) = \inf \left\{ \widetilde{\Phi}(X) \big/ \widetilde{\Phi} \geq \Phi, \frac{\partial \widetilde{\Phi}}{\partial X} \geq 1/k \text{ on } \mathbb{R} \right\};$$

and one can show that $\widehat{\Phi}$ is given by

$$\widehat{\Phi}(X) = \sup_{X \geq Z \geq 0} \left\{ \Phi(X - Z) - \frac{1}{k} Z \right\}.$$

Then, as λ goes to $+\infty$, the indifference price converges to the solution of

$$\frac{\partial \pi}{\partial t} + \frac{1}{2} \sigma^2 \left((1 + k \frac{\partial \pi}{\partial X}) X \right) X^2 \frac{\partial^2 \pi}{\partial X^2} = 0 \qquad (29')$$

with $\pi\big|_{t=0} = \widehat{\Phi}$.

We conclude by stating the analogue of (35) namely

$$\sigma_{\text{eff}} = \sigma \left\{ 1 + k\epsilon S \frac{\partial^2 \pi_{BS}}{\partial S^2} + O(\epsilon^2) \right\}, \qquad (35')$$

where π_{BS} is the Black-Scholes price namely $\pi_{BS}(S, t) = E\left[\Phi(S_T)/S_t = S\right]$ with S_t solution of

$$\frac{dS_t}{S_t} = \sigma(S_t) \, dB_t.$$

4 Variants and Extensions

We mention briefly in this section some variants and extensions of the preceding results in several directions namely i) interest rates, ii) more general influence models and iii) multi-dimensonnal assets.

The first one concerns interest rates and we simply observe that incorporating simple interest rates models is a straightforward matter that creates no additional difficulty and we skip this easy adaptation in order to restrict the length of this paper.

Concerning other influence models, we mention that we studied many other models where one replaces the term $(-ku_t dt)$ in (1) or (1') by other terms of a similar nature. Our general approach can be adapted to those situations and we shall not attempt here to describe numerous possibilities we looked at. For instance, one may incorporate memory or relaxation effects. One may also consider the case when k is no longer a constant but a function of S_t and possibly even of α_t. Once again, the adaptations are straightforward. However, one variant is slightly more delicate to handle namely the case when ku_t is replaced by a nonlinear term $k(u_t)$ where k is, for instance, a Lipschitz function on $\mathbb{R}$ such that we have

$$k(x) - kx \quad \text{is bounded on } \mathbb{R} \tag{43}$$

where $k \geq 0$. An example of such a k is provided by

$$k(x) = \begin{cases} 0 & \text{if } x \in [-a, b], \\ k(x-b) & \text{if } x \geq b, \\ k(x+a) & \text{if } x \leq -a \end{cases}$$

where $k \geq 0$, $a \geq 0$, $b \geq 0$, example that corresponds to an influence which is perceptible only when the rate of buying or selling orders exceeds some given threshold. In such a nonlinear situation, one has to adapt the results (and methods of proofs of [12] and we briefly sketch here, at a formal level, how this can be done "in the case of (1)" that is

$$dS_t = \sigma(S_t) \, dB_t - k(u_t) dt \,. \tag{44}$$

As we did in Section 2, we introduce $X_t = S_t - k\alpha_t$ and $P_t = C_t + \alpha_t S_t - k\dfrac{\alpha_t^2}{2}$ and we have

$$\begin{cases} dX_t = \sigma\left(X_t + k\alpha_t\right) dB_t - \ell(u_t) dt \\ dP_t = \alpha_t \sigma\left(X_t + k\alpha_t\right) dB_t - \alpha_t \ell(u_t) dt = \alpha_t dX_t \end{cases} \tag{45}$$

with $\ell(u) = k(u) - ku$ on $\mathbb{R}$. And one can show that the value function is equal (for all $S, \alpha, t < T$) to

$$V(X, P, t) = \sup_{\alpha_t, u_t \text{ adapted}} \mathbb{E}\left[U(P_T - \Phi(X_T)) \,\big|\, P_t = P, \, X_t = X\right] \tag{46}$$

which solves the following HJB equation

$$\begin{cases} \dfrac{\partial V}{\partial t} + \sup_{\alpha, u}\left\{\dfrac{1}{2}\sigma^2(X+kx)\left[\dfrac{\partial^2 V}{\partial X^2} + \alpha^2 \dfrac{\partial^2 V}{\partial P^2} + 2\alpha \dfrac{\partial^2 V}{\partial X \partial t}\right] + \right. \\ \left. - \ell(u)\dfrac{\partial V}{\partial X} - \alpha\ell(u)\dfrac{\partial V}{\partial P}\right\} = 0 \quad \text{for all } X, \, t < T \end{cases}$$

with $V\big|_{t=T} = U(P - \phi(X))$. At this stage, it is then straightforward to adapt the results obtained in the preceding section. For instance, if σ is constant, $\Phi \equiv 0$ ("fair value") and $U = 1 - e^{-\lambda z}$, then we have

$$V = U\left(C + \alpha S - \dfrac{k}{2}\alpha^2 + \dfrac{L^2}{\lambda \sigma^2}(T-t)\right) \tag{47}$$

where $L = \sup\limits_{u \in \mathbb{R}} |\ell(u)|$.

Our final extension concerns situations involving many assets, namely when the dynamics and the controls are given by stochastic differential equations of the form

$$dS_t = \sigma(S_t).dB_t - K.u_t\, dt, \qquad (48)$$

$$d\alpha_t = -u_t\, dt, \qquad (49)$$

$$dC_t = u_t \cdot S_t\, dt, \qquad (50)$$

where $S_t \in \mathbb{R}^N$, σ is a Lipschitz $N \times m$ matrix such that $\sigma\sigma^T$ is bounded and positive (uniformly on $\mathbb{R}^N$), B_t is a standard m-dimensional Brownian matrix, $\alpha_t \in \mathbb{R}^N$ and K is a $N \times N$ matrix (not necessarily diagonal).

We emphasize the fact that the "influence" matrix K is not necessarily diagonal, a fact which means that buying or selling significant quantities of one of the assets may influence the price of the others (or some of the others).

One can then follow and adapt trivially all the arguments made in the previous sections and the results obtained above are extended in a very natural way to this multi-dimensional setting. For instance, the value of a quantity α of the assets S is

$$\alpha.S - \frac{1}{2}K\alpha.\alpha \qquad (51)$$

(or $\dfrac{1 - e^{-K_{ij}\mu_j}}{K_{ik}\mu_k}\mu_i$ if dS_t^i in (48) is replaced by $\dfrac{dS_t^i}{S_t^i}$),

and the analogue of the asymptotic formula (35) is given by

$$\sigma_{\text{eff}} = \left(I + \epsilon K.\frac{\partial^2 \pi_{BS}}{\partial S^2} + 0(\epsilon^2)\right).\sigma. \qquad (52)$$

Let us observe finally that, even if we consider assets that are a priori independent and a diagonal matrix K, our results show that hedging an option whose pay-off depends upon all the assets will create correlations between those assets.

References

1. M. Avellaneda and M.D. Lipkin: A market induced mechanism for stock pinning. (preprint)
2. K. Back: Insider trading in continuous time. *Review of Financial Studies*, **5** (1992) 387-409.
3. K. Back, C.H. Cao and G. Willard: Imperfect condition among informed traders. *J. Finance*, **LV** (2000) 2117-2155.
4. M. Bardi and I. Capuzzo-Dolcetta: *Optimal control and viscosity solutions of Hamilton-Jacobi-Bellman equations*. Birkhäusser, Boston (1997).
5. F. Black and M. Scholes: The pricing of options and corporate liabilities. *J. Political Economy*, **81** (1973) 637-659.

6. G. Constantinides and Th. Zariphopoulou: Bounds on prices of contingent claims in an intertemporal economy with proportional transaction costs and general preferences. *Finance Stoch.*, **3** (1999) 345-369.
7. W.H. Fleming and M.H. Soner: *Controlled Markow processes and Viscosity Solutions.* Springer, Berlin (1993).
8. H. Föllmer: Stock price fluctuation as a diffusion in a random environment. In *Mathematical Models in Finance*, eds. S.D. Howison, F.P. Kelly and P. Wilmott, Chapman Hall, London (1995).
9. R. Frey and A. Stremme: Portfolio insurance and volatility. Department of Economics, Univ. of Bonn (Discussion paper B-256).
10. A.S. Kyle: Continuous auctions and insider trading. *Econometrica*, **53** (1985) 1315-1335.
11. J.-M. Lasry and P.-L. Lions: Towards a self-consistent theory of volatility. (preprint)
12. J.-M. Lasry and P.-L. Lions: Une classe nouvelle de problèmes singuliers de contrôle stochastique. *C.R. Acad. Sci. Paris*, **331** (2000) 879-885.
13. G. Lasserre: Asymmetric information and imperfect competition in a continuous time multivariate secutiry model. *Finance and Stochastics*, **8** (2004) 285-309.
14. R. Merton: Theory of rational option pricing. *Bull. J. Econom. Manag. Sci.*, **4** (1973) 141-183.
15. M. Musiela and Th. Zariphopoulou: Indifference prices and related measures. (preprint)

Some Applications and Methods of Large Deviations in Finance and Insurance *

Huyên Pham

Laboratoire de Probabilités et Modèles Aléatoires
CNRS, UMR 7599
Université Paris 7 and Institut Universitaire de France
e-mail: pham@math.jussieu.fr

Summary. In these notes, we present some methods and applications of large deviations to finance and insurance. We begin with the classical ruin problem related to the Cramer's theorem and give en extension to an insurance model with investment in stock market. We then describe how large deviation approximation and importance sampling are used in rare event simulation for option pricing. We finally focus on large deviations methods in risk management for the estimation of large portfolio losses in credit risk and portfolio performance in market investment.

Keywords: large deviations, ruin problem, importance sampling, rare event simulation, exit probability, credit risk, portfolio performance

MSC Classification (2000) : 60F10, 62P05, 65C05, 91B28, 91B30.

1 Introduction

The area of large deviations is a set of asymptotic results on rare event probabilities and a set of methods to derive such results. Large deviations is a very active area in applied probability, and questions related to extremal events both in finance and insurance applications, play an increasingly important role. For instance, recent applications of interest concern ruin probabilities in risk theory, risk management, option pricing, and simulation of quantities involved in this context.

Large deviations appear historically in insurance mathematics with the ruin probability estimation problem within the classical Cramer-Lundberg model. The problem was then subsequently extended to more general models involving for example Lévy risk processes. In finance, large deviations arise in various contexts.

---

* Lectures from a Bachelier course at Institut Henri Poincaré, Paris, February 2007. I would like to thank the participants for their interest and comments.

They occur in risk management for the computation of the probability of large losses of a portfolio subject to market risk or the default probabilities of a portfolio under credit risk. Large deviations methods are largely used in rare events simulation and so appear naturally in the approximation of option pricing, in particular for barrier options and far from the money options.

We illustrate our purpose with the following toy example. Let X be a (real-valued) random variable, and consider the problem of computing or estimating $\mathbb{P}[X > \ell]$, the probability that X exceeds some level ℓ. In finance, we may think of X as the loss of a portfolio subject to market or credit risk, and we are interested in the probability of large loss or default probability. The r.v. X may also correspond to the terminal value of a stock price, and the quantity $\mathbb{P}[X > \ell]$ appears typically in the computation of a call or barrier option, with a small probability of payoff when the option is far from the money or the barrier ℓ is large. To estimate $p = \mathbb{P}[X > \ell]$, a basic technique is Monte Carlo simulation : generate n independent copies $X_1, \ldots, X_n$ of X, and use the sample mean :

$$\bar{S}_n = \frac{1}{n} \sum_{i=1}^{n} Y_i, \quad \text{with } Y_i = 1_{X_i > \ell}.$$

The convergence of this estimate (when $n \to \infty$) follows from the law of large numbers, while the standard rate of convergence is given, via the central limit theorem, in terms of the variance $v = p(1-p)$ of Y_i :

$$\mathbb{P}\left[|\bar{S}_n - p| \geq \frac{a}{\sqrt{n}}\right] \to 2\Phi\left(-\frac{a}{\sqrt{v}}\right),$$

where Φ is the cumulative distribution function of the standard normal law. Furthermore, the convergence of the estimator $\bar{S}_n$ is precised with the large deviation result, known here as the Cramer's theorem, which is concerned with approximation of rare event probabilities $\mathbb{P}[\bar{S}_n \in A]$, and typically states that

$$\mathbb{P}\left[|\bar{S}_n - p| \geq a\right] \simeq C e^{-\gamma n},$$

for some constants C and γ.

Let us now turn again to the estimation of $p = \mathbb{P}[X > \ell]$. As mentioned above, the rate of convergence of the naive estimator $\bar{S}_n$ is determined by :

$$\text{Var}(\bar{S}_n) = \frac{\text{Var}(1_{X > \ell})}{n} = \frac{p(1-p)}{n},$$

and the relative error is

$$\text{relative error} = \frac{\text{standard deviation of } \bar{S}_n}{\text{mean of } \bar{S}_n} = \frac{\sqrt{p(1-p)}}{\sqrt{np}}.$$

Hence, if $p = \mathbb{P}[X > \ell]$ is small, and since $\sqrt{p - p^2}/p \to \infty$ as p goes to zero, we see that a large sample size (i.e. n) is required for the estimator to achieve a reasonable relative error bound. This is a common occurence when estimating rare events. In order to improve the estimate of the tail probability $\mathbb{P}[X > \ell]$, one is tempted to use importance sampling to reduce variance, and hence speed up the computation by requiring fewer samples. This consists basically in changing measures to try to give more weight to "important" outcomes, (increase the default probability). Since large deviations theory also deals with rare events, we can see its strong link with importance sampling.

To make the idea concrete, consider again the problem of estimating $p = \mathbb{P}[X > \ell]$, and suppose that X has distribution $\mu(dx)$. Let us look at an alternative sampling distribution $\nu(dx)$ absolutely continuous with respect to μ, with density $f(x) = d\nu/d\mu(x)$. The tail probability can then be written as:

$$p = \mathbb{P}[X > \ell] = \int 1_{x>\ell} \mu(dx) = \int 1_{x>\ell} \phi(x) \nu(dx) = \mathbb{E}_\nu[1_{X>\ell} \phi(X)],$$

where $\phi = 1/f$, and $\mathbb{E}_\nu$ denotes the expectation under the measure ν. By generating i.i.d. samples $\tilde{X}_1, \ldots, \tilde{X}_n, \ldots$ with distribution ν, we have then an alternative unbiased and convergent estimate of p with

$$\tilde{S}_n = \frac{1}{n} \sum_{i=1}^{n} 1_{\tilde{X}_i > \ell} \phi(\tilde{X}_i),$$

and whose rate of convergence is determined by

$$\mathrm{Var}_\nu(\bar{S}_n) = \frac{1}{n} \int \left(1_{x>\ell} - pf(x)\right)^2 \phi^2(x) \nu(dx).$$

The minimization of this quantity over all possible ν (or f) leads to a zero variance with the choice of a density $f(x) = 1_{x>\ell}/p$. This is of course only a theoretical result since it requires the knowledge of p, the very thing we want to estimate! However, by noting that in this case $\nu(dx) = f(x)\mu(dx) = 1_{x>\ell}\mu(dx)/\mathbb{P}[X>\ell]$ is nothing else than the conditional distribution of X given $\{X > \ell\}$, this suggests to use an importance sampling change of measure that makes the rare event $\{X > \ell\}$ more likely. This method of suitable change of measure is also the key step in proving large deviation results.

The plan of these lectures is the following. In Section 2, we give some basic tools and results on large deviations, in particular the most classical result on large deviations, Cramer's theorem. We illustrate in Section 3 the first applications of large deviations to ruin problems in insurance industry, and give some extension to an insurance model with financial investment opportunity. Section 4 is concerned with the large deviations approximation for rare event simulation in option pricing, and we shall use asymptotic results from large deviations theory : Freidlin-Wentzell theory on sample path large deviations, and Varadhan's integral principle. Finally, Section 5 is devoted to applications of large deviations in risk management, where we use conditional and control variants of the Ellis-Gartner theorem.

2 Basic Tools and Results on Large Deviations

2.1 Laplace Functional and Exponential Change of Measures

If X is a (real-valued) random variable on $(\Omega, \mathcal{F})$ with probability distribution $\mu(dx)$, the cumulant generating function (c.g.f.) of μ is the logarithm of the Laplace function of X, i.e. :

$$\Gamma(\theta) = \ln \mathbb{E}[e^{\theta X}] = \ln \int e^{\theta x} \mu(dx) \in (-\infty, \infty], \quad \theta \in \mathbb{R}.$$

Notice that $\Gamma(0) = 0$, and Γ is convex by Hölder inequality. We denote $\mathcal{D}(\Gamma) = \{\theta \in \mathbb{R} : \Gamma(\theta) < \infty\}$, and for any $\theta \in \mathcal{D}(\Gamma)$, we define a probability measure μ_θ on $\mathbb{R}$ by :

$$\mu_\theta(dx) = \exp(\theta x - \Gamma(\theta))\mu(dx). \tag{1}$$

Suppose that $X_1, \ldots, X_n, \ldots$, is an i.i.d. sequence of random variables with distribution μ and consider the new probability measure $\mathbb{P}_\theta$ on $(\Omega, \mathcal{F})$ with likelihood ratio evaluated at $(X_1, \ldots, X_n)$, $n \in \mathbb{N}^*$, by :

$$\frac{d\mathbb{P}_\theta}{d\mathbb{P}}(X_1, \ldots, X_n) = \prod_{i=1}^n \frac{d\mu_\theta}{d\mu}(X_i) = \exp\left(\theta \sum_{i=1}^n X_i - n\Gamma(\theta)\right). \tag{2}$$

By denoting $\mathbb{E}_\theta$ the corresponding expectation under $\mathbb{P}_\theta$, formula (2) means that for all $n \in \mathbb{N}^*$,

$$\mathbb{E}\Big[f(X_1, \ldots, X_n)\Big] = \mathbb{E}_\theta\Big[f(X_1, \ldots, X_n) \exp\Big(-\theta \sum_{i=1}^n X_i + n\Gamma(\theta)\Big)\Big], \tag{3}$$

for all Borel functions f for which the expectation on the l.h.s. of (3) is finite. Moreover, the random variables $X_1, \ldots, X_n$, $n \in \mathbb{N}^*$, are i.i.d. with probability distribution μ_θ under $\mathbb{P}_\theta$. Actually, the relation (3) extends from a fixed number of steps n to a random number of steps, provided the random horizon is a stopping time. More precisely, if τ is a stopping time in $\mathbb{N}$ for $X_1, \ldots, X_n, \ldots$, i.e. the event $\{\tau < n\}$ is measurable with respect to the algebra generated by $\{X_1, \ldots, X_n\}$ for all n, then

$$\mathbb{E}\Big[f(X_1, \ldots, X_\tau)1_{\tau < \infty}\Big]$$
$$= \mathbb{E}_\theta\Big[f(X_1, \ldots, X_\tau) \exp\Big(-\theta \sum_{i=1}^\tau X_i + \tau\Gamma(\theta)\Big)1_{\tau < \infty}\Big], \tag{4}$$

for all Borel functions f for which the expectation on the l.h.s. of (4) is finite.

The cumulant generating function Γ records some useful information on the probability distributions μ_θ. For example, $\Gamma'(\theta)$ is the mean of μ_θ. Indeed, for any θ in the interior of $\mathcal{D}(\Gamma)$, differentiation yields by dominated convergence :

$$\Gamma'(\theta) = \frac{\mathbb{E}[Xe^{\theta X}]}{\mathbb{E}[e^{\theta X}]} = \mathbb{E}\big[X \exp\big(\theta X - \Gamma(\theta)\big)\big] = \mathbb{E}_\theta[X]. \tag{5}$$

A similar calculation shows that $\Gamma''(\theta)$ is the variance of μ_θ. Notice in particular that if 0 lies in the interior of $\mathcal{D}(\Gamma)$, then $\Gamma'(0) = \mathbb{E}[X]$ and $\Gamma''(0) = Var(X)$.

Bernoulli distribution

Let μ the Bernoulli distribution of parameter p. Its c.g.f. is given by

$$\Gamma(\theta) = \ln(1 - p + pe^\theta).$$

A direct simple algebra calculation shows that μ_θ is the Bernoulli distribution of parameter $pe^\theta/(1 - p + pe^\theta)$.

Poisson distribution

Let μ the Poisson distribution of intensity λ. Its c.g.f. is given by

$$\Gamma(\theta) = \lambda(e^\theta - 1).$$

A direct simple algebra calculation shows that μ_θ is the Poisson distribution of intensity λe^θ. Hence, the effect of the change of probability measure $\mathbb{P}_\theta$ is to multiply the intensity by a factor e^θ.

Normal distribution

Let μ the normal distribution $\mathcal{N}(0, \sigma^2)$, whose c.g.f. is given by :

$$\Gamma(\theta) = \frac{\theta^2 \sigma^2}{2}.$$

A direct simple algebra calculation shows that μ_θ is the normal distribution $\mathcal{N}(\theta\sigma^2, \sigma^2)$. Hence, if $X_1, \ldots, X_n$ are i.i.d. with normal distribution $\mathcal{N}(0, \sigma^2)$, then under the change of measure $\mathbb{P}_\theta$ with likelihood ratio :

$$\frac{d\mathbb{P}_\theta}{d\mathbb{P}}(X_1, \ldots, X_n) = \exp\left(\theta \sum_{i=1}^n X_i - n\frac{\theta^2 \sigma^2}{2}\right),$$

the random variables $X_1, \ldots, X_n$ are i.i.d. with normal distribution $\mathcal{N}(\theta\sigma^2, \sigma^2)$: the effect of $\mathbb{P}_\theta$ is to change the mean of X_i from 0 to $\theta\sigma^2$. This result can be interpreted as the finite-dimensional version of Girsanov's theorem.

Exponential distribution

Let μ the exponential distribution of intensity λ. Its c.g.f. is given by

$$\Gamma(\theta) = \begin{cases} \ln\left(\frac{\lambda}{\lambda-\theta}\right), & \theta < \lambda \\ \infty, & \theta \geq \lambda \end{cases}$$

A direct simple algebra calculation shows that for $\theta < \lambda$, μ_θ is the exponential distribution of intensity $\lambda - \theta$. Hence, the effect of the change of probability measure $\mathbb{P}_\theta$ is to shift the intensity from λ to $\lambda - \theta$.

2.2 Cramer's Theorem

The most classical result in large deviations area is Cramer's theorem. This concerns large deviations associated with the empirical mean of i.i.d. random variables valued in a finite-dimensional space. We do not state the Cramer's theorem in whole generality. Our purpose is to put emphasis on the methods used to derive such result. For simplicity, we consider the case of real-valued i.i.d. random variables X_i with (nondegenerate) probability distribution μ of finite mean $\mathbb{E}X_1 = \int x\mu(dx) < \infty$, and we introduce the random walk $S_n = \sum_{i=1}^n X_i$. It is well-known by the law of large numbers that the empirical mean S_n/n converges in probability to $\bar{x} = \mathbb{E}X_1$, i.e. $\lim_n \mathbb{P}[S_n/n \in (\bar{x} - \varepsilon, \bar{x} + \varepsilon)] = 1$ for all $\varepsilon > 0$. Notice also, by the central limit theorem that $\lim_n \mathbb{P}[S_n/n \in [\bar{x}, \bar{x} + \varepsilon)] = 1/2$ for all $\varepsilon > 0$. Large deviations results focus on asymptotics for probabilities of rare events, for example of the form $\mathbb{P}[\frac{S_n}{n} \geq x]$ for $x > \mathbb{E}X_1$, and state that

$$\mathbb{P}\left[\frac{S_n}{n} \geq x\right] \simeq Ce^{-\gamma x},$$

for some constants C and γ to be precised later. The symbol $\simeq$ means that the ratio is one in the limit (here when n goes to infinity). The rate of convergence is characterized by the Fenchel-Legendre transform of the c.g.f. Γ of X_1:

$$\Gamma^*(x) = \sup_{\theta \in \mathbb{R}} \left[\theta x - \Gamma(\theta)\right] \in [0, \infty], \quad x \in \mathbb{R}.$$

As supremum of affine functions, Γ^* is convex. The sup in the definition of Γ^* can be evaluated by differentiation : for $x \in \mathbb{R}$, if $\theta = \theta(x)$ is solution to the saddle-point equation, $x = \Gamma'(\theta)$, then $\Gamma^*(x) = \theta x - \Gamma(\theta)$. Notice, from (5), that the exponential change of measure $\mathbb{P}_\theta$ puts the expectation of X_1 to x. Actually, exponential change of measure is a key tool in large deviations methods. The idea is to select a measure under which the rare event is no longer rare, so that the rate of decrease of the original probability is given by the rate of decrease of the likelihood ratio. This particular change of measure is intended to approximate the most likely way for the rare event to occur.

By Jensen's inequality, we show that $\Gamma^*(\mathbb{E}X_1) = 0$. This implies that for all $x \geq \mathbb{E}X_1$, $\Gamma^*(x) = \sup_{\theta \geq 0} \left[\theta x - \Gamma(\theta)\right]$, and so Γ^* is nondecreasing on $[\mathbb{E}X_1, \infty)$.

Theorem 2.1 *(Cramer's theorem)*
For any $x \geq \mathbb{E}X_1$, we have

$$\lim_{n \to \infty} \frac{1}{n} \ln \mathbb{P}\left[\frac{S_n}{n} \geq x\right] = -\Gamma^*(x) = -\inf_{y \geq x} \Gamma^*(y). \tag{6}$$

Proof. 1) *Upper bound.* The main step in the upper bound $\leq$ of (6) is based on Chebichev inequality combined with the i.i.d. assumption on the X_i :

$$\mathbb{P}\left[\frac{S_n}{n} \geq x\right] = \mathbb{E}\left[1_{\frac{S_n}{n} \geq x}\right] \leq \mathbb{E}\left[e^{\theta(S_n - nx)}\right] = \exp\left(n\Gamma(\theta) - \theta nx\right), \quad \forall \theta \geq 0.$$

By taking the infimum over $\theta \geq 0$, and since $\Gamma^*(x) = \sup_{\theta \geq 0}[\theta x - \Gamma(\theta)]$ for $x \geq \mathbb{E} X_1$, we then obtain

$$\mathbb{P}\Big[\frac{S_n}{n} \geq x\Big] \leq \exp\big(-n\Gamma^*(x)\big),$$

and so in particular the upper bound $\leq$ of (6).

2) *Lower bound.* Since $\mathbb{P}\big[\frac{S_n}{n} \geq x\big] \geq \mathbb{P}\big[\frac{S_n}{n} \in [x, x+\varepsilon)\big]$, for all $\varepsilon > 0$, it suffices to show that

$$\lim_{\varepsilon \to 0} \liminf_{n \to \infty} \frac{1}{n} \ln \mathbb{P}\Big[\frac{S_n}{n} \in [x, x+\varepsilon)\Big] \geq -\Gamma^*(x). \tag{7}$$

Suppose that μ is supported on a bounded support so that Γ is finite everywhere. Suppose first that there exists a solution $\theta = \theta(x) > 0$ to the saddle-point equation: $\Gamma'(\theta) = x$, i.e. attaining the supremum in $\Gamma^*(x) = \theta(x)x - \Gamma(\theta(x))$. The key step is now to introduce the new probability distribution μ_θ as in (1) and $\mathbb{P}_\theta$ the corresponding probability measure on $(\Omega, \mathcal{F})$ with likelihood ratio :

$$\frac{d\mathbb{P}_\theta}{d\mathbb{P}} = \prod_{i=1}^{n} \frac{d\mu_\theta}{d\mu}(X_i) = \exp\big(\theta S_n - n\Gamma(\theta)\big).$$

Then, we have by (3) and for all $\varepsilon > 0$:

$$\mathbb{P}\Big[\frac{S_n}{n} \in [x, x+\varepsilon)\Big] = \mathbb{E}_\theta\Big[\exp\big(-\theta S_n + n\Gamma(\theta)\big) 1_{\frac{S_n}{n} \in [x, x+\varepsilon)}\Big]$$

$$= e^{-n(\theta x - \Gamma(\theta))} \mathbb{E}_\theta\Big[\exp\big(-n\theta(\frac{S_n}{n} - x)\big) 1_{\frac{S_n}{n} \in [x, x+\varepsilon)}\Big]$$

$$\geq e^{-n(\theta x - \Gamma(\theta))} e^{-n|\theta|\varepsilon} \mathbb{P}_\theta\Big[\frac{S_n}{n} \in [x, x+\varepsilon)\Big],$$

and so

$$\frac{1}{n} \ln \mathbb{P}\Big[\frac{S_n}{n} \in [x, x+\varepsilon)\Big]$$

$$\geq -[\theta x - \Gamma(\theta)] - |\theta|\varepsilon + \frac{1}{n} \ln \mathbb{P}_\theta\Big[\frac{S_n}{n} \in [x, x+\varepsilon)\Big]. \tag{8}$$

Now, since $\Gamma'(\theta) = x$, we have $\mathbb{E}_\theta[X_1] = x$, and by the law of large numbers and CLT : $\lim_n \mathbb{P}_\theta\big[\frac{S_n}{n} \in [x, x+\varepsilon)\big] = 1/2 \,(>0)$. We also have $\Gamma^*(x) = \theta x - \Gamma(\theta)$. Therefore, by sending n to infinity and then ε to zero in (8), we get (7).

Now, if the supremum in $\Gamma^*(x)$ is not attained, we can find a sequence $(\theta_k)_k \nearrow \infty$, such that $\theta_k x - \Gamma(\theta_k) \to \Gamma^*(x)$. Since $\mathbb{E}[e^{\theta_k(X_1-x)} 1_{X_1 < x}] \to 0$, we then get

$$\mathbb{E}[e^{\theta_k(X_1-x)} 1_{X_1 \geq x}] \to e^{-\Gamma^*(x)},$$

as k goes to infinity. This is possible only if $\mathbb{P}[X_1 > x] = 0$ and $\mathbb{P}[X_1 = x] = e^{-\Gamma^*(x)}$. By the i.i.d. assumption on the X_i, this implies $\mathbb{P}[S_n/n \geq x] \geq (\mathbb{P}[X_1 \geq x])^n = e^{-n\Gamma^*(x)}$, which proves (7).

Suppose now that μ is of unbounded support, and fix M large enough s.t. $\mu([-M, M]) > 0$. By the preceding proof, the lower bound (7) holds with the law of S_n/n conditional on $\{|X_i| \leq M, i = 1, \ldots, n\}$, and with a c.g.f. equal to the c.g.f. of the conditional law of X_1 given $|X_1| \leq M$:

$$\lim_{\varepsilon \to 0} \liminf_{n \to \infty} \frac{1}{n} \ln \mathbb{P}\left[\frac{S_n}{n} \in [x, x+\varepsilon) \Big| |X_i| \leq M, i = 1, \ldots, n\right]$$
$$\geq -\tilde{\Gamma}_M^*(x) := -\sup_{\theta \in \mathbb{R}}[\theta x - \tilde{\Gamma}_M(\theta)], \tag{9}$$

with $\tilde{\Gamma}_M(\theta) = \ln \mathbb{E}[e^{\theta X_1}||X_1| \leq M] = \Gamma_M(\theta) - \ln \mu([-M, M])$, $\Gamma_M(\theta) = \ln \mathbb{E}[e^{\theta X_1} 1_{|X_1| \leq M}]$. Now, by writing from Bayes formula that $\mathbb{P}\left[\frac{S_n}{n} \in [x, x+\varepsilon)\right]$
$= \mathbb{P}\left[\frac{S_n}{n} \in [x, x+\varepsilon) \Big| |X_i| \leq M, i = 1, \ldots, n\right] \cdot (\mu([-M, M]))^n$, we get with (9)

$$\lim_{\varepsilon \to 0} \liminf_{n \to \infty} \frac{1}{n} \ln \mathbb{P}\left[\frac{S_n}{n} \in [x, x+\varepsilon)\right]$$
$$\geq \lim_{\varepsilon \to 0} \liminf_{n \to \infty} \frac{1}{n} \ln \mathbb{P}\left[\frac{S_n}{n} \in [x, x+\varepsilon) \Big| |X_i| \leq M, i = 1, \ldots, n\right] + \ln \mu([-M, M])$$
$$\geq -\Gamma_M^*(x) := -\sup_{\theta \in \mathbb{R}}[\theta x - \Gamma_M(\theta)].$$

The required result is obtained by sending M to infinity. Notice also finally that $\inf_{y \geq x} \Gamma^*(y) = \Gamma^*(x)$ since Γ^* is nondecreasing on $[\mathbb{E}X_1, \infty)$. □

Examples

1) Bernoulli distribution : for $X_1 \sim \mathcal{B}(p)$, we have $\Gamma^*(x) = x \ln\left(\frac{x}{p}\right) + (1-x) \ln\left(\frac{1-x}{1-p}\right)$ for $x \in [0, 1]$ and ∞ otherwise.
2) Poisson distribution : for $X_1 \sim \mathcal{P}(\lambda)$, we have $\Gamma^*(x) = x \ln\left(\frac{x}{\lambda}\right) + \lambda - x$ for $x \geq 0$ and ∞ otherwise.
3) Normal distribution : for $X_1 \sim \mathcal{N}(0, \sigma^2)$, we have $\Gamma^*(x) = \frac{x^2}{2\sigma^2}$, $x \in \mathbb{R}$.
2) Exponential distribution : for $X_1 \sim \mathcal{E}(\lambda)$, we have $\Gamma^*(x) = \lambda x - 1 - \ln(\lambda x)$ for $x > 0$ and $\Gamma^*(x) = \infty$ otherwise.

Remark 2.1 Cramer's theorem possesses a multivariate counterpart dealing with the large deviations of the empirical means of i.i.d. random vectors in $\mathbb{R}^d$.

Remark 2.2 (Relation with importance sampling)
Fix n and let us consider the estimation of $p_n = \mathbb{P}[S_n/n \geq x]$. A standard estimator for p_n is the average with N independent copies of $X = 1_{S_n/n \geq x}$. However, as shown in the introduction, for large n, p_n is small, and the relative error of this estimator is large. By using an exponential change of measure $\mathbb{P}_\theta$ with likelihood ratio

$$\frac{d\mathbb{P}_\theta}{d\mathbb{P}} = \exp\left(\theta S_n - n\Gamma(\theta)\right),$$

so that

$$p_n = \mathbb{E}_\theta\Big[\exp\big(-\theta S_n + n\Gamma(\theta)\big)1_{\frac{S_n}{n}\geq x}\Big],$$

we have an importance sampling (IS) (unbiased) estimator of p_n, by taking the average of independent replications of

$$\exp\big(-\theta S_n + n\Gamma(\theta)\big)1_{\frac{S_n}{n}\geq x}.$$

The parameter θ is chosen in order to minimize the variance of this estimator, or equivalently its second moment :

$$M_n^2(\theta, x) = \mathbb{E}_\theta\Big[\exp\big(-2\theta S_n + 2n\Gamma(\theta)\big)1_{\frac{S_n}{n}\geq x}\Big]$$
$$\leq \exp\big(-2n(\theta x - \Gamma(\theta))\big). \tag{10}$$

By noting from Cauchy-Schwarz's inequality that $M_n^2(\theta, x) \geq p_n^2 = \mathbb{P}[S_n/n \geq x] \simeq Ce^{-2n\Gamma^*(x)}$ as n goes to infinity, from Cramer's theorem, we see that the fastest possible exponential rate of decay of $M_n^2(\theta, x)$ is twice the rate of the probability itself, i.e. $2\Gamma^*(x)$. Hence, from (10), and with the choice of $\theta = \theta_x$ s.t. $\Gamma^*(x) = \theta_x x - \Gamma(\theta_x)$, we get an asymptotic optimal IS estimator in the sense that :

$$\lim_{n\to\infty} \frac{1}{n} \ln M_n^2(\theta_x, x) = 2 \lim_{n\to\infty} \frac{1}{n} \ln p_n.$$

This parameter θ_x is such that $\mathbb{E}_{\theta_x}[S_n/n] = x$ so that the event $\{S_n/n \geq x\}$ is no more rare under $\mathbb{P}_{\theta_x}$, and is precisely the parameter used in the derivation of the large deviations result in Cramer's theorem.

2.3 Some General Principles in the Theory of Large Deviations

In this section, we give some general principles in large deviations theory. We refer to the classical references [13] or [15] for a detailed treatment on the subject.

We first give the formal definition of a large deviation principle (LDP). Consider a sequence $\{Z^\varepsilon\}_\varepsilon$ on $(\Omega, \mathcal{F}, \mathbb{P})$ valued in some topological space $\mathcal{X}$. The LDP characterizes the limiting behaviour as $\varepsilon \to 0$ of the family of probability measures $\{\mathbb{P}[Z^\varepsilon \in dx]\}_\varepsilon$ on $\mathcal{X}$ in terms of a *rate function*. A rate function I is a lower semi-continuous function mapping $I : \mathcal{X} \to [0, \infty]$. It is a *good* rate function if the level sets $\{x \in \mathcal{X} : I(x) \leq M\}$ are compact for all $M < \infty$.

The sequence $\{Z^\varepsilon\}_\varepsilon$ satisfies a LDP on $\mathcal{X}$ with rate function I (and speed ε) if :

(i) *Upper bound* : for any closed subset F of $\mathcal{X}$

$$\limsup_{\varepsilon \to 0} \varepsilon \ln \mathbb{P}[Z^\varepsilon \in F] \leq - \inf_{x \in F} I(x).$$

(ii) *Lower bound* : for any open subset G of $\mathcal{X}$

$$\liminf_{\varepsilon \to 0} \varepsilon \ln \mathbb{P}[Z^\varepsilon \in G] \geq - \inf_{x \in G} I(x).$$

If F is a subset of $\mathcal{X}$ s.t. $\inf_{x \in F^\circ} I(x) = \inf_{x \in \bar{F}} I(x) := I_F$, then

$$\lim_{\varepsilon \to 0} \varepsilon \ln \mathbb{P}[Z^\varepsilon \in F] = -I_F,$$

which formally means that $\mathbb{P}[Z^\varepsilon \in F] \simeq C e^{-I_F/\varepsilon}$ for some constant C. The classical Cramer's theorem considered the case of the empirical mean $Z^\varepsilon = S_n/n$ of i.i.d. random variables in $\mathbb{R}^d$, with $\varepsilon = 1/n$. Further main results in large deviations theory are the Gärtner-Ellis theorem, which is a version of Cramer's theorem where independence is weakened to the existence of

$$\Gamma(\theta) := \lim_{\varepsilon \to 0} \varepsilon \ln \mathbb{E}\big[e^{\frac{\theta \cdot Z^\varepsilon}{\varepsilon}}\big], \quad \theta \in \mathbb{R}^d.$$

LDP is then stated for the sequence $\{Z^\varepsilon\}_\varepsilon$ with a rate function equal to the Fenchel-Legendre transform of Γ:

$$\Gamma^*(x) = \sup_{\theta \in \mathbb{R}^d} [\theta.x - \Gamma(\theta)], \quad x \in \mathbb{R}^d.$$

Other results in large deviations theory include Sanov's theorem, which gives rare events asymptotics for empirical distributions. In many problems, the interest is in rare events that depend on random process, and the corresponding asymptotics probabilities, usually called sample path large deviations, were developed by Freidlin-Wentzell and Donsker-Varadhan. For instance, the problem of diffusion exit from a domain is an important application of Freidlin-Wentzell theory, and occurs naturally in finance, see Section 4.1. We briefly summarize these results. Let $\varepsilon > 0$ a (small) positive parameter and consider the stochastic differential equation in $\mathbb{R}^d$ on some interval $[0, T]$,

$$dX^\varepsilon_s = b_\varepsilon(s, X^\varepsilon_s)ds + \sqrt{\varepsilon}\sigma(s, X^\varepsilon_s)dW_s, \tag{11}$$

and suppose that there exists a Lipschitz function b on $[0, T] \times \mathbb{R}^d$ s.t.

$$\lim_{\varepsilon \to 0} b_\varepsilon = b,$$

uniformly on compact sets. Given an open set Γ of $[0, T] \times \mathbb{R}^d$, we consider the exit time from Γ,

$$\tau^\varepsilon_{t,x} = \inf\big\{s \geq t \,:\, X^{\varepsilon, t, x}_s \notin \Gamma\big\},$$

and the corresponding exit probability

$$v_\varepsilon(t, x) = \mathbb{P}[\tau^\varepsilon_{t,x} \leq T], \quad (t, x) \in [0, T] \times \mathbb{R}^d.$$

Here $X^{\varepsilon,t,x}$ denotes the solution to (11) starting from x at time t. It is well-known that the process $X^{\varepsilon,t,x}$ converge to $X^{0,t,x}$ the solution to the ordinary differential equation

$$dX_s^0 = b(s, X_s^0)ds, \quad X_t^0 = x.$$

In order to ensure that v_ε goes to zero, we assume that for all $t \in [0, T]$,

$$(\mathbf{H}) \quad x \in \Gamma \Longrightarrow X_s^{0,t,x} \in \Gamma, \quad \forall s \in [t, T].$$

Indeed, under **(H)**, the system (11) tends, when ε is small, to stay inside Γ, so that the event $\{\tau_{t,x}^\varepsilon \leq T\}$ is rare. The large deviations asymptotics of $v_\varepsilon(t, x)$, when ε goes to zero, was initiated by Varadhan and Freidlin-Wentzell by probabilistic arguments. An alternative approach, introduced by Fleming, connects this theory with optimal control and Bellman equation, and is developed within the theory of viscosity solutions, see e.g. [9]. We sketch here this approach. It is well-known that the function v_ε satisfies the linear PDE

$$\frac{\partial v_\varepsilon}{\partial t} + b_\varepsilon(t, x).D_x v_\varepsilon + \frac{\varepsilon}{2}\mathrm{tr}(\sigma\sigma'(t, x)D_x^2 v_\varepsilon) = 0, \quad (t, x) \in [0, T) \times \Gamma, \quad (12)$$

together with the boundary conditions

$$v_\varepsilon(t, x) = 1, \quad (t, x) \in [0, T) \times \partial\Gamma \quad (13)$$
$$v_\varepsilon(T, x) = 0, \quad x \in \Gamma. \quad (14)$$

Here $\partial\Gamma$ is the boundary of Γ. We now make the logarithm transformation

$$V_\varepsilon = -\varepsilon \ln v_\varepsilon.$$

Then, after some straightforward derivation, (12) becomes the nonlinear PDE

$$-\frac{\partial V_\varepsilon}{\partial t} - b_\varepsilon(t, x).D_x V_\varepsilon - \frac{\varepsilon}{2}\mathrm{tr}(\sigma\sigma'(t, x)D_x^2 V_\varepsilon)$$
$$+ \frac{1}{2}(D_x V_\varepsilon)'\sigma\sigma'(t, x)D_x V_\varepsilon = 0, (t, x) \in [0, T) \times \Gamma, \quad (15)$$

and the boundary data (13)-(14) become

$$V_\varepsilon(t, x) = 0, \quad (t, x) \in [0, T) \times \partial\Gamma \quad (16)$$
$$V_\varepsilon(T, x) = \infty, \quad x \in \Gamma. \quad (17)$$

At the limit $\varepsilon = 0$, the PDE (15) becomes a first-order PDE

$$-\frac{\partial V_0}{\partial t} - b(t, x).D_x V_0$$
$$+ \frac{1}{2}(D_x V_0)'\sigma\sigma'(t, x)D_x V_0 = 0, \quad (t, x) \in [0, T) \times \Gamma, \quad (18)$$

with the boundary data (16)-(17). By PDE-viscosity solutions methods and comparison results, we can prove (see e.g. [9] or [19]) that V_ε converges uniformly on compact subsets of $[0, T) \times \Gamma$, as ε goes to zero, to V_0 the unique viscosity solution to (18) with the boundary data (16)-(17). Moreover, V_0 has a representation in terms of control problem. Consider the Hamiltonian function

$$H(t, x, p) = -b(t, x).p + \frac{1}{2} p' \sigma \sigma'(t, x) p, \quad (t, x, p) \in [0, T] \times \Gamma \times \mathbb{R}^d,$$

which is quadratic and in particular convex in p. Then, using the Legendre transform, we may rewrite

$$H(t, x, p) = \sup_{q \in \mathbb{R}^d} \big[- q.p - H^*(t, x, q) \big],$$

where

$$\begin{aligned} H^*(t, x, q) &= \sup_{p \in \mathbb{R}^d} \big[- p.q - H(t, x, p) \big] \\ &= \frac{1}{2} (q - b(t, x))' (\sigma \sigma'(t, x))^{-1} (q - b(t, x)), \quad (t, x, q) \in [0, T] \times \Gamma \times \mathbb{R}^d. \end{aligned}$$

Hence, the PDE (18) is rewritten as

$$\frac{\partial V_0}{\partial t} + \inf_{q \in \mathbb{R}^d} \big[q.D_x V_0 + H^*(t, x, q) \big] = 0, \quad (t, x) \in [0, T) \times \Gamma,$$

which, together with the boundary data (16)-(17), is associated to the value function for the following calculus of variations problem : for an absolutely continuous function $x(.)$ on $[0, T)$ and valued in $\mathbb{R}^d$, i.e. $x \in H^1_{loc}([0, T], \mathbb{R}^d)$, we denote $\dot{x}(u) = q_u$ its time derivative, and $\tau(x)$ the exit time of $x(.)$ from Γ. Then,

$$\begin{aligned} V_0(t, x) \\ &= \inf_{x(.) \in \mathcal{A}(t,x)} \int_t^T H^*(u, x(u), \dot{x}(u)) du, \quad (t, x) \in [0, T) \times \Gamma, \\ &= \inf_{x(.) \in \mathcal{A}(t,x)} \int_t^T \frac{1}{2} (\dot{x}(u) - b(u, x(u)))' (\sigma \sigma'(u, x(u)))^{-1} (\dot{x}(u) - b(u, x(u))) du \end{aligned}$$

where

$$\mathcal{A}(t, x) = \big\{ x(.) \in H^1_{loc}([0, T], \mathbb{R}^d) \ : \ x(t) = x \text{ and } \tau(x) \leq T \big\}.$$

The large deviations result is then stated as

$$\lim_{\varepsilon \to 0} \varepsilon \ln v_\varepsilon(t, x) = -V_0(t, x), \tag{19}$$

and the above limit holds uniformly on compact subsets of $[0, T) \times \Gamma$. A more precise result may be obtained, which allows to remove the above log estimate. This type of result is developed in [17], and is called sharp large deviations estimate. It states asymptotic expansion (in ε) of the exit probability for points (t,x) belonging to a set N of $[0,T'] \times \Gamma$ for some $T' < T$, open in the relative topology, and s.t. $V_0 \in C^\infty(N)$. Then, under the condition that

$$b_\varepsilon = b + \varepsilon b_1 + 0(\varepsilon^2),$$

one has

$$v_\varepsilon(t,x) = \exp\left(-\frac{V_0(t,x)}{\varepsilon} - w(t,x)\right)(1 + O(\varepsilon)),$$

uniformly on compact sets of N, where w is solution to the PDE problem

$$-\frac{\partial W}{\partial t} - (b - \sigma\sigma' D_x V_0).D_x w = \frac{1}{2}\text{tr}(\sigma\sigma' D_x^2 V_0) + b_1.D_x V_0 \quad \text{in } N$$

$$w(t,x) = 0 \quad \text{on} \quad \left([0,T] \times \partial\Gamma\right) \cup \bar{N}.$$

The function w may be represented as

$$w(t,x) = \int_t^\rho \left(\frac{1}{2}\text{tr}(\sigma\sigma' D_x^2 V_0) + b_1.D_x V_0\right)(s,\xi(s))ds,$$

where ξ is the solution to

$$\dot{\xi}(s) = (b - \sigma\sigma' D_x V_0)(s,\xi(s)), \quad \xi(t) = x,$$

and ρ is the exit time (after t) of $(s,\xi(s))$ from N.

We shall develop more in detail in the next sections some applications of the Gärtner-Ellis and Freidlin-Wentzell theories in finance.

We end this paragraph by stating the important Varadhan's integral formula, which involves the asymptotics behavior of certain expectations. It extends the well-known method of Laplace for studying the asymptotics of certain integrals on $\mathbb{R}$: given a continuous function φ from $[0,1]$ into $\mathbb{R}$, Laplace's method states that

$$\lim_{n\to\infty} \frac{1}{n} \ln \int_0^1 e^{n\varphi(x)} dx = \max_{x\in[0,1]} \varphi(x).$$

Varadhan result's is formulated as follows :

Theorem 2.2 *(Varadhan) Suppose that $\{Z^\varepsilon\}_\varepsilon$ satisfies a LDP on $\mathcal{X}$ with good rate function I, and let $\varphi : \mathcal{X} \to \mathbb{R}$ be any continuous function s.t. the following moment condition holds for some $\gamma > 1$:*

$$\limsup_{\varepsilon\to 0} \varepsilon \ln \mathbb{E}\left[e^{\gamma\varphi(Z^\varepsilon)/\varepsilon}\right] < \infty.$$

Then,

$$\lim_{\varepsilon\to 0} \varepsilon \ln \mathbb{E}\left[e^{\varphi(Z^\varepsilon)/\varepsilon}\right] = \sup_{x\in\mathcal{X}} \left[\varphi(x) - I(x)\right]. \qquad (20)$$

Proof. (a) For simplicity, we show the inequality $\leq$ in (20) when φ is bounded on $\mathcal{X}$. Hence, there exists $M \in (0, \infty)$ s.t. $-M \leq \varphi(x) \leq M$ for all $x \in \mathcal{X}$. For N positive integer, and $j \in \{1, \ldots, N\}$, we consider the closed subsets of $\mathcal{X}$

$$F_{N,j} = \left\{ x \in \mathcal{X} : -M + \frac{2(j-1)M}{N} \leq \varphi(x) \leq -M + \frac{2jM}{N} \right\},$$

so that $\cup_{j=1}^{N} F_{N,j} = \mathcal{X}$. We then have from the large deviations upper bound on (Z^ε),

$$\limsup_{\varepsilon \to 0} \varepsilon \ln \mathbb{E}\left[e^{\varphi(Z^\varepsilon)/\varepsilon}\right]$$

$$= \limsup_{\varepsilon \to 0} \varepsilon \ln \int_\mathcal{X} e^{\varphi(Z^\varepsilon)/\varepsilon} \mathbb{P}[Z^\varepsilon \in dx]$$

$$\leq \limsup_{\varepsilon \to 0} \varepsilon \ln \Big(\sum_{j=1}^{N} \int_{F_{N,j}} e^{\varphi(Z^\varepsilon)/\varepsilon} \mathbb{P}[Z^\varepsilon \in dx] \Big)$$

$$\leq \limsup_{\varepsilon \to 0} \varepsilon \ln \Big(\sum_{j=1}^{N} e^{(-M+2jM/N)/\varepsilon} \mathbb{P}[Z^\varepsilon \in F_{N,j}] \Big)$$

$$\leq \limsup_{\varepsilon \to 0} \varepsilon \ln \Big(\max_{j=1,\ldots,N} e^{(-M+2jM/N)/\varepsilon} \mathbb{P}[Z^\varepsilon \in F_{N,j}] \Big)$$

$$\leq \max_{j=1,\ldots,N} \Big(-M + \frac{2jM}{N} + \limsup_{\varepsilon \to 0} \varepsilon \ln \mathbb{P}[Z^\varepsilon \in F_{N,j}] \Big)$$

$$\leq \max_{j=1,\ldots,N} \Big(-M + \frac{2jM}{N} + \sup_{x \in F_{N,j}} [-I(x)] \Big)$$

$$\leq \max_{j=1,\ldots,N} \Big(-M + \frac{2jM}{N} + \sup_{x \in F_{N,j}} [\varphi(x) - I(x)] - \inf_{x \in F_{N,j}} \varphi(x) \Big)$$

$$\leq \sup_{x \in F_{N,j}} [\varphi(x) - I(x)] + \frac{2M}{N}.$$

By sending N to infinity, we get the inequality $\leq$ in (20).

(b) To prove the reverse inequality, we fix an arbitrary point $x_0 \in \mathcal{X}$, an arbitrary $\delta > 0$, and we consider the open set $G = \{x \in \mathcal{X} : \varphi(x) > \varphi(x_0) - \delta\}$. Then, we have from the large deviations lower bound on (Z^ε),

$$\liminf_{\varepsilon \to 0} \varepsilon \ln \mathbb{E}\left[e^{\varphi(Z^\varepsilon)/\varepsilon}\right] \geq \liminf_{\varepsilon \to 0} \varepsilon \ln \mathbb{E}\left[e^{\varphi(Z^\varepsilon)/\varepsilon} 1_{Z^\varepsilon \in G}\right]$$

$$\geq \varphi(x_0) - \delta + \liminf_{\varepsilon \to 0} \varepsilon \ln \mathbb{P}[Z^\varepsilon \in G]$$

$$\geq \varphi(x_0) - \delta - \inf_{x \in G} I(x)$$

$$\geq \varphi(x_0) - I(x_0) - \delta.$$

Since $x_0 \in \mathcal{X}$ and $\delta > 0$ are arbitrary, we get the required result. □

Remark 2.3 The relation (20) has the following interpretation. By writing formally the LDP for (Z^ε) with rate function I as $\mathbb{P}[Z^\varepsilon \in dx] \simeq e^{-I(x)/\varepsilon} dx$, we can write

$$\mathbb{E}\left[e^{\varphi(Z^{\varepsilon})/\varepsilon}\right] = \int e^{\varphi(x)/\varepsilon}\mathbb{P}[Z^{\varepsilon} \in dx] \simeq \int e^{(\varphi(x)-I(x))/\varepsilon}dx$$
$$\simeq C\exp\left(\frac{\sup_{x\in\mathcal{X}}(\varphi(x)-I(x))}{\varepsilon}\right).$$

As in Laplace's method, Varadhan's formula states that to exponential order, the main contribution to the integral is due to the largest value of the exponent.

3 Ruin Probabilities in Risk Theory

3.1 The Classical Ruin Problem

The Insurance Model

We consider an insurance company earning premiums at a constant rate p per unit of time, and paying claims that arrive at the jumps of a Poisson process with intensity λ. We denote by N_t the number of claims arriving in $[0,t]$, by T_n, $n \geq 1$, the arrival times of the claim, and by $\xi_1 = T_1$, $\xi_n = T_n - T_{n-1}$, $n \geq 2$, the interarrival times, which are then i.i.d. exponentially distributed with finite mean $E\xi_1 = 1/\lambda$. The size of the n-th claim is denoted Y_n, and we assume that the claim sizes Y_n, $n \in \mathbb{N}^*$, are (positive) i.i.d., and independent of the Poisson process. Starting from an initial reserve $x > 0$, the risk reserve process $X_t = X_t^x$, $t \geq 0$, of the insurance company is then given by :

$$X_t^x = x + pt - \sum_{i=1}^{N_t} Y_i. \tag{21}$$

The probability of ruin with infinite horizon is

$$\psi(x) = \mathbb{P}[\tau_x < \infty],$$

where $\tau_x = \inf\{t \geq 0 : X_t^x < 0\}$ is the time of ruin. We are interested in the estimation of the ruin probability, in particular for large values of the initial reserve.

The Cramer-Lundberg Estimate

The Cramer-Lundberg approximation concerns the estimation of the ruin probability $\psi(x)$, and is one of the most celebrated result of risk theory. There are several approaches for deriving such a result. We follow in this paragraph a method based on large deviations arguments and change of probability measures.

First, we easily see, by the strong law of large numbers, that

$$\frac{1}{t}\sum_{i=1}^{N_t} Y_i \to \rho \quad a.s., \quad t \to \infty,$$

where $\rho = \lambda \mathbb{E}[Y_1] > 0$ is interpreted as the average amount of claim per unit of time. The safety loading η plays a key role in ruin probability. It is defined as the relative amount by which the premium rate exceeds ρ :

$$\eta = \frac{p-\rho}{\rho} \iff p = (1+\eta)\rho.$$

Hence $X_t^x/t \to p - \rho = \rho\eta$ when t goes to infinity. Therefore, if $\eta < 0$, $X_t^x \to -\infty$, and we clearly have $\psi(x) = 1$ for all x. For $\eta = 0$, we can also show that $\limsup X_t^x = -\infty$ so that $\psi(x) = 1$. In the sequel, we make the net profit assumption :

$$\eta = \frac{p - \lambda \mathbb{E}[Y_1]}{\lambda \mathbb{E}[Y_1]} > 0, \tag{22}$$

which ensures that the probability of ruin is less than 1.

Since ruin may occur only at the arrival of a claim, i.e. when X jumps downwards, it suffices to consider the discrete-time process embedded at the jumps of the Poisson process. We then define the discrete-time process $X_{T_n}^x$, $n \geq 1$, so that

$$\psi(x) = \mathbb{P}[\sigma_x < \infty],$$

where $\sigma_x = \inf\{n \geq 1 : X_{T_n}^x < 0\} = \inf\{n \geq 1 : S_n > x\}$, and $S_n = x - X_{T_n}^x$ is the net payout up to the n-th claim and given by the random walk :

$$S_n = Z_1 + \ldots + Z_n, \quad Z_i = Y_i - p\xi_i, \quad i \in \mathbb{N}^*.$$

The r.v. Z_i are i.i.d. and satisfy under the net profit condition, $\mathbb{E}[Z_1] < 0$. We denote by Γ_Z the c.g.f. of the Z_i, and we see that by independence of Y_i and ξ_i :

$$\Gamma_Z(\theta) = \Gamma_Y(\theta) + \Gamma_\xi(-p\theta)$$
$$= \Gamma_Y(\theta) + \ln\left(\frac{\lambda}{\lambda + p\theta}\right), \quad \theta > -\frac{\lambda}{p},$$

where Γ_Y (resp. Γ_ξ) is the c.g.f. of the Y_i (resp. ξ_i). For any θ in the domain of Γ_Z, we consider the exponential change of measure with parameter θ, and since σ_x is a stopping time in the filtration of $(Z_1, \ldots, Z_n)$, we apply formula (4) to write the ruin probability as an $\mathbb{E}_\theta$ expectation :

$$\psi(x) = \mathbb{P}[\sigma_x < \infty] = \mathbb{E}\left[1_{\sigma_x < \infty}\right]$$
$$= \mathbb{E}_\theta\left[1_{\sigma_x < \infty} \exp\left(-\theta S_{\sigma_x} + \sigma_x \Gamma_Z(\theta)\right)\right]. \tag{23}$$

We now assume Y has a light-tailed distribution, i.e. : there exists $\bar{\theta} \in (0, \infty]$ s.t. $\Gamma_y(\theta) < \infty$ for $\theta < \bar{\theta}$, and $\Gamma_Y(\theta) \to \infty$ as $\theta \nearrow \bar{\theta}$. In this case, the c.g.f. Γ_Z of the Z_i is finite on $(-\lambda/p, \bar{\theta})$, it is differentiable in 0 with $\Gamma'_Z(0) = \mathbb{E}[Z_1] < 0$ under the net profit condition (22). Moreover, since $\mathbb{E}[Y_1] > 0$ and Y_1 is independent of ξ_1, we see that $\mathbb{P}[Z_1 > 0] > 0$, which implies that $\Gamma_Z(\theta)$ goes to infinity as θ goes to $\bar{\theta}$. By convexity of Γ_Z and recalling that $\Gamma_Z(0) = 0$, we deduce the existence of an unique $\theta_L > 0$ s.t. $\Gamma_Z(\theta_L) = 0$. This unique positive θ_L is the solution to the so-called *Cramer-Lundberg* equation :

$$\Gamma_Y(\theta_L) + \ln\left(\frac{\lambda}{\lambda + p\theta_L}\right) = 0,$$

which is also written equivalently in :

$$\gamma_Y(\theta_L) = \frac{p\theta_L}{\lambda}, \qquad (24)$$

where $\gamma_Y = \exp(\Gamma_Y(\theta)) - 1 = \int e^{\theta y} F_Y(dy) - 1$ is the shifted ($\gamma_Y(0) = 0$) moment generating function of Y_i, and F_Y is the distribution function of the claim sizes Y_i. θ_L is called *adjustment coefficient* (or sometimes *Lundberg exponent*). Notice also that by convexity of Γ_Z, we have $\Gamma'_Z(\theta_L) > 0$. Hence, under $\mathbb{P}_{\theta_L}$, the random walk has positive drift $\mathbb{E}_{\theta_L}[Z_n] = \Gamma'_Z(\theta_L) > 0$, and this implies $\mathbb{P}_{\theta_L}[\sigma_x < \infty] = 1$. For this choice of $\theta = \theta_L$, (23) becomes

$$\psi(x) = \mathbb{E}_{\theta_L}\left[e^{-\theta_L S_{\sigma_x}}\right] = e^{-\theta_L x} \mathbb{E}_{\theta_L}\left[e^{-\theta_L(S_{\sigma_x} - x)}\right]. \qquad (25)$$

By noting that the overshoot $S_{\sigma_x} - x$ is nonnegative, we obtain the Lundberg's inequality on the ruin probability :

$$\psi(x) \leq e^{-\theta_L x}, \quad \forall x > 0. \qquad (26)$$

Moreover, by renewal's theory, the overshoot $R^x = S_{\sigma_x} - x$ has a limit R^∞ (in the sense of weak convergence with respect to $\mathbb{P}_{\theta_L}$), when x goes to infinity, and therefore $\mathbb{E}_{\theta_L}[e^{-\theta_L R^x}]$ converges to some positive constant C. We then get the classical approximation for large values of the initial reserve :

$$\psi(x) \simeq C e^{-\theta_L x},$$

as $x \to \infty$, which implies a large deviation type estimation

$$\lim_{x \to \infty} \frac{1}{x} \ln \psi(x) = -\theta_L. \qquad (27)$$

Further details and extensions can be found in [3] or [16]. They concern more general processes (e.g. Levy processes) for the risk reserve, heavy-tailed distribution for the claim size ... In the next paragraph, we study an extension of the classical ruin problem, developed by [23] and [31], where the insurer has the additional opportunity to invest in a stock market.

Application: Importance sampling for the ruin probability estimation

From the perspective of estimation of the ruin probability $\psi(x)$, and by choosing the Lundberg exponent θ_L, we have an unbiased estimator with the associated importance sampling estimator based on Monte-Carlo simulations of

$$\psi(x) = \mathbb{E}_{\theta_L}\left[e^{-\theta_L S_{\sigma_x}}\right].$$

Since, obviously, $S_{\sigma_x} > x$, the second order moment of this estimator satisfies

$$M^2(\theta_L, x) = \mathbb{E}_{\theta_L}\left[e^{-2\theta_L S_{\sigma_x}}\right] \leq e^{-2\theta_L x}.$$

On the other hand, by Cauchy-Schwarz's inequality, the second moment of any unbiased estimator must be as large as the square of the ruin probability, and we have seen that this probability is $O(e^{-\theta_L x})$. Therefore, the IS estimator based on θ_L is asymptotically optimal as $x \to \infty$:

$$\lim_{x \to \infty} \frac{1}{x} \ln M^2(\theta_L, x) = 2 \lim_{x \to \infty} \frac{1}{x} \ln \psi(x) \ (= 2\theta_L).$$

Remark 3.4 For any $x > 0$, $\theta > 0$, consider the process

$$M_t(x, \theta) = \exp(-\theta X_t^x), \quad t \geq 0.$$

where X_t^x is the risk reserve process defined in (21). A standard calculation shows that for all $t \geq 0$,

$$\mathbb{E}\left[M_t(0, \theta)\right] = e^{(\lambda \gamma_Y(\theta) - p\theta)t}.$$

Moreover, since X_t^x has stationary independent increments, and by denoting $\mathbb{F} = (\mathcal{F}_t)_{t \geq 0}$ the filtration generated by the risk reserve X, we have for all $0 \leq t \leq T$

$$\begin{aligned}
\mathbb{E}\left[M_T(x, \theta) | \mathcal{F}_t\right] &= \mathbb{E}\left[e^{-\theta X_T^x} | \mathcal{F}_t\right] \\
&= M_t(x, \theta) \mathbb{E}\left[e^{-\theta(X_T^x - X_t^x)} | \mathcal{F}_t\right] \\
&= M_t(x, \theta) \mathbb{E}\left[M_{T-t}(0, \theta)\right] \\
&= M_t(x, \theta) \ e^{(\lambda \gamma_Y(\theta) - p\theta)(T-t)}. \quad (28)
\end{aligned}$$

Hence, for the choice of $\theta = \theta_L$: the adjustment coefficient, the process $M_t(x, \theta_L)$, $t \geq 0$, is a $(\mathbb{P}, \mathbb{F})$-martingale. The use of this martingale property in the derivation of ruin estimate was initiated by Gerber [24]. We show in the next paragraph how to extend this idea for a ruin problem with investment in a stock market.

3.2 Ruin Probabilities and Optimal Investment

The Insurance-Finance Model

In the setting of the classical model described in the previous paragraph, we consider the additional feature that the insurance company is also allowed to invest in some stock market, modeled by a geometric brownian motion:

$$dS_t = bS_t dt + \sigma S_t dW_t,$$

where b, σ are constants, $\sigma > 0$, and W is a standard brownian motion, independent of the risk reserve X as defined in (21). We denote by $\mathbb{F} = (\mathcal{F}_t)_{t \geq 0}$ the filtration generated by X and S. The insurer may invest at any time t an amount of money α_t in the stock, and the rest in the bond (which in the present model yields no interest). The set $\mathcal{A}$ of admissible investment strategies is defined as the set of $\mathbb{F}$-adapted processes $\alpha = (\alpha_t)$ s.t. $\int_0^t \alpha_s^2 ds < \infty$ a.s. Given an initial capital $x \geq 0$, and an admissible investment control α, the insurer's wealth process can then be written as

$$V_t^{x,\alpha} = X_t^x + \int_0^t \frac{\alpha_u}{S_u} dS_u$$

$$= x + pt - \sum_{i=1}^{N_t} Y_i + \int_0^t \alpha_u (b du + \sigma dW_u), \quad t \geq 0.$$

We define the infinite time ruin probability

$$\psi(x, \alpha) = \mathbb{P}[\tau_{x,\alpha} < \infty],$$

where $\tau_{x,\alpha} = \inf\{t \geq 0 : V_t^{x,\alpha} < 0\}$ is the time of ruin, depending on the initial wealth x and the investment strategy α. We are interested in the minimal ruin probability of the insurer

$$\psi^*(x) = \inf_{\alpha \in \mathcal{A}} \psi(x, \alpha).$$

Asymptotic Ruin Probability Estimate

The main result is an asymptotic large deviation estimation for the minimal ruin probability when the initial reserve goes to infinity:

Theorem 3.3 *We have*

$$\lim_{x \to \infty} \frac{1}{x} \ln \psi^*(x) = -\theta^*, \tag{29}$$

where $\theta^* > 0$ is he unique positive solution to the equation

$$\gamma_Y(\theta) = p\frac{\theta}{\lambda} + \frac{b^2}{2\sigma^2 \lambda}. \tag{30}$$

Here $\gamma_Y(\theta) = \mathbb{E}[e^{\theta Y_1}] - 1$ is the shifted moment generating function of the claim size. Moreover, the constant strategy $\alpha^* = \frac{b}{\sigma^2 \theta^*}$ is asymptotically optimal in the sense that

$$\lim_{x \to \infty} \frac{1}{x} \ln \psi(x, \alpha^*) = -\theta^*.$$

Finally, if $b \neq 0$, $\theta^* > \theta_L$ the Lundberg exponent.

Remark 3.5 The estimate (29) is analogue to the classical Lundberg estimate (27) without investment. The exponent is larger than the Lundberg one, and thus one gets a sharper bound on the minimal ruin probability. Moreover, the trading strategy yielding the optimal asymptotic exponential decay consists in holding a fixed (explicit) amount in the risky asset. This surprising result, in apparent contradiction with the common believe that 'rich' companies should invest more than 'poor' ones, is explained by the fact that minimization of ruin probability is an extremely conservative criterion.

We follow the martingale approach of Gerber for stating this theorem. We emphasize the main steps of the proof. Let us introduce, for fixed $x, \theta \in \mathbb{R}_+$, and $\alpha \in \mathcal{A}$, the process

$$M_t(x, \theta, \alpha) = \exp(-\theta V_t^{x,\alpha}), \quad t \geq 0.$$

Then, a straightforward calculation shows that for any constant process $\alpha = a$, and $t \geq 0$,

$$\begin{aligned} \mathbb{E}\big[M_t(0, \theta, a)\big] &= \mathbb{E}\big[e^{-\theta(pt - \sum_{i=1}^{N_t} Y_i + abt + a\sigma W_t)}\big] \\ &= e^{-\theta(p+ab)t} \mathbb{E}\big[e^{\theta \sum_{i=1}^{N_t} Y_i}\big] \mathbb{E}\big[e^{-\theta a \sigma W_t}\big] \\ &= e^{-\theta(p+ab)t} e^{\gamma_Y(\theta) \lambda t} e^{\frac{\theta^2 a^2 \sigma^2}{2} t} \\ &= e^{f(\theta, a) t}, \end{aligned}$$

where

$$f(\theta, a) = \lambda \gamma_Y(\theta) - p\theta - ab\theta + \frac{1}{2} a^2 \theta^2 \sigma^2.$$

We recall that under the assumption of light-tailed distribution on the claim size Y_i, the shifted moment generating function γ_Y is finite and convex on $(-\infty, \bar{\theta})$ for some $\bar{\theta} \in (0, \infty]$, and $\gamma_Y \to \infty$ as θ goes to $\bar{\theta}$. Moreover, recalling that $\mathbb{E}[Y_1] \geq 0$, then γ_Y is increasing on $[0, \bar{\theta})$ since $\gamma_Y'(\theta) = \mathbb{E}[Y_1 e^{\theta Y_1}] > \mathbb{E}[Y_1]$ for $0 < \theta < \bar{\theta}$. Now, we see that for all $\theta > 0$,

$$\bar{f}(\theta) := \inf_{a \in \mathbb{R}} f(\theta, a) = \lambda \gamma_Y(\theta) - p\theta - \frac{b^2}{2\sigma^2},$$

with an infimum attained for $\hat{a}(\theta) = b/(\theta \sigma^2)$. From the properties of γ_Y, we clearly have the existence and uniqueness of θ^* solution to $\bar{f}(\theta^*) = 0$, i.e. (30). Since the r.h.s. of (30) is just the r.h.s. of (24), but shifted by the positive constant $b^2/2\sigma^2 \lambda$ (if $b \neq 0$), it is also obvious that $\theta^* > \theta_L$. By choosing $\alpha^* = \hat{a}(\theta^*) = b^2/(\theta^* \sigma^2)$, we then have

$$\bar{f}(\theta^*) = f(\theta^*, \alpha^*) = 0.$$

A straightforward calculation also shows that for all $a \in \mathbb{R}$,

$$f(\theta^*, a) = \frac{1}{2}(\theta^*)^2 \sigma^2 (a - \alpha^*)^2 \geq 0. \tag{31}$$

Hence, since V_t^{x,α^*} has independent stationary increments, we obtain similarly as in (28), for all $0 \leq t \leq T$,

$$\mathbb{E}[M_T(x,\theta^*,\alpha^*)|\mathcal{F}_t] = M_t(x,\theta^*,\alpha^*)\mathbb{E}[M_{T-t}(0,\theta^*,\alpha^*)]$$
$$= M_t(x,\theta,a),$$

which shows that the process $M(x,\theta^*,\alpha^*)$ is a $(\mathbb{P},\mathbb{F})$-martingale. Therefore, from the optional sampling theorem at the (bounded) stopping time $\tau_{x,\alpha^*} \wedge T$, we have

$$e^{-\theta^* x} = M_0(x,\theta^*,\alpha^*) = \mathbb{E}[M_{\tau_{x,\alpha^*}\wedge T}(x,\theta^*,\alpha^*)]$$
$$= \mathbb{E}[M_{\tau_{x,\alpha^*}}(x,\theta^*,\alpha^*)1_{\tau_{x,\alpha^*}\leq T}] + \mathbb{E}[M_T(x,\theta^*,\alpha^*)1_{\tau_{x,\alpha^*}> T}]$$
$$\geq \mathbb{E}[M_{\tau_{x,\alpha^*}}(x,\theta^*,\alpha^*)1_{\tau_{x,\alpha^*}\leq T}],$$

since the process M is nonnegative. By the monotone convergence theorem, we then get by sending T to infinity into the previous inequality

$$e^{-\theta^* x} \geq \mathbb{E}[M_{\tau_{x,\alpha^*}}(x,\theta^*,\alpha^*)1_{\tau_{x,\alpha^*}<\infty}]$$
$$= \mathbb{E}[M_{\tau_{x,\alpha^*}}(x,\theta^*,\alpha^*)|\tau_{x,\alpha^*}<\infty]\mathbb{P}[\tau_{x,\alpha^*}<\infty],$$

from Bayes formula. Thus, we get

$$\psi(x,\alpha^*) = \mathbb{P}[\tau_{x,\alpha^*}<\infty] \leq \frac{e^{-\theta^* x}}{\mathbb{E}[M_{\tau_{x,\alpha^*}}(x,\theta^*,\alpha^*)|\tau_{x,\alpha^*}<\infty]}.$$

Now, by definition of the time of ruin, $V_{\tau_{x,\alpha^*}}^{x,\alpha^*}$ is nonpositive and so $M_{\tau_{x,\alpha^*}}(x,\theta^*,\alpha^*) \geq 1$ a.s. on $\{\tau_{x,\alpha^*}<\infty\}$. We deduce that

$$\psi^*(x) \leq \psi(x,\alpha^*) \leq e^{-\theta^* x}. \tag{32}$$

In order to state a lower bound on the minimal ruin probability, we proceed as follows. We apply Itô's formula to the process $M(x,\theta^*,\alpha)$ for arbitrary $\alpha \in \mathcal{A}$:

$$\frac{dM_t(x,\theta^*,\alpha)}{M_{t-}(x,\theta^*,\alpha)} = \left(-\theta^*(p+b\alpha_t) + \frac{1}{2}(\theta^*)^2\sigma^2\alpha_t^2\right)dt - \theta^*\sigma dW_t + (e^{\theta^* Y_{N_t}}-1)dN_t.$$

Observing that $\gamma_Y(\theta) = \mathbb{E}[e^{\theta Y_{N_t}}-1]$, we rewrite as

$$\frac{dM_t(x,\theta^*,\alpha)}{M_{t-}(x,\theta^*,\alpha)} = \left(-\theta^*(p+b\alpha_t) + \frac{1}{2}(\theta^*)^2\sigma^2\alpha_t^2 + \lambda\gamma_Y(\theta^*)\right)dt - \theta^*\sigma dW_t$$
$$+(e^{\theta^* Y_{N_t}}-1)dN_t - \lambda\mathbb{E}[e^{\theta^* Y_{N_t}}-1]dt$$
$$= f(\theta^*,\alpha_t)dt - \theta^*\sigma dW_t + d\tilde{N}_t,$$

where $\tilde{N}_t = \int_0^t (e^{\theta^* Y_{N_u}}-1)dN_u - \int_0^t \lambda\mathbb{E}[e^{\theta^* Y_{N_u}}-1]du$. By using the martingale property of $N_t - \lambda t$, we can check that $\tilde{N}$ is a martingale. Since $f(\theta^*,\alpha_t) \geq 0$ a.s. for all t (see (31)), we deduce that $M(x,\theta^*,\alpha)$ is a (local) submartingale. To go to a true submartingale, we need some additional assumption on the distribution of the claim size. Actually, we can prove that under the following uniform exponential tail distribution

$$\sup_{z\geq 0} \mathbb{E}[e^{-\theta^*(z-Y_1)}|Y_1 > z] = \sup_{z\geq 0} \frac{\int_z^\infty e^{-\theta^*(z-y)} dF_Y(y)}{\int_z^\infty dF_Y} < \infty, \quad (33)$$

the process $M(x, \theta^*, \alpha)$ is an uniformly integrable submartingale. Therefore, from the optional sampling theorem at the (bounded) stopping time $\tau_{x,\alpha} \wedge T$, we have

$$\begin{aligned} e^{-\theta^* x} &= M_0(x, \theta^*, \alpha) \\ &\leq \mathbb{E}\big[M_{\tau_{x,\alpha}\wedge T}(x, \theta^*, \alpha)\big] \\ &= \mathbb{E}\big[M_{\tau_{x,\alpha}}(x, \theta^*, \alpha)1_{\tau_{x,\alpha}\leq T}\big] + \mathbb{E}\big[M_T(x, \theta^*, \alpha)1_{\tau_{x,\alpha}>T}\big]. \quad (34) \end{aligned}$$

We now claim that $M_T(x, \theta^*, \alpha)$ converges a.s. on $\{\tau_{x,\alpha} = \infty\}$ to zero as T goes to infinity. First, we know from Doob's submartingale convergence theorem that $\lim_{T\to\infty} M_T(x, \theta^*, \alpha)$ exists a.s., hence also $\lim_{T\to\infty} V_T(x, \alpha)$. Since the expectation of a jump size $E[Y_1]$ is positive, there exists $y > 0$ s.t. $\mathbb{P}[Y_1 > y] > 0$. By independence of the jump sizes in the compound Poisson process of the risk reserve, it is then an easy exercice to see that with probability 1, a jump of size greater than y occurs infinitely often on $[0, \infty)$. On the other hand, the stochastic integral due to the invesment strategy α in the stock price, is a.s. continuous, and so cannot compensate the jumps of the compound Poisson process, greater than y, which will occur infinitely often a.s. It follows that on $\{\tau_{x,\alpha} = \infty\}$ (where ruin does not occur), $V_T^{x,\alpha}$ cannot converge to a finite value with positive probability. Therefore on $\{\tau_{x,\alpha} = \infty\}$, we have $\lim_{T\to\infty} V_T(x, \alpha) = \infty$ and thus, since $\theta^* > 0$, $\lim_{T\to\infty} M_T(x, \theta^*, \alpha) = 0$ a.s. As $T \to \infty$, we have then from the dominated convergence theorem ($M_T(x, \theta^*, \alpha) \leq 1$ on $\{\tau_{x,\alpha} > T\}$) for the second term in (34), and by the monotone convergence theorem for the first term,

$$\begin{aligned} e^{-\theta^* x} &\leq \mathbb{E}\big[M_{\tau_{x,\alpha}}(x, \theta^*, \alpha)1_{\tau_{x,\alpha}<\infty}\big] \\ &= \mathbb{E}\big[M_{\tau_{x,\alpha}}(x, \theta^*, \alpha)\big|\tau_{x,\alpha} < \infty\big]\mathbb{P}[\tau_{x,\alpha} < \infty], \end{aligned}$$

and so

$$\psi(x, \alpha) = \mathbb{P}[\tau_{x,\alpha} < \infty] \geq \frac{e^{-\theta^* x}}{\mathbb{E}\big[M_{\tau_{x,\alpha}}(x, \theta^*, \alpha)\big|\tau_{x,\alpha} < \infty\big]}. \quad (35)$$

We finally prove that $\mathbb{E}\big[M_{\tau_{x,\alpha}}(x, \theta^*, \alpha)\big|\tau_{x,\alpha} < \infty\big]$ is bounded by a constant independent of $\alpha \in \mathcal{A}$. Fix some arbitrary $\alpha \in \mathcal{A}$ and set for shorthand notation $\tau = \tau_{x,\alpha}$ the time of ruin of the wealth process $V^{x,\alpha}$. First observe that ruin $\{\tau < \infty\}$ occurs either through the brownian motion, i.e. on $\{\tau < \infty, V_{\tau-}^{x,\alpha} = 0\}$, and in this case $V_\tau^{x,\alpha} = 0$ and so $M_\tau(x, \theta^*, \alpha) = 1$, or through a jump, i.e. on $\{\tau < \infty, V_{\tau-}^{x,\alpha} > 0\}$, and in this case $V_\tau^{x,\alpha} < 0$ and so $M_\tau(x, \theta^*, \alpha) > 1$. Hence,

$$\begin{aligned} \mathbb{E}\big[M_\tau(x, \theta^*, \alpha)\big|\tau < \infty\big] &\leq \mathbb{E}\big[M_\tau(x, \theta^*, \alpha)\big|\tau < \infty, V_{\tau-}^{x,\alpha} > 0\big] \\ &= \mathbb{E}\big[e^{-\theta^*(V_{\tau-}^{x,\alpha}-Y_{N_\tau})}\big|\tau < \infty, V_{\tau-}^{x,\alpha} > 0\big]. \quad (36) \end{aligned}$$

Let $H^{x,\alpha}(dt,dz)$ denote the joint distribution of τ and $V_{\tau-}^{x,\alpha}$ conditional on the event $\{\tau < \infty, V_{\tau-}^{x,\alpha} > 0\}$ that ruin occurs through a jump. Given $\tau = t$ and $V_{\tau-}^{x,\alpha} = z > 0$, a claim Y_{N_τ} occurs at time t and has distribution $dF_Y(y)/\int_z^\infty dF_Y$ for $y > z$. Hence

$$\mathbb{E}\left[e^{-\theta^*(V_{\tau-}^{x,\alpha}-Y_{N_\tau})}\big|\tau < \infty, V_{\tau-}^{x,\alpha} > 0\right]$$

$$= \int_0^\infty \int_0^\infty H^{x,\alpha}(dt,dz) \int_z^\infty e^{-\theta^*(z-y)} \frac{dF_Y(y)}{\int_z^\infty dF_Y}$$

$$\leq \sup_{z\geq 0} \int_z^\infty e^{-\theta^*(z-y)} \frac{dF_Y(y)}{\int_z^\infty dF_Y} < \infty, \qquad (37)$$

by assumption (33). By setting

$$C = \frac{1}{\sup_{z\geq 0} \int_z^\infty e^{-\theta^*(z-y)} \frac{dF_Y(y)}{\int_z^\infty dF_Y}} = \inf_{z\geq 0} \frac{\int_z^\infty dF_Y}{\int_z^\infty e^{-\theta^*(z-y)} dF_Y(y)} \in (0,1],$$

we then have from (35)-(36)-(37), for all $x \geq 0$,

$$\psi(x,\alpha) \geq Ce^{-\theta^* x}, \quad \forall \alpha \in \mathcal{A}.$$

Together with the upper bound (32), this completes the proof of Theorem 3.3.

4 Large Deviations and Rare Event Simulation in Option Pricing

4.1 Importance Sampling and Large Deviations Approximations

In this paragraph, we show how to use large deviations approximation via importance sampling for Monte-carlo computation of expectations arising in option pricing. In the context of continuous-time models, we are interested in the computation of

$$I_g = \mathbb{E}\left[g(S_t, 0 \leq t \leq T)\right],$$

where S is the underlying asset price, and g is the payoff of the option, eventually path-dependent, i.e. depending on the path process S_t, $0 \leq t \leq T$. The Monte-Carlo approximation technique consists in simulating N independent sample paths $(S_t^i)_{0\leq t\leq T}$, $i = 1, \ldots, N$, in the distribution of $(S_t)_{0\leq t\leq T}$, and approximating the required expectation by the sample mean estimator:

$$I_g^N = \frac{1}{N}\sum_{i=1}^N g(S^i).$$

The consistency of this estimator is ensured by the law of large numbers, while the error approximation is given by the variance of this estimator from the central limit theorem : the lower is the variance of $g(S)$, the better is the approximation for a given number N of simulations. As already mentioned in the introduction, the basic principle of importance sampling is to reduce variance by changing probability measure from which paths are generated. Here, the idea is to change the distribution of the price process to be simulated in order to take into account the specificities of the payoff function g. We focus in this section in the importance sampling technique within the context of diffusion models, and then show how to obtain an optimal change of measure by a large deviation approximation of the required expectation.

Importance Sampling for Diffusions via Girsanov's Theorem

We briefly describe the importance sampling variance reduction technique for diffusions. Let X be a d-dimensional diffusion process governed by

$$dX_s = b(X_s)ds + \Sigma(X_s)dW_s, \tag{38}$$

where $(W_t)_{t\geq 0}$ is a n-dimensional brownian motion on a filtered probability space $(\Omega, \mathcal{F}, \mathbb{F} = (\mathcal{F}_t)_{t\geq 0}, \mathbb{P})$, and the Borel functions b, Σ satisfy the usual Lipschitz condition ensuring the existence of a strong solution to the s.d.e. (38). We denote by $X_s^{t,x}$ the solution to (38) starting fom x at time t, and we define the function :

$$v(t,x) = \mathbb{E}\Big[g(X_s^{t,x}, t \leq s \leq T)\Big], \quad (t,x) \in [0,T] \times \mathbb{R}^d.$$

Let $\phi = (\phi_t)_{0\leq t\leq T}$ be an $\mathbb{R}^d$-valued adapted process such that the process

$$M_t = \exp\left(-\int_0^t \phi_u' dW_u - \frac{1}{2}\int_0^t |\phi_u|^2 du\right), \quad 0 \leq t \leq T,$$

is a martingale, i.e. $\mathbb{E}[M_T] = 1$. This is ensured for instance under the Novikov criterion : $\mathbb{E}\big[\exp\big(\frac{1}{2}\int_0^T |\phi_u|^2 du\big)\big] < \infty$. In this case, one can define a probability measure $\mathbb{Q}$ equivalent to $\mathbb{P}$ on $(\Omega, \mathcal{F}_T)$ by :

$$\frac{d\mathbb{Q}}{d\mathbb{P}} = M_T.$$

Moreover, by Girsanov's theorem, the process $\hat{W}_t = W_t + \int_0^t \phi_u du, 0 \leq t \leq T$, is a brownian motion under $\mathbb{Q}$, and the dynamics of X under $\mathbb{Q}$ is given by

$$dX_s = \big(b(X_s) - \Sigma(X_s)\phi_s\big)ds + \Sigma(X_s)d\hat{W}_s. \tag{39}$$

From Bayes formula, the expectation of interest can be written as

$$v(t,x) = \mathbb{E}^{\mathbb{Q}}\Big[g(X_s^{t,x}, t \leq s \leq T)L_T\Big], \tag{40}$$

where L is the $\mathbb{Q}$-martingale

$$L_t = \frac{1}{M_t} = \exp\left(\int_0^t \phi_u' d\hat{W}_u - \frac{1}{2}\int_0^t |\phi_u|^2 du\right), \quad 0 \le t \le T. \quad (41)$$

The expression (40) suggests, for any choice of ϕ, an alternative Monte-Carlo estimator for $v(t,x)$ with

$$I_{g,\phi}^N(t,x) = \frac{1}{N}\sum_{i=1}^N g(X^{i,t,x})L_T^i,$$

by simulating N independent sample paths $(X^{i,t,x})$ and L_T^i of $(X^{t,x})$ and L_T under $\mathbb{Q}$ given by (39)-(41). Hence, the change of probability measure through the choice of ϕ leads to a modification of the drift process in the simulation of X. The variance reduction technique consists in determining a process ϕ, which induces a smaller variance for the corresponding estimator $I_{g,\phi}$ than the initial one $I_{g,0}$. The two next paragraphs present two approaches leading to the construction of such processes ϕ. In the first approach developed in [22], the process ϕ is stochastic, and requires an approximation of the expectation of interest. In the second approach due to [25], the process ϕ is deterministic and derived through a simple optimization problem. Both approaches rely on asymptotic results from the theory of large deviations.

Option Pricing Approximation with a Freidlin-Wentzell Large Deviation Principle

We are looking for a stochastic process ϕ, which allows to reduce (possibly to zero!) the variance of the corresponding estimator. The heuristics for achieving this goal is based on the following argument. Suppose for the moment that the payoff g depends only on the terminal value X_T. Then, by applying Itô's formula to the $\mathbb{Q}$-martingale $v(s, X_s^{t,x})L_s$ between $s = t$ and $s = T$, we obtain:

$$g(X_T^{t,x})L_T = v(t,x)L_t + \int_t^T L_s\big(D_x v(s, X_s^{t,x})'\Sigma(X_s^{t,x}) + v(x, X_s^{t,x})\phi_s'\big)d\hat{W}_s.$$

Hence, the variance of $I_{g,\phi}^N(t,x)$ is given by

$$Var_{\mathbb{Q}}(I_{g,\phi}^N(t,x)) = \frac{1}{N}\mathbb{E}^{\mathbb{Q}}\left[\int_t^T L_s^2 \big|D_x v(s, X_s^{t,x})'\Sigma(X_s^{t,x}) + v(x, X_s^{t,x})\phi_s'\big|^2 ds\right].$$

The choice of the process ϕ is motivated by the following remark. If the function v were known, then one could vanish the variance by choosing

$$\phi_s = \phi_s^* = -\frac{1}{v(s, X_s^{t,x})}\Sigma'(X_s^{t,x})D_x v(s, X_s^{t,x}), \quad t \le s \le T. \quad (42)$$

Of course, the function v is unknown (this is precisely what we want to compute), but this suggests to use a process ϕ from the above formula with an approximation of the function v. We may then reasonably hope to reduce the variance, and also to use such a method for more general payoff functions, possibly path-dependent. We shall use a large deviations approximation for the function v.

The basic idea for the use of large deviations approximation to the expectation function v is the following. Suppose the option of interest, characterized by its payoff function g, has a low probability of exercice, e.g. it is deeply out the money. Then, a large proportion of simulated paths end up out of the exercice domain, giving no contribution to the Monte-carlo estimator but increasing the variance. In order to reduce the variance, it is interesting to change of drift in the simulation of price process to make the domain exercice more likely. This is achieved with a large deviations approximation of the process of interest in the asymptotics of small diffusion term : such a result is known in the literature as Freidlin-Wentzell sample path large deviations principle. Equivalently, by time-scaling, this amounts to large deviation approximation of the process in small time, studied by Varadhan.

To illustrate our purpose, let us consider the case of an up-in bond, i.e. an option that pays one unit of numéraire iff the underlying asset reached a given up-barrier K. Within a stochastic volatility model $X = (S, Y)$ as in (38) and given by :

$$dS_t = \sigma(Y_t) S_t dW_t^1 \tag{43}$$
$$dY_t = \eta(Y_t)dt + \gamma(Y_t)dW_t^2, \quad d<W_1, W_2>_t = \rho dt, \tag{44}$$

its price is then given by

$$v(t,x) = \mathbb{E}\big[1_{\max_{t \leq u \leq T} S_u^{t,x} \geq K}\big]$$
$$= \mathbb{P}[\tau_{t,x} \leq T], \quad t \in [0,T], \quad x = (s,y) \in (0,\infty) \times \mathbb{R},$$

where

$$\tau_{t,x} = \inf\{u \geq t : X_u^{t,x} \notin \Gamma\}, \quad \Gamma = (0,K) \times \mathbb{R}.$$

The event $\{\max_{t \leq u \leq T} S_u^{t,x} \geq K\} = \{\tau_{t,x} \leq T\}$ is rare when $x = (s,y) \in \Gamma$, i.e. $s < K$ (out the money option) and the time to maturity $T - t$ is small. The large deviations asymptotics for the exit probability $v(t,x)$ in small time to maturity $T - t$ is provided by the Freidlin-Wentzell and Varadhan theories. Indeed, we see from the time-homogeneity of the coefficients of the diffusion and by time-scaling that we may write $v(t,x) = w_{T-t}(0,x)$, where for $\varepsilon > 0$, w_ε is the function defined on $[0,1] \times (0,\infty) \times \mathbb{R}$ by

$$w_\varepsilon(t,x) = \mathbb{P}[\tau_{t,x}^\varepsilon \leq 1],$$

and $X^{\varepsilon,t,x}$ is the solution to

$$dX_s^\varepsilon = \varepsilon b(X_s^\varepsilon)ds + \sqrt{\varepsilon}\Sigma(X_s^\varepsilon)dW_s, \quad X_t^\varepsilon = x.$$

and $\tau_{t,x}^\varepsilon = \inf\{s \geq t : X_s^{\varepsilon,t,x} \notin \Gamma\}$. From the large deviations result (19) stated in paragraph 2.3, we have :

$$\lim_{t \nearrow T} -(T-t)\ln v(t,x) = V_0(0,x),$$

where

$$V_0(t,x) = \inf_{x(.)\in\mathcal{A}(t,x)} \int_t^1 \frac{1}{2}\dot{x}(u)'M(x(u))\dot{x}(u)du, \quad (t,x) \in [0,1) \times \Gamma,$$

$\Sigma(x)$ is the diffusion matrix of $X = (S,Y)$, $M(x) = (\Sigma\Sigma'(x))^{-1}$, and

$$\mathcal{A}(t,x) = \{x(.) \in H^1_{loc}([0,1],(0,\infty) \times \mathbb{R}) \;:\; x(t) = x \text{ and } \tau(x) \leq 1\}.$$

Here, for an absolutely continuous function $x(.)$ on $[0,1)$ and valued in $(0,\infty) \times \mathbb{R}$, we denote $\dot{x}(u)$ its time derivative, and $\tau(x)$ the exit time of $x(.)$ from Γ.

We also have another interpretation of the positive function V_0 in terms of Riemanian distance on $\mathbb{R}^d$ associated to the metric $M(x) = (\Sigma\Sigma'(x))^{-1}$. By denoting $L_0(x) = \sqrt{2V_0(0,x)}$, one can prove (see [34]) that L_0 is the unique viscosity solution to the eikonal equation

$$(D_x L_0)' \Sigma\Sigma'(x) D_x L_0 = 1, \quad x \in \Gamma$$
$$L_0(x) = 0, \quad x \in \partial\Gamma$$

and that it may be represented as

$$L_0(x) = \inf_{z \in \partial\Gamma} L_0(x,z), \quad x \in \Gamma, \tag{45}$$

where

$$L_0(x,z) = \inf_{x(.)\in\mathcal{A}(x,z)} \int_0^1 \sqrt{\dot{x}(u)'M(x(u))\dot{x}(u)}du,$$

and $\mathcal{A}(x,z) = \{x(.) \in H^1_{loc}([0,1],(0,\infty) \times \mathbb{R}) \;:\; x(0) = x \text{ and } x(1) = z\}$. Hence, the function L_0 can be computed either by the numerical resolution of the eikonal equation or by using the representation (45). $L_0(x)$ is interpreted as the minimal length (according to the metric M) of the path $x(.)$ allowing to reach the boundary $\partial\Gamma$ from x. From the above large deviations result, which is written as

$$\ln v(t,x) \simeq -\frac{L_0^2(x)}{2(T-t)}, \quad \text{as } T - t \to 0,$$

and the expression (42) for the optimal theoretical ϕ^*, we use a change of probability measure with

$$\phi(t,x) = \frac{L_0(x)}{T-t} \Sigma'(x) D_x L_0(x).$$

Such a process ϕ may also appear interesting to use in a more general framework than up-in bond : one can use it for computing any option whose exercice domain looks similar to the up and in bond. We also expect that the variance reduction is more significant as the exercice probability is low, i.e. for deep out the money options. In the particular case of the Black-Scholes model, i.e. $\sigma(y) = \sigma$ constant, we have

$$L_0(x) = \frac{1}{\sigma}\left|\ln\left(\frac{s}{K}\right)\right|,$$

and so

$$\phi(t,x) = \frac{1}{\sigma(T-t)}\ln\left(\frac{s}{K}\right), \quad s < K.$$

Change of Drift via Varadhan-Laplace Principle

We describe here a method due to [25], which, in contrast with the above approach, does not require the knowledge of the option price. This method restricts to deterministic changes of drift over discrete time steps. Hence, the diffusion model X for state variables (stock price, volatility) is simulated (eventually using an Euler scheme if needed) on a discrete time grid $0 = t_0 < t_1 < \ldots < t_m = T$: the increment of the brownian motion from t_{i-1} to t_i is simulated as $\sqrt{t_i - t_{i-1}} Z_i$, where $Z_1, \ldots, Z_m$ are i.i.d. n-dimensional standard normal random vectors. We denote by Z the concatenation of the Z_i into a single vector of lengh $l = mn$. Each outcome of Z determines a path of state variables. Let $G(Z)$ denote the payoff derived from Z, and our aim is to compute the (path-dependent) option price $\mathbb{E}[G(Z)]$. For example, in the case of the Black-Scholes model for the stock price S, we have

$$S_{t_i} = S_{t_{i-1}} \exp\left(-\frac{\sigma^2}{2}(t_i - t_{i-1}) + \sigma\sqrt{t_i - t_{i-1}}Z_i\right),$$

and the payoff of the Asian option is

$$G(Z) = G(Z_1, \ldots, Z_m) = \left(\frac{1}{m}\sum_{i=1}^{m} S_{t_i} - K\right)_+$$

We apply importance sampling by changing the mean of Z from 0 to some vector $\mu = (\mu_1, \ldots, \mu_m)$. We denote $\mathbb{P}_\mu$ and $\mathbb{E}_\mu$ the probability and expectation when $Z \sim \mathcal{N}(\mu, I_m)$. Notice that with the notations of paragraph 4.1, this corresponds to a piecewise constant process ϕ s.t. $\phi_t = -\mu_i/\sqrt{t_i - t_{i-1}}$ on $[t_{i-1}, t_i)$. By Girsanov's theorem or here more simply from the likelihood ratio for normal random vectors, the corresponding unbiased estimator is then obtained by taking the average of independent replications of

$$\vartheta_\mu = G(Z)e^{-\mu' Z + \frac{1}{2}\mu'\mu},$$

where Z is sampled from $\mathcal{N}(\mu, I_m)$. We call ϑ_μ a μ-IS estimator. In order to minimize over μ the variance of this estimator, it suffices to minimize its second moment, which is given by :

$$M^2(\mu) = \mathbb{E}_\mu\left[G(Z)^2 e^{-2\mu' Z + \mu'\mu}\right] = \mathbb{E}\left[G(Z)^2 e^{-\mu' Z + \frac{1}{2}\mu'\mu}\right]$$

We are then looking for an optimal μ solution to

$$\inf_\mu M^2(\mu) := \inf_\mu \mathbb{E}\Big[G(Z)^2 e^{-\mu' Z + \frac{1}{2}\mu'\mu}\Big]. \tag{46}$$

This minimization problem is, in general, a well-posed problem. Indeed, it is shown in [2] that if $\mathbb{P}[G(Z) > 0] > 0$, and $\mathbb{E}[G(Z)^{2+\delta}] < \infty$ for some $\delta > 0$, then $M^2(.)$ is a strictly convex function, and thus μ^* solution to (46) exists and is unique. This μ^* can be computed by solving (numerically) $\nabla M^2(\mu) = 0$. This is the method adopted in [2] with a Robbins-Monro stochastic algorithm. We present here an approximate resolution of (46) by means of large deviations approximation. For this, assume that G takes only nonnegative values, so that it is written as $G(z) = \exp(F(z))$, with the convention that $F(z) = -\infty$ if $G(z) = 0$, and let us consider the more general estimation problem where Z is replaced by $Z^\varepsilon = \sqrt{\varepsilon} Z$ and we simultaneously scale the payoff by raising it to the power of $1/\varepsilon$:

$$v_\varepsilon = \mathbb{E}[e^{\frac{1}{\varepsilon} F(Z^\varepsilon)}]$$

The quantity of interest $\mathbb{E}[G(Z)] = \mathbb{E}[e^{F(Z)}]$ is v_ε for $\varepsilon = 1$. We embed the problem of estimating v_1 in the more general problem of estimating v_ε and analyze the second moment of corresponding IS estimators as ε is small, by means of Varadhan-Laplace principle. For any μ, we consider μ_ε-IS estimator of v_ε with $\mu^\varepsilon = \mu/\sqrt{\varepsilon}$:

$$\vartheta^\varepsilon_\mu = e^{\frac{1}{\varepsilon} F(\sqrt{\varepsilon} Z)} e^{-\mu'_\varepsilon Z + \frac{1}{2}\mu'_\varepsilon \mu_\varepsilon} = e^{\frac{1}{\varepsilon}(F(Z^\varepsilon) - \mu' Z^\varepsilon + \frac{1}{2}\mu'\mu)}$$

where Z is sampled from $\mathcal{N}(\mu_\varepsilon, I_m) = \mathcal{N}(\mu/\sqrt{\varepsilon}, I_m)$. Its second moment is

$$M^2_\varepsilon(\mu) = \mathbb{E}_{\mu_\varepsilon}\Big[e^{\frac{1}{\varepsilon}(2F(Z^\varepsilon) - 2\mu' Z^\varepsilon + \mu'\mu)}\Big] = \mathbb{E}\Big[e^{\frac{1}{\varepsilon}(2F(Z^\varepsilon) - \mu' Z^\varepsilon + \frac{1}{2}\mu'\mu)}\Big]$$

Now, from Cramer's theorem, $(Z^\varepsilon)_\varepsilon$ satisfies a LDP with rate function $I(z) = \frac{1}{2}z'z$. Hence, under the condition that $F(z) \leq c_1 + c_2 z'z$ for some $c_2 < 1/4$, one can apply Varadhan's integral principle (see Theorem 2.2) to the function $z \to 2F(z) - \mu'z + \frac{1}{2}\mu'\mu$, and get

$$\lim_{\varepsilon \to 0} \varepsilon \ln M^2_\varepsilon(\mu) = \sup_z \Big[2F(z) - \mu'z + \frac{1}{2}\mu'\mu - \frac{1}{2}z'z\Big]. \tag{47}$$

This suggests to search for a μ solution to the problem:

$$\inf_\mu \sup_z \Big[2F(z) - \mu'z + \frac{1}{2}\mu'\mu - \frac{1}{2}z'z\Big]. \tag{48}$$

This min-max problem may be reduced to a simpler one. Indeed, assuming that the conditions of the min/max theorem hold, then the inf and sup can be permuted, and we find

$$\inf_\mu \sup_z \Big[2F(z) - \mu'z + \frac{1}{2}\mu'\mu - \frac{1}{2}z'z\Big]$$
$$= \sup_z \Big[\inf_\mu \big(-\mu'z + \frac{1}{2}\mu'\mu\big) + 2F(z) - \frac{1}{2}z'z\Big]$$
$$= 2\sup_z \Big[F(z) - \frac{1}{2}z'z\Big]. \tag{49}$$

Actually, under suitable convexity conditions on F and its domain, one can show (see [25]) that (49) holds. Furthermore, if $\hat{z}$ is solution to

$$\sup_z \left(F(z) - \frac{1}{2} z'z \right), \tag{50}$$

then a solution $\hat{\mu}$ solution to (48) should be identified with the conjugate of $\hat{z}$, via $\inf_\mu \left(-\mu' \hat{z} + \frac{1}{2}\mu'\mu \right)$ that is $\hat{\mu} = \hat{z}$. The solution to problem (50) has also the following interpretation. From heuristic arguments of importance sampling (see the introduction), an optimal effective importance sampling density should assign high probability to regions on which the product of the integrand payoff and the original density is large. For our problem, this product is proportional to

$$e^{F(z) - \frac{1}{2} z'z},$$

since $\exp(-z'z/2)$ is proportional to the standard normal density. This suggests to choose $\mu = \hat{\mu}$ solution to (50). Another heuristics indication for the choice of (50) is based on the following argument. Assume that F is C^1 on its domain, and the maximum $\hat{\mu}$ in (50) is attained in the interior of the domain, so that it solves the fixed point equation:

$$\nabla F(\hat{\mu}) = \hat{\mu}. \tag{51}$$

By using a first-order Taylor approximation of F around the mean $\hat{\mu}$ of Z under $\mathbb{P}_{\hat{\mu}}$, we may approximate the estimator as

$$\vartheta_{\hat{\mu}} = e^{F(Z) - \hat{\mu}'Z + \frac{1}{2}\hat{\mu}'\hat{\mu}} \simeq e^{F(\hat{\mu}) + \nabla F(\hat{\mu})'(Z - \hat{\mu}) - \hat{\mu}'Z + \frac{1}{2}\hat{\mu}'\hat{\mu}}. \tag{52}$$

Hence, for the choice of $\hat{\mu}$ satisfying (51), the expression of the r.h.s. of (52) collapses to a constant with no dependence on Z. Thus, applying importance sampling with such a $\hat{\mu}$ would produce a zero-variance estimator if (52) holds exactly, e.g. if F is linear, and it should produce a low-variance estimator if (52) holds only approximately.

The choice of $\mu = \hat{\mu}$ solution to (50) leads also to an asymptotically optimal IS-estimator in the following sense. First, notice that for any μ, we have by Cauchy-Schwarz's inequality: $M_\varepsilon^2(\mu) \geq (v_\varepsilon)^2$, and so

$$\lim_{\varepsilon \to 0} \varepsilon \ln M_\varepsilon^2(\mu) \geq 2 \lim_{\varepsilon \to 0} \varepsilon \ln v_\varepsilon = 2 \lim_{\varepsilon \to 0} \varepsilon \ln \mathbb{E}[e^{\frac{1}{\varepsilon} F(Z^\varepsilon)}].$$

From Varadhan's integral principle applied to the function $z \to F(z)$, we thus deduce for any μ,

$$\lim_{\varepsilon \to 0} \varepsilon \ln M_\varepsilon^2(\mu) \geq 2 \sup_z [F(z) - \frac{1}{2} z'z] = 2[F(\hat{\mu}) - \frac{1}{2}\hat{\mu}'\hat{\mu}].$$

Hence, $2[F(\hat{\mu}) - \frac{1}{2}\hat{\mu}'\hat{\mu}]$ is the best-possible exponential decay rate for a μ_ε-IS estimator $\vartheta_\mu^\varepsilon$. Now, by choosing $\mu = \hat{\mu}$, and from (47), (49), we have

$$\lim_{\varepsilon \to 0} \varepsilon \ln M_\varepsilon^2(\hat{\mu}) = 2[F(\hat{\mu}) - \frac{1}{2}\hat{\mu}'\hat{\mu}],$$

which shows that the $\hat{\mu}_\varepsilon$-IS estimator $\vartheta_{\hat{\mu}}^\varepsilon$ is asymptotically optimal.

Remark 4.6 From the first-order approximation of F in (52), we see that in order to obtain further variance reduction, it is natural to address the quadratic component of F. This can be achieved by a method of stratified sampling as developed in [25].

Recently, the above approach of [25] was extended in [29] to a continuous-time setting, where the optimal deterministic drift in the Black-Scholes model is characterized as the solution to a classical one-dimensional variational problem.

4.2 Computation of Barrier Crossing Probabilities and Sharp Large Deviations

In this paragraph, we present a simulation procedure for computing the probability that a diffusion process crosses pre-specified barriers in a given time interval $[0, T]$. Let $(X_t)_{t \in [0,T]}$ be a diffusion process in $\mathbb{R}^d$,

$$dX_t = b(X_t)dt + \sigma(X_t)dW_t$$

and τ is the exit time of X from some domain Γ of $\mathbb{R}^d$, eventually depending on time:

$$\tau = \inf\{t \in [0, T] : X_t \notin \Gamma(t)\},$$

with the usual convention that $\inf \emptyset = \infty$. Such a quantity appears typically in finance in the computation of barrier options, for example with a knock-out option:

$$C_0 = \mathbb{E}\big[e^{-rT} g(X_T) 1_{\tau > T}\big], \tag{53}$$

with $\Gamma(t) = (-\infty, U(t))$ in the case of single barrier options, and $\Gamma(t) = (L(t), U(t))$, for double barrier options. Here, L, U are real functions : $[0, \infty) \to (0, \infty)$ s.t. $L < U$.

The direct naive approach would consist first of simulating the process X on $[0, T]$ through a discrete Euler scheme of step size $\varepsilon = T/n = t_{i+1} - t_i$, $i = 0, \ldots, n$:

$$\bar{X}^\varepsilon_{t_{i+1}} = \bar{X}^\varepsilon_{t_i} + b(\bar{X}^\varepsilon_{t_i})\varepsilon + \sigma(\bar{X}^\varepsilon_{t_i})(W_{t_{i+1}} - W_{t_i}),$$

and the exit time τ is approximated by the first time the discretized process reaches the barrier :

$$\bar{\tau}^\varepsilon = \inf\big\{t_i : \bar{X}^\varepsilon_{t_i} \notin \Gamma(t_i)\big\}.$$

Then, the barrier option price C_0 in (53) is approximated by Monte-Carlo simulations of the quantity

$$\bar{C}^\varepsilon_0 = \mathbb{E}\big[e^{-rT} g(\bar{X}^\varepsilon_T) 1_{\bar{\tau}^\varepsilon > T}\big].$$

In this procedure, one considers that the price diffusion is killed if there exists a value $\bar{X}^\varepsilon_{t_i}$, which is out of the domain $\Gamma(t_i)$. Hence, such an approach is suboptimal since it does not control the diffusion path between two successive dates t_i and t_{i+1} : the diffusion path could have crossed the barriers and come back to the domain without being detected. In this case, one over-estimates the exit time probability of $\{\tau > T\}$. This suboptimality is confirmed by the property that the error between C_0 and $\bar{C}_0^\varepsilon$ is of order $\sqrt{\varepsilon}$, as shown in [28], instead of the usual order ε obtained for standard vanilla options.

In order to improve the above procedure, we need to determine the probability that the process X crosses the barrier between discrete simulation times. We then consider the continuous Euler scheme

$$\bar{X}_t^\varepsilon = \bar{X}_{t_i}^\varepsilon + b(\bar{X}_{t_i}^\varepsilon)(t - t_i) + \sigma(\bar{X}_{t_i}^\varepsilon)(W_t - W_{t_i}), \quad t_i \leq t \leq t_{i+1},$$

which evolves as a Brownian with drift between two time discretizations $t_i, t_{i+1} = t_i + \varepsilon$. Given a simulation path of $(\bar{X}_{t_i}^\varepsilon)_i$, and values $\bar{X}_{t_i}^\varepsilon = x_i$, $\bar{X}_{t_{i+1}}^\varepsilon = x_{i+1}$, we denote

$$p_i^\varepsilon(x_i, x_{i+1}) = \mathbb{P}\Big[\exists t \in [t_i, t_{i+1}] : \bar{X}_t^\varepsilon \notin \Gamma(t_i) \big| (\bar{X}_{t_i}^\varepsilon, \bar{X}_{t_{i+1}}^\varepsilon) = (x_i, x_{i+1})\Big],$$

the exit probability of the Euler scheme conditionally on the simulated path values. The correction Monte-Carlo procedure works then as follows : with probability $p_i^\varepsilon = p_i^\varepsilon(\bar{X}_{t_i}^\varepsilon, \bar{X}_{t_{i+1}}^\varepsilon)$, we stop the simulation by considering that the diffusion is killed, and we set $\tau^\varepsilon = t_i$; with probability $1 - p_i^\varepsilon$, we carry on the simulation. The approximation of (53) is thus computed by Monte-Carlo simulations of

$$C_0^\varepsilon = \mathbb{E}\big[e^{-rT}(\bar{X}_T^\varepsilon - K)_+ 1_{\tau^\varepsilon > T}\big].$$

We then recover a rate of convergence of order ε for $C_0^\varepsilon - C_0$, see [28].

The effective implementation of this corrected procedure requires the calculation of p_i^ε. Notice that on the interval $[t_i, t_{i+1}]$, the diffusion $\bar{X}^\varepsilon$ conditionned to $\bar{X}_{t_i}^\varepsilon = x_i$, $\bar{X}_{t_{i+1}}^\varepsilon = x_{i+1}$, is a brownian bridge : it coincides in distribution with the process

$$\tilde{B}_t^{i,\varepsilon} = x_i + \frac{t}{\varepsilon}(x_{i+1} - x_i) + \sigma(x_i)\left(W_t - \frac{t}{\varepsilon}W_\varepsilon\right), \quad 0 \leq t \leq \varepsilon,$$

and so by time change $t \to t/\varepsilon$, with the process

$$Y_t^{i,\varepsilon} := \tilde{B}_{\varepsilon t}^{i,\varepsilon} = x_i + t(x_{i+1} - x_i) + \sqrt{\varepsilon}\sigma(x_i)(W_t - tW_1), \quad 0 \leq t \leq 1.$$

It is known that the process $Y^{i,\varepsilon}$ is solution to the s.d.e.

$$dY_t^{i,\varepsilon} = -\frac{Y_t^{i,\varepsilon} - x_{i+1}}{1 - t}dt + \sqrt{\varepsilon}\sigma(x_i)dW_t, \quad 0 \leq t < 1,$$

$$Y_0^{i,\varepsilon} = x_i.$$

The probability p_i^ε can then be expressed as

$$p_i^\varepsilon(x_i, x_{i+1}) = \mathbb{P}[\tau^{i,\varepsilon} \leq 1], \qquad (54)$$

where $\tau^{i,\varepsilon} = \inf\{t \geq 0 : Y_t^{i,\varepsilon} \notin \Gamma(t_i + \varepsilon t)\}$. In the case of a half-space, i.e. single constant barrier, one has an explicit expression for the exit probability of a Brownian bridge. For example, if $\Gamma(t) = (-\infty, U)$, we have

$$p_i^\varepsilon(x_i, x_{i+1}) = \exp\left(-\frac{I_U(x_i, x_{i+1})}{\varepsilon}\right),$$

with

$$I_U(x_i, x_{i+1}) = \frac{2}{\sigma^2(x_i)}(U - x_i)(U - x_{i+1}).$$

In the general case, we do not have analytical expressions for p_i^ε, and one has to rely on simulation techniques or asymptotic approximations. We shall here consider asymptotic techniques based on large deviations and Freidlin-Wentzell theory. Let us illustrate this point in the case of two time-dependent barriers, i.e. $\Gamma(t) = (L(t), U(t))$ for smooth barriers functions $L < U$. Problem (54) does not exactly fit into the Freidlin-Wentzell framework considered in paragraph 2.3, but was adapted for Brownian bridges with time-dependent barriers in [7]. We then have the large deviation estimate for p_i^ε:

$$\lim_{\varepsilon \to 0} \varepsilon \ln p_i^\varepsilon(x_i, x_{i+1}) = -I_{L,U}(x_i, x_{i+1}),$$

where $I_{L,U}(x_i, x_{i+1})$ is the infimum of the functional

$$y(.) \longrightarrow \frac{1}{2\sigma(x_i)^2} \int_0^1 \left|\dot{y}(t) + \frac{y(t) - x_{i+1}}{1-t}\right|^2 dt,$$

over all absolutely continuous paths $y(.)$ on $[0,1]$ s.t. $y(0) = x_i$, and there exists some $t \in [0,1]$ for which $y(t) \leq L(t_i)$ or $y(t) \geq U(t_i)$. This infimum is a classical problem of calculus of variations, which is explicitly solved and gives for any $x_i, x_{i+1} \in (L(t_i), U(t_i))$ (otherwise $I_{L,U}(x_i, x_{i+1}) = 0$):

$$I_{L,U}(x_i, x_{i+1})$$
$$= \begin{cases} \frac{2}{\sigma^2(x_i)}(U(t_i) - x_i)(U(t_i) - x_{i+1}) & \text{if } x_i + x_{i+1} > L(t_i) + U(t_i) \\ \frac{2}{\sigma^2(x_i)}(x_i - L(t_i))(x_{i+1} - L(t_i)) & \text{if } x_i + x_{i+1} < L(t_i) + U(t_i). \end{cases}$$

In order to remove the log estimate on p_i^ε, we need a sharper large deviation estimate, and this is analyzed by the results of [17] recalled in paragraph 2.3. More precisely, we have

$$p_i^\varepsilon(x_i, x_{i+1}) = \exp\left(-\frac{I_{L,U}(x_i, x_{i+1})}{\varepsilon} - w_{L,U}(x_i, x_{i+1})\right)(1 + O(\varepsilon)),$$

where $w_{L,U}(x_i, x_{i+1})$ is explicited in [7] as

$$w_{L,U}(x_i, x_{i+1}) = \begin{cases} \frac{2}{\sigma^2(x_i)}(U(t_i) - x_i)U'(t_i) & \text{if } x_i + x_{i+1} > L(t_i) + U(t_i) \\ \frac{2}{\sigma^2(x_i)}(x_i - L(t_i))L'(t_i) & \text{if } x_i + x_{i+1} < L(t_i) + U(t_i). \end{cases}$$

Some recent extensions of this large deviations approach to the computation of exit probabilities for multivariate Brownian bridge are studied in [32], which also gives applications for the estimation of default probabilities in credit risk models, and the pricing of credit default swaps.

5 Large Deviations in Risk Management

5.1 Large Portfolio Losses in Credit Risk

Portfolio Credit Risk in a Single Factor Normal Copula Model

A basic problem in measuring portfolio credit risk is determining the distribution of losses from default over a fixed horizon. Credit portfolios are often large, including exposure to thousands of obligors, and the default probabilities of high-quality credits are extremely small. These features in credit risk context lead to consider rare but significant large loss events, and emphasis is put on the small probabilities of large losses, as these are relevant for calculation of value at risk and related risk measures.

We use the following notation:

n = number of obligors to which portfolio is exposed,
Y_k = default indicator (= 1 if default, 0 otherwise) for k-th obligor,
p_k = marginal probability that k-th obligor defaults, i.e. $p_k = \mathbb{P}[Y_k = 1]$,
c_k = loss resulting from default of the k-th obligor,
$L_n = c_1 Y_1 + \ldots + c_n Y_n$ = total loss from defaults.

We are interested in estimating tail probabilities $\mathbb{P}[L_n > \ell_n]$ in the limiting regime at increasingly high loss thresholds ℓ_n, and rarity of large losses resulting from a large number n of obligors and multiple defaults.

For simplicity, we consider a homogeneous portfolio where all p_k are equal to p, and all c_k are equal constant to 1. An essential feature for credit risk management is the mechanism used to model the dependence across sources of credit risk. The dependence among obligors is modelled by the dependence among the default indicators Y_k. This dependence is introduced through a normal copula model as follows: each default indicator is represented as

$$Y_k = 1_{\{X_k > x_k\}}, \quad k = 1, \ldots, n,$$

where $(X_1, \ldots, X_n)$ is a multivariate normal vector. Without loss of generality, we take each X_k to have a standard normal distribution, and we choose x_k to match the marginal default probability p_k, i.e. $x_k = \Phi^{-1}(1 - p_k) = -\Phi^{-1}(p_k)$, with Φ cumulative normal distribution. We also denote $\varphi = \Phi'$ the density of the normal distribution. The correlations along the X_k, which determine the dependence among the Y_k, are specified through a single factor model of the form:

$$X_k = \rho Z + \sqrt{1 - \rho^2} \varepsilon_k, \quad k = 1, \ldots, n. \tag{55}$$

where Z has the standard normal distribution $\mathcal{N}(0,1)$, ε_k are independent $\mathcal{N}(0,1)$ distribution, and Z is independent of ε_k, $k = 1, \ldots, n$. Z is called systematic risk factor (industry, regional risk factors for example ...), and ε_k is an idiosyncratic risk associated with the k-th obligor. The constant ρ in $[0,1)$ is a factor loading on the single factor Z, and assumed here to be identical for all obligors. We shall distinguish the case of independent obligors ($\rho = 0$), and dependent obligors ($\rho > 0$). More general multivariate factor models with inhomogeneous obligors are studied in [26].

Independent Obligors

In this case, $\rho = 0$, the default indicators Y_k are i.i.d. with Bernoulli distribution of parameter p, and L_n is a binomial distribution of parameters n and p. By the law of large numbers, L_n/n converges to p. Hence, in order that the loss event $\{L_n \geq l_n\}$ becomes rare (without being trivially impossible), we let l_n/n approach $q \in (p, 1)$. It is then appropriate to specify $l_n = nq$ with $p < q < 1$. From Cramer's theorem and the expressions of the c.g.f. of the Bernoulli distribution and its Fenchel-Legendre transform, we obtain the large deviation result for the loss probability:

$$\lim_{n \to \infty} \frac{1}{n} \ln \mathbb{P}[L_n \geq nq] = -q \ln \left(\frac{q}{p}\right) - (1-q) \ln \left(\frac{1-q}{1-p}\right) < 0.$$

Remark 5.7 By denoting $\Gamma(\theta) = \ln(1 - p + pe^\theta)$ the c.g.f. of Y_k, we have an IS (unbiased) estimator of $\mathbb{P}[L_n \geq nq]$ by taking the average of independent replications of

$$\exp(-\theta L_n + n\Gamma(\theta))1_{L_n \geq nq}$$

where L_n is sampled with a default probability $p(\theta) = \mathbb{P}_\theta[Y_k = 1] = pe^\theta/(1 - p + pe^\theta)$. Moreover, see Remark 2.2, this estimator is asymptotically optimal, as n goes to infinity, for the choice of parameter $\theta_q \geq 0$ attaining the argmax in $\theta q - \Gamma(\theta)$.

Dependent Obligors

We consider the case where $\rho > 0$. Then, conditionally on the factor Z, the default indicators Y_k are i.i.d. with Bernoulli distribution of parameter:

$$p(Z) = \mathbb{P}[Y_k = 1 | Z] = \mathbb{P}[\rho Z + \sqrt{1-\rho^2}\varepsilon_k > -\Phi^{-1}(p)|Z]$$
$$= \Phi\left(\frac{\rho Z + \Phi^{-1}(p)}{\sqrt{1-\rho^2}}\right). \tag{56}$$

Hence, by the law of large numbers, L_n/n converges in law to the random variable $p(Z)$ valued in $(0,1)$. In order that $\{L_n \geq l_n\}$ becomes a rare event (without being impossible) as n increases, we therefore let l_n/n approach 1 from below. We then set

$$l_n = nq_n, \quad \text{with } q_n < 1, \ q_n \nearrow 1 \text{ as } n \to \infty. \tag{57}$$

Actually, we assume that the rate of increase of q_n to 1 is of order n^{-a} with $a \leq 1$:

$$1 - q_n = O(n^{-a}), \quad \text{with } 0 < a \leq 1. \tag{58}$$

We then state the large deviations result for the large loss threshold regime.

Theorem 5.4 *In the single-factor homogeneous portfolio credit risk model* (55), *and with large threshold l_n as in* (57)-(58), *we have*

$$\lim_{n \to \infty} \frac{1}{\ln n} \ln \mathbb{P}[L_n \geq nq_n] = -a \frac{1-\rho^2}{\rho^2}.$$

Observe that in the above theorem, we normalize by $\ln n$, indicating that the probability decays like $n^{-\gamma}$, with $\gamma = a(1-\rho^2)/\rho^2$. We find that the decay rate is determined by the effect of the dependence structure in the Gaussian copula model. When ρ is small (weak dependence between sources of credit risk), large losses occur very rarely, which is formalized by a high decay rate. In the opposite case, this decay rate is small when ρ tends to one, which means that large losses are most likely to result from systematic risk factors.

Proof. 1) We first prove the lower bound :

$$\liminf_{n \to \infty} \frac{1}{\ln n} \ln \mathbb{P}[L_n \geq nq_n] \geq -a \frac{1-\rho^2}{\rho^2}. \tag{59}$$

From Bayes formula, we have

$$\begin{aligned}\mathbb{P}[L_n \geq nq_n] &\geq \mathbb{P}[L_n \geq nq_n, p(Z) \geq q_n] \\ &= \mathbb{P}[L_n \geq nq_n | p(Z) \geq q_n] \, \mathbb{P}[p(Z) \geq q_n].\end{aligned} \tag{60}$$

For any $n \geq 1$, we define $z_n \in \mathbb{R}$ the solution to

$$p(z_n) = q_n, \quad n \geq 1.$$

Since $p(.)$ is an increasing one to one function, we have $\{p(Z) \geq q_n\} = \{Z \geq z_n\}$. Moreover, observing that L_n is an increasing function of Z, we get

$$\begin{aligned}\mathbb{P}[L_n \geq nq_n | p(Z) \geq q_n] &= \mathbb{P}[L_n \geq nq_n | Z \geq z_n] \\ &\geq \mathbb{P}[L_n \geq nq_n | Z = z_n] = \mathbb{P}[L_n \geq nq_n | p(Z) = q_n],\end{aligned}$$

so that from (60)

$$\mathbb{P}[L_n \geq nq_n] \geq \mathbb{P}[L_n \geq nq_n | p(Z) = q_n] \mathbb{P}[Z \geq z_n]. \tag{61}$$

Now given $p(Z) = q_n$, L_n is binomially distributed with parameters n and q_n, and thus

$$\mathbb{P}[L_n \geq nq_n | p(Z) = q_n] \geq 1 - \Phi(0) = \frac{1}{2} (>0). \tag{62}$$

We focus on the tail probability $\mathbb{P}[Z \geq z_n]$ as n goes to infinity. First, observe that since q_n goes to 1, we have z_n going to infinity as n tends to infinity. Furthermore, from the expression (56) of $p(z)$, the rate of decrease (58), and using the property that $1 - \Phi(x) \simeq \varphi(x)/x$ as $x \to \infty$, we have

$$O(n^{-a}) = 1 - q_n = 1 - p(z_n) = 1 - \Phi\left(\frac{\rho z_n + \Phi^{-1}(p)}{\sqrt{1-\rho^2}}\right)$$

$$\simeq \frac{\sqrt{1-\rho^2}}{\rho z_n + \Phi^{-1}(p)} \exp\left(-\frac{1}{2}\left(\frac{\rho z_n + \Phi^{-1}(p)}{\sqrt{1-\rho^2}}\right)^2\right),$$

as $n \to \infty$, so that by taking logarithm :

$$a \ln n - \frac{1}{2}\frac{\rho^2 z_n^2}{1-\rho^2} - \ln z_n = O(1).$$

This implies

$$\lim_{n \to \infty} \frac{z_n^2}{\ln n} = 2a\frac{1-\rho^2}{\rho^2}. \tag{63}$$

By writing

$$\mathbb{P}[Z \geq z_n] = \mathbb{P}[z_n \leq Z \leq z_n + 1]$$
$$\geq \frac{1}{\sqrt{2\pi}} \exp\left(-\frac{1}{2}(z_n+1)^2\right),$$

we deduce with (63)

$$\liminf_{n \to \infty} \frac{1}{\ln n} \ln \mathbb{P}[Z \geq z_n] \geq a\frac{1-\rho^2}{\rho^2}.$$

Combining with (61) and (62), we get the required lower bound (59).

2) We now focus on the upper bound

$$\limsup_{n \to \infty} \frac{1}{\ln n} \ln \mathbb{P}[L_n \geq nq_n] \leq -a\frac{1-\rho^2}{\rho^2}. \tag{64}$$

We introduce the conditional c.g.f. of Y_k :

$$\Gamma(\theta, z) = \ln \mathbb{E}\left[e^{\theta Y_k}|Z = z\right] \tag{65}$$
$$= \ln(1 - p(z) + p(z)e^\theta). \tag{66}$$

Then, for any $\theta \geq 0$, we get by Chebichev's inequality,

$$\mathbb{P}[L_n \geq nq_n|Z] \leq \mathbb{E}\left[e^{\theta(L_n - nq_n)}|Z\right] = e^{-n(\theta q_n - \Gamma(\theta, Z))},$$

so that
$$\mathbb{P}[L_n \geq nq_n|Z] \leq e^{-n\Gamma^*(q_n,Z)}, \qquad (67)$$

where
$$\Gamma^*(q,z) = \sup_{\theta \geq 0}[\theta q - \Gamma(\theta,z)]$$
$$= \begin{cases} 0, & \text{if } q \leq p(z) \\ q\ln\left(\frac{q}{p(z)}\right) + (1-q)\ln\left(\frac{1-q}{1-p(z)}\right), & \text{if } p(z) < q \leq 1. \end{cases}$$

By taking expectation on both sides on (67), we get
$$\mathbb{P}[L_n \geq nq_n] \leq \mathbb{E}\big[e^{F_n(Z)}\big], \qquad (68)$$

where we set $F_n(z) = -n\Gamma^*(q_n, z)$. Since $\rho > 0$, the function $p(z)$ is increasing in z, so $\Gamma(\theta, z)$ is an increasing function of z for all $\theta \geq 0$. Hence, $F_n(z)$ is an increasing function of z, which is nonpositive and attains its maximum value 0, for all z s.t. $q_n = p(z_n) \leq p(z)$, i.e. $z \geq z_n$. Moreover, by differentiation, we can check that F_n is a concave function of z. We now introduce a change of measure. The idea is to shift the factor mean to reduce the variance of the term $e^{F_n(Z)}$ in the r.h.s. of (68). We consider the change of measure $\mathbb{P}_\mu$ that puts the distribution of Z to $\mathcal{N}(\mu, 1)$. Its likelihood ratio is given by

$$\frac{d\mathbb{P}_\mu}{d\mathbb{P}} = \exp\left(\mu Z - \frac{1}{2}\mu^2\right),$$

so that
$$\mathbb{E}\big[e^{F_n(Z)}\big] = \mathbb{E}_\mu\big[e^{F_n(Z)-\mu Z + \frac{1}{2}\mu^2}\big],$$

where $\mathbb{E}_\mu$ denotes the expectation under $\mathbb{P}_\mu$. By concavity of F_n, we have $F_n(Z) \leq F_n(\mu) + F'_n(\mu)(Z - \mu)$, so that
$$\mathbb{E}\big[e^{F_n(Z)}\big] \leq \mathbb{E}_\mu\big[e^{F_n(\mu)+(F'_n(\mu)-\mu)Z-\mu F'_n(\mu)+\frac{1}{2}\mu^2}\big]. \qquad (69)$$

We now choose $\mu = \mu_n$ solution to
$$F'_n(\mu_n) = \mu_n, \qquad (70)$$

so that the term in the expectation in the r.h.s. of (69) does not depend on Z, and is therefore a constant term (with zero-variance). Such a μ_n exists, since, by strict concavity of the function $z \to F_n(z) - \frac{1}{2}z^2$, equation (70) is the first-order equation associated to the optimization problem :

$$\mu_n = \arg\max_{\mu \in \mathbb{R}}[F_n(\mu) - \frac{1}{2}\mu^2].$$

With this choice of factor mean μ_n, and by inequalities (68), (69), we get

$$\mathbb{P}[L_n \geq nq_n] \leq e^{F_n(\mu_n) - \frac{1}{2}\mu_n^2}. \tag{71}$$

We now prove that μ_n/z_n converges to 1 as n goes to infinity. Actually, we show that for all $\varepsilon > 0$, there is n_0 large enough so that for all $n \geq n_0$, $z_n(1-\varepsilon) < \mu_n < z_n$. Since $F'_n(\mu_n) - \mu_n = 0$, and the function $F'_n(z) - z$ is decreasing by concavity $F_n(z) - z^2/2$, it suffices to show that

$$F'_n(z_n(1-\varepsilon)) - z_n(1-\varepsilon) > 0 \text{ and } F'_n(z_n) - z_n < 0. \tag{72}$$

We have

$$F'_n(z) = n\left(\frac{p(z_n)}{p(z)} - \frac{1-p(z_n)}{1-p(z)}\right)\varphi\left(\frac{\rho z + \Phi^{-1}(p)}{\sqrt{1-\rho^2}}\right)\frac{\rho}{\sqrt{1-\rho^2}}.$$

The second inequality in (72) holds since $F'_n(z_n) = 0$ and $z_n > 0$ for $q_n > p$, hence for n large enough. Actually, z_n goes to infinity as n goes to infinity from (63). For the first inequality in (72), we use the property that $1 - \Phi(x) \simeq \varphi(x)/x$ as $x \to \infty$, so that

$$\lim_{n \to \infty} \frac{p(z_n)}{p(z_n(1-\varepsilon))} = 1, \text{ and } \lim_{n \to \infty} \frac{1-p(z_n)}{1-p(z_n(1-\varepsilon))} = 0.$$

From (63), we have

$$\varphi\left(\frac{\rho z_n(1-\varepsilon) + \Phi^{-1}(p)}{\sqrt{1-\rho^2}}\right) = 0(n^{-a(1-c)^2}),$$

and therefore

$$F'_n(z_n(1-\varepsilon)) = 0(n^{1-a(1-\varepsilon)^2}).$$

Moreover, from (63) and as $a \leq 1$, we have

$$z_n(1-\varepsilon) = 0(\sqrt{\ln n}) = o(n^{1-a(1-\varepsilon)^2})$$

We deduce that for n large enough $F'_n(z_n(1-\varepsilon)) - z_n(1-\varepsilon) > 0$ and so (72). Finally, recalling that F_n is nonpositive, and from (71), we obtain :

$$\limsup_{n \to \infty} \frac{1}{\ln n} \ln \mathbb{P}[L_n \geq nq_n] \leq -\frac{1}{2}\lim_{n \to} \frac{\mu_n^2}{\ln n} = -\frac{1}{2}\lim_{n \to} \frac{z_n^2}{\ln n} = -a\frac{1-\rho^2}{\rho^2}. \tag{73}$$

□

Application: asymptotic optimality of two-step importance sampling estimator

Consider the estimation problem of $\mathbb{P}[L_n \geq nq]$. We apply a two-step importance sampling (IS) by using IS conditional on the common factors Z and IS to the distribution of the factors Z. Observe that conditioning on Z reduces to the problem of the independent case studied in the previous paragraph, with default probability $p(Z)$ as defined in (56), and c.g.f. $\Gamma(.,Z)$ in (65). Choose $\theta_{q_n}(Z) \geq 0$ attaining the argmax in $\theta q_n - \Gamma(\theta, Z)$, and consider the estimator

$$\exp(-\theta_{q_n}(Z)L_n + n\Gamma(\theta_{q_n}(Z), Z))1_{L_n \geq nq_n},$$

where L_n is sampled with a default probability $p(\theta_q(Z), Z) = p(Z)e^{\theta_{q_n}(Z)}/(1 - p(Z) + p(Z)e^{\theta_{q_n}(Z)})$. This provides an unbiased conditional estimator of $\mathbb{P}[L_n \geq nq_n|Z]$ and an asymptotically optimal conditional variance. We further apply IS to the factor $Z \sim \mathcal{N}(0,1)$ under $\mathbb{P}$, by shifting the factor mean to μ, and then considering the estimator

$$\exp(-\mu Z + \frac{1}{2}\mu^2)\exp(-\theta_{q_n}(Z)L_n + n\Gamma(\theta_{q_n}(Z), Z))1_{L_n \geq nq_n}, \quad (74)$$

where Z is sampled from $\mathcal{N}(\mu, 1)$. To summarize, the two-step IS estimator is generated as follows :

- Sample Z from $\mathcal{N}(\mu, 1)$
- Compute $\theta_{q_n}(Z)$ and $p(\theta_{q_n}(Z), Z)$
- Return the estimator (74) where L_n is sampled with default probability $p(\theta_{q_n}(Z), Z)$.

By construction, this provides an unbiaised estimator of $\mathbb{P}[L_n \geq nq_n]$, and the key point is to specify the choice of μ in order to reduce the global variance or equivalently the second moment $M_n^2(\mu, q_n)$ of this estimator. First, recall from Cauchy-Schwarz's inequality : $M_n^2(\mu, q_n) \geq (\mathbb{P}[L_n \geq nq])^2$, so that the fastest possible rate of decay of $M_n^2(\mu, q_n)$ is twice the probability itself :

$$\liminf_{n\to\infty} \frac{1}{\ln n} \ln M_n^2(q_n, \mu) \geq 2 \lim_{n\to\infty} \frac{1}{\ln n} \ln \mathbb{P}[L_n \geq nq_n]. \quad (75)$$

To achieve this twice rate, we proceed as follows. Denoting by $\bar{\mathbb{E}}$ the expectation under the IS distribution, we have

$$M_n^2(\mu, q_n) = \bar{\mathbb{E}}\Big[\exp(-2\mu Z + \mu^2)\exp(-2\theta_{q_n}(Z)L_n + 2n\Gamma(\theta_{q_n}(Z), Z))1_{L_n \geq nq_n}\Big]$$
$$\leq \bar{\mathbb{E}}\Big[\exp(-2\mu Z + \mu^2)\exp(-2n\theta_{q_n}(Z)q_n + 2n\Gamma(\theta_{q_n}(Z), Z))\Big]$$
$$= \bar{\mathbb{E}}\Big[\exp(-2\mu Z + \mu^2 + 2F_n(Z))\Big],$$

by definition of $\theta_{q_n}(Z)$ and $F_n(z) = -n\sup_{\theta \geq 0}[\theta q_n - \Gamma(\theta, z)]$ introduced in the proof of the upper bound in Theorem 5.4. As in (69), (71), by choosing $\mu = \mu_n$ solution to $F_n'(\mu_n) = \mu_n$, we then get

$$M_n^2(\mu_n, q_n) \leq \exp(2F_n(\mu_n) - \mu_n^2) \leq \exp(-\mu_n^2),$$

Some Applications and Methods of Large Deviations in Finance and Insurance   231

since F_n is nonpositive. From (73), this yields

$$\limsup_{n \to \infty} \frac{1}{\ln n} \ln M_n^2(\mu_n, q_n) \leq -2a \frac{1-\rho^2}{\rho^2} = 2 \lim_{n \to \infty} \frac{1}{\ln n} \ln \mathbb{P}[L_n \geq nq_n],$$

which proves together with (75) that

$$\lim_{n \to \infty} \frac{1}{\ln n} \ln M_n^2(\mu_n, q_n) = -2a \frac{1-\rho^2}{\rho^2} = 2 \lim_{n \to \infty} \frac{1}{\ln n} \ln \mathbb{P}[L_n \geq nq_n],$$

and thus the estimator (74) for the choice $\mu = \mu_n$ is asymptotically optimal. The choice of $\mu = z_n$ also leads to an asymptotically optimal estimator.

Remark 5.8 We also prove by similar methods large deviation results for the loss distribution in the limiting regime where individual loss probabilities decrease toward zero, see [26] for the details. This setting is relevant to portfolios of highly-rated obligors, for which one-year default probabilities are extremely small. This is also relevant to measuring risk over short time horizons. In this limiting regime, we set

$$l_n = nq, \quad \text{with } 0 < q < 1, \quad p = p_n = O(e^{-na}), \text{ with } a > 0.$$

Then,

$$\lim_{n \to \infty} \frac{1}{n} \ln \mathbb{P}[L_n \geq nq] = -\frac{a}{\rho^2},$$

and we may construct similarly as in the case of large losses, a two-step IS asymptotically optimal estimator.

5.2 A Large Deviations Approach to Optimal Long Term Investment

An Asymptotic Outperforming Benchmark Criterion

A popular approach for institutional managers is concerned about the performance of their portfolio relative to the achievement of a given benchmark. This means that investors are interested in maximizing the probability that their wealth exceed a predetermined index. Equivalently, this may be also formulated as the problem of minimizing the probability that the wealth of the investor falls below a specified value. This target problem was studied by several authors for a goal achievement in finite time horizon, see e.g. [11] or [20]. Recently, and in a static framework, the paper [40] considered an asymptotic version of this outperformance criterion when time horizon goes to infinity, which leads to a large deviations portfolio criterion. To illustrate the purpose, let us consider the following toy example. Suppose that an investor trades a number α of shares in stock of price S, and keep it until time T. Her wealth at time T is then $X_T^\alpha = \alpha S_T$. For simplicity, we take a Bachelier model for the stock price :

$S_t = \mu t + \sigma W_t$, where W is a brownian motion. We now look at the behavior of the average wealth when time horizon T goes to infinity. By the law of large numbers, for any $\alpha \in \mathbb{R}$, the average wealth converges a.s. to :

$$\bar{X}_T^\alpha := \frac{X_T^\alpha}{T} = \alpha\mu + \alpha\sigma\frac{W_T}{T} \longrightarrow \alpha\mu,$$

when T goes to infinity. When considering positive stock price, as in the Black-Scholes model, the relevant ergodic mean is the average of the growth rate, i.e. the logarithm of the wealth. Fix some benchmark level $x \in \mathbb{R}$. Then, from Cramer's theorem, the probability of outperforming x decays exponentially fast as :

$$\mathbb{P}[\bar{X}_T^\alpha \geq x] \simeq e^{-I(x,\alpha)T},$$

in the sense that $\lim_{T \to \infty} \frac{1}{T} \ln \mathbb{P}[\bar{X}_T^\alpha \geq x] = -I(x,\alpha)$, where

$$I(x,\alpha) = \sup_{\theta \in \mathbb{R}}[\theta x - \Gamma(\theta,\alpha)]$$

$$\Gamma(\theta,\alpha) = \frac{1}{T} \ln \mathbb{E}[e^{\theta X_T^\alpha}].$$

Thus, the lower is the decay rate $I(x,\alpha)$, the more chance there is of realizing a portfolio performance above x. The asymptotic version of the outperforming benchmark criterion is then formulated as :

$$\sup_{\alpha \in \mathbb{R}} \lim_{T \to \infty} \frac{1}{T} \ln \mathbb{P}[\bar{X}_T^\alpha \geq x] = -\inf_{\alpha \in \mathbb{R}} I(x,\alpha). \tag{76}$$

In this simple example, the quantities involved are all explicit :

$$\Gamma(\theta,\alpha) = \theta\alpha\mu + \frac{(\theta\alpha\sigma)^2}{2}$$

$$I(x,\alpha) = \begin{cases} \frac{1}{2}\left(\frac{\alpha\mu - x}{\alpha\sigma}\right)^2, & \alpha \neq 0 \\ 0, & \alpha = 0, \ x = 0 \\ \infty, & \alpha = 0, \ x \neq 0. \end{cases}$$

The solution to (76) is then given by $\alpha^* = x/\mu$, which means that the associated expected wealth $\mathbb{E}[\bar{X}_T^{\alpha^*}]$ is equal to the target x.

We now develop an asymptotic dynamic version of the outperformance management criterion due to [36]. Such a problem corresponds to an ergodic objective of beating a given benchmark, and may be of particular interest for institutional managers with long term horizon, like mutual funds. On the other hand, stationary long term horizon problems are expected to be more tractable than finite horizon problems, and should provide some good insight for management problems with long, but finite, time horizon.

We formulate the problem in a rather abstract setting. Let $Z = (X,Y)$ be a process valued in $\mathbb{R} \times \mathbb{R}^d$, controlled by α, a control process valued in some subset A

of $\mathbb{R}^q$. We denote by $\mathcal{A}$ the set of control processes. As usual, to alleviate notations, we omitted the dependence of $Z = (X, Y)$ in $\alpha \in \mathcal{A}$. We shall then study the large deviations control problem :

$$v(x) = \sup_{\alpha \in \mathcal{A}} \limsup_{T \to \infty} \frac{1}{T} \ln \mathbb{P}[\bar{X}_T \geq x], \quad x \in \mathbb{R}, \tag{77}$$

where $\bar{X}_T = X_T/T$. The variable X should typically be viewed in finance as the (logarithm) of the wealth process, Y are factors on market (stock, volatility ...), and α represents the trading portfolio.

Duality to the Large Deviations Control Problem

The large deviations control problem (77) is a non standard stochastic control problem, where the objective is usually formulated as an expectation of some functional to optimize. In particular, in a Markovian continuous-time setting, we do not know if there is a dynamic programming principle and a corresponding Hamilton-Jacobi-Bellman equation for our problem. We shall actually adopt a duality approach based on the relation relating rate function of a LDP and cumulant generating function. The formal derivation is the following. Given $\alpha \in \mathcal{A}$, if there is a LDP for $\bar{X}_T = X_T/T$, its rate function $I(., \alpha)$ should be related by the Fenchel-Legendre transform :

$$I(x, \alpha) = \sup_{\theta}[\theta x - \Gamma(\theta, \alpha)],$$

to the c.g.f.

$$\Gamma(\theta, \alpha) = \limsup_{T \to \infty} \frac{1}{T} \ln \mathbb{E}[e^{\theta X_T}]. \tag{78}$$

In this case, we would get

$$v(x) = \sup_{\alpha \in \mathcal{A}} \limsup_{T \to \infty} \frac{1}{T} \ln \mathbb{P}[\bar{X}_T \geq x] = -\inf_{\alpha \in \mathcal{A}} I(x, \alpha)$$
$$= -\inf_{\alpha \in \mathcal{A}} \sup_{\theta}[\theta x - \Gamma(\theta, \alpha)],$$

and so, provided that one could intervert infinum and supremum in the above relation (actually, the minmax theorem does not apply since $\mathcal{A}$ is not necessarily compact and $\alpha \to \theta x - \Gamma(\theta, \alpha)$ is not convex) :

$$v(x) = -\sup_{\theta}[\theta x - \Gamma(\theta)], \tag{79}$$

where

$$\Gamma(\theta) = \sup_{\alpha \in \mathcal{A}} \Gamma(\theta, \alpha) = \sup_{\alpha \in \mathcal{A}} \limsup_{T \to \infty} \frac{1}{T} \ln \mathbb{E}[e^{\theta X_T}]. \tag{80}$$

Problem (80) is the dual problem via (79) to the original problem (77). We shall see in the next section that (80) can be reformulated as a risk-sensitive ergodic control problem, which is more tractable than (77) and is studied by dynamic programming methods leading in some cases to explicit calculations.

First, we show rigorously the duality relation between the large deviations control problem and the risk-sensitive control problem and how the optimal controls to the former one are related to the latter one. This result may be viewed as an extension of the Gärtner-Ellis theorem with control components.

Theorem 5.5 *Suppose that there exists $\bar{\theta} \in (0,\infty]$ such that for all $\theta \in [0,\bar{\theta})$, there exists a solution $\hat{\alpha}(\theta) \in \mathcal{A}$ to the dual problem $\Gamma(\theta)$, with a limit in (78), i.e.*

$$\Gamma(\theta) = \lim_{T\to\infty} \frac{1}{T} \ln \mathbb{E}\left[\exp\left(\theta X_T^{\hat{\alpha}(\theta)}\right)\right]. \tag{81}$$

Suppose also that $\Gamma(\theta)$ is continuously differentiable on $[0,\bar{\theta})$. Then for all $x < \Gamma'(\bar{\theta}) := \lim_{\lambda \nearrow \bar{\theta}} \Gamma'(\theta)$, we get

$$v(x) = - \sup_{\theta \in [0,\bar{\theta})} [\theta x - \Gamma(\theta)]. \tag{82}$$

Moreover, the sequence of controls

$$\alpha_t^{*,n} = \begin{cases} \hat{\alpha}_t\left(\theta\left(x + \frac{1}{n}\right)\right), & \Gamma'(0) < x < \Gamma'(\bar{\theta}) \\ \hat{\alpha}_t\left(\theta\left(\Gamma'(0) + \frac{1}{n}\right)\right), & x \leq \Gamma'(0), \end{cases}$$

with $\theta(x) \in (0,\bar{\theta})$ s.t. $\Gamma'(\theta(x)) = x \in (\Gamma'(0), \Gamma'(\bar{\theta}))$, is nearly optimal in the sense that

$$\lim_{n\to\infty} \limsup_{T\to\infty} \frac{1}{T} \ln \mathbb{P}\left[\bar{X}_T^{\alpha^{*,n}} \geq x\right] = v(x).$$

Proof.
Step 1. Let us consider the Fenchel-Legendre transform of the convex function Γ on $[0,\bar{\theta})$:

$$\Gamma^*(x) = \sup_{\theta \in [0,\bar{\theta})} [\theta x - \Gamma(\theta)], \quad x \in \mathbb{R}. \tag{83}$$

Since Γ is C^1 on $[0,\bar{\theta})$, it is well-known (see e.g. Lemma 2.3.9 in [13]) that the function Γ^* is convex, nondecreasing and satisfies:

$$\Gamma^*(x) = \begin{cases} \theta(x)x - \Gamma(\theta(x)), & \text{if } \Gamma'(0) < x < \Gamma'(\bar{\theta}) \\ 0, & \text{if } x \leq \Gamma'(0), \end{cases} \tag{84}$$

$$\theta(x)x - \Gamma^*(x) > \theta(x)x' - \Gamma^*(x'), \quad \forall \Gamma'(0) < x < \Gamma'(\bar{\theta}), \forall x' \neq x, \tag{85}$$

where $\theta(x) \in (0,\bar{\theta})$ is s.t. $\Gamma'(\theta(x)) = x \in (\Gamma'(0), \Gamma'(\bar{\theta}))$. Moreover, Γ^* is continuous on $(-\infty, \Gamma'(\bar{\theta}))$.

Step 2 : Upper bound. For all $x \in \mathbb{R}$, $\alpha \in \mathcal{A}$, an application of Chebycheff's inequality yields :

$$\mathbb{P}[\bar{X}_T \geq x] \leq \exp(-\theta x T)\mathbb{E}[\exp(\theta X_T)], \quad \forall\, \theta \in [0, \bar{\theta}),$$

and so

$$\limsup_{T \to \infty} \frac{1}{T} \ln \mathbb{P}[\bar{X}_T \geq x] \leq -\theta x + \limsup_{T \to \infty} \frac{1}{T} \ln \mathbb{E}[\exp(\theta X_T)], \quad \forall\, \theta \in [0, \bar{\theta}).$$

By definitions of Γ and Γ^*, we deduce :

$$\sup_{\alpha \in \mathcal{A}} \limsup_{T \to \infty} \frac{1}{T} \ln \mathbb{P}[\bar{X}_T^\alpha \geq x] \leq -\Gamma^*(x). \tag{86}$$

Step 3 : Lower bound. Given $x < \Gamma'(\bar{\theta})$, let us define the probability measure $\mathbb{Q}_T^n$ on $(\Omega, \mathcal{F}_T)$ via :

$$\frac{d\mathbb{Q}_T^n}{d\mathbb{P}} = \exp\left[\theta(x_n) X_T^{\alpha^{*,n}} - \Gamma_T(\theta(x_n), \alpha^{*,n})\right], \tag{87}$$

where $x_n = x + 1/n$ if $x > \Gamma'(0)$, $x_n = \Gamma'(0) + 1/n$ otherwise, $\alpha^{*,n} = \hat{\alpha}(\theta(x_n))$, and

$$\Gamma_T(\theta, \alpha) = \ln \mathbb{E}[\exp(\theta X_T^\alpha)], \quad \theta \in [0, \bar{\theta}),\ \alpha \in \mathcal{A}.$$

Here n is large enough so that $x + 1/n < \Gamma'(\bar{\theta})$. We now take $\varepsilon > 0$ small enough so that $x \leq x_n - \varepsilon$ and $x_n + \varepsilon < \Gamma'(\bar{\theta})$. We then have :

$$\frac{1}{T} \ln \mathbb{P}[\bar{X}_T^{\alpha^{*,n}} \geq x] \geq \frac{1}{T} \ln \mathbb{P}\left[x_n - \varepsilon < \bar{X}_T^{\alpha^{*,n}} < x_n + \varepsilon\right]$$

$$= \frac{1}{T} \ln \left(\int \frac{d\mathbb{P}}{d\mathbb{Q}_T^n} \mathbf{1}_{\{x_n - \varepsilon < \bar{X}_T^{\alpha^{*,n}} < x_n + \varepsilon\}} d\mathbb{Q}_T^n \right)$$

$$\geq -\theta(x_n)(x_n + \varepsilon) + \frac{1}{T} \Gamma_T(\theta(x_n), \alpha^{*,n})$$

$$+ \frac{1}{T} \ln \mathbb{Q}_T^n \left[x_n - \varepsilon < \bar{X}_T^{\alpha^{*,n}} < x_n + \varepsilon \right],$$

where we use (87) in the last inequality. By definition of the dual problem, this yields :

$$\liminf_{T \to \infty} \frac{1}{T} \ln \mathbb{P}[\bar{X}_T^{\alpha^{*,n}} \geq x] \geq -\theta(x_n)(x_n + \varepsilon) + \Gamma(\theta(x_n))$$

$$+ \liminf_{T \to \infty} \frac{1}{T} \ln \mathbb{Q}_T^n \left[x_n - \varepsilon < \bar{X}_T^{\alpha^{*,n}} < x_n + \varepsilon \right]$$

$$\geq -\Gamma^*(x_n) - \theta(x_n)\varepsilon \tag{88}$$

$$+ \liminf_{T \to \infty} \frac{1}{T} \ln \mathbb{Q}_T^n \left[x_n - \varepsilon < \bar{X}_T^{\alpha^{*,n}} < x_n + \varepsilon \right],$$

where the second inequality follows by the definition of Γ^* (and actually holds with equality due to (84)). We now show that :

$$\liminf_{T\to\infty} \frac{1}{T} \ln Q_T^n \left[x_n - \varepsilon < \bar{X}_T^{\alpha^*,n} < x_n + \varepsilon \right] = 0. \tag{89}$$

Denote by $\tilde{\Gamma}_T^n$ the c.g.f. under Q_T^n of $X_T^{\alpha^*,n}$. For all $\zeta \in \mathbb{R}$, we have by (87) :

$$\begin{aligned}\tilde{\Gamma}_T^n(\zeta) &:= \ln E^{Q_T^n}[\exp(\zeta X_T^{\alpha^*,n})] \\ &= \Gamma_T(\theta(x_n) + \zeta, \alpha^{*,n}) - \Gamma_T(\theta(x_n), \alpha^{*,n}).\end{aligned}$$

Therefore, by definition of the dual problem and (81), we have for all $\zeta \in [-\theta(x_n), \bar{\theta} - \theta(x_n))$:

$$\limsup_{T\to\infty} \frac{1}{T} \tilde{\Gamma}_T^n(\zeta) \leq \Gamma(\theta(x_n) + \zeta) - \Gamma(\theta(x_n)). \tag{90}$$

As in part 1) of this proof, by Chebycheff's inequality, we have for all $\zeta \in [0, \bar{\theta} - \theta(x_n))$:

$$\begin{aligned}\limsup_{T\to\infty} \frac{1}{T} \ln Q_T^n \left[\bar{X}_T^{\alpha^*,n} \geq x_n + \varepsilon \right] &\leq -\zeta(x_n + \varepsilon) + \limsup_{T\to\infty} \frac{1}{T} \tilde{\Gamma}_T^n(\zeta) \\ &\leq -\zeta(x_n + \varepsilon) + \Gamma(\zeta + \theta(x_n)) - \Gamma(\theta(x_n)),\end{aligned}$$

where the second inequality follows from (90). We deduce

$$\begin{aligned}\limsup_{T\to\infty} \frac{1}{T} \ln Q_T^n \left[\bar{X}_T^{\alpha^*,n} \geq x_n + \varepsilon \right] &\leq -\sup\{\zeta(x_n + \varepsilon) - \Gamma(\zeta) : \zeta \in [\theta(x_n), \bar{\theta})\} \\ &\quad - \Gamma(\theta(x_n)) + \theta(x_n)(x_n + \varepsilon) \\ &\leq -\Gamma^*(x_n + \varepsilon) - \Gamma(\theta(x_n)) + \theta(x_n)(x_n + \varepsilon) \\ &= -\Gamma^*(x_n + \varepsilon) + \Gamma^*(x_n) + \varepsilon\theta(x_n), \tag{91}\end{aligned}$$

where the second inequality and the last equality follow from (84). Similarly, we have for all $\zeta \in [-\theta(x_n), 0]$:

$$\begin{aligned}\limsup_{T\to\infty} \frac{1}{T} \ln Q_T^n \left[\bar{X}_T^{\alpha^*,n} \leq x_n - \varepsilon \right] &\leq -\zeta(x_n - \varepsilon) + \limsup_{T\to\infty} \frac{1}{T} \tilde{\Gamma}_T^n(\zeta) \\ &\leq -\zeta(x_n - \varepsilon) + \Gamma(\theta(x_n) + \zeta) - \Gamma(\theta(x_n)),\end{aligned}$$

and so :

$$\begin{aligned}\limsup_{T\to\infty} \frac{1}{T} \ln Q_T^n \left[\bar{X}_T^{\alpha^*,n} \leq x_n - \varepsilon \right] &\leq -\sup\{\zeta(x_n - \varepsilon) - \Gamma(\zeta) : \zeta \in [0, \theta(x_n)]\} \\ &\quad - \Gamma(\theta(x_n)) + \theta(x_n)(x_n - \varepsilon) \\ &\leq -\Gamma^*(x_n - \varepsilon) + \Gamma^*(\theta(x_n)) - \varepsilon\theta(x_n). \tag{92}\end{aligned}$$

By (91)-(92), we then get :

$$\limsup_{T\to\infty} \frac{1}{T} \ln \mathbb{Q}_T^n \left[\left\{ \bar{X}_T^{\alpha^*,n} \leq x_n - \varepsilon \right\} \cup \left\{ \bar{X}_T^{\alpha^*,n} \geq x_n + \varepsilon \right\} \right]$$

$$\leq \max \left\{ \limsup_{T\to\infty} \frac{1}{T} \ln \mathbb{Q}_T^n \left[\bar{X}_T^{\alpha^*,n} \geq x_n + \varepsilon \right] ; \right.$$

$$\left. \limsup_{T\to\infty} \frac{1}{T} \ln \mathbb{Q}_T^n \left[\bar{X}_T^{\alpha^*,n} \leq x_n - \varepsilon \right] \right\}$$

$$\leq \max \left\{ -\Gamma^*(x_n + \varepsilon) + \Gamma^*(x_n) + \varepsilon \theta(x_n); \right.$$

$$\left. -\Gamma^*(x_n - \varepsilon) + \Gamma^*(\theta(x_n)) - \varepsilon \theta(x_n) \right\} < 0,$$

where the strict inequality follows from (85). This implies that $\mathbb{Q}_T^n [\{\bar{X}_T^{\alpha^*,n} \leq x_n - \varepsilon\} \cup \{\bar{X}_T^{\alpha^*,n} \geq x_n + \varepsilon\}] \to 0$ and hence $\mathbb{Q}_T^n [x_n - \varepsilon < \bar{X}_T^{\alpha^*,n} < x_n + \varepsilon] \to 1$ as T goes to infinity. In particular (89) is satisfied, and by sending ε to zero in (88), we get :

$$\liminf_{T\to\infty} \frac{1}{T} \ln \mathbb{P}[\bar{X}_T^{\alpha^*,n} \geq x] \geq -\Gamma^*(x_n).$$

By continuity of Γ^* on $(-\infty, \Gamma'(\bar{\theta}))$, we obtain by sending n to infinity and recalling that $\Gamma^*(x) = 0 = \Gamma^*(\Gamma'(0))$ for $x \leq \Gamma'(0)$:

$$\liminf_{n\to\infty} \liminf_{T\to\infty} \frac{1}{T} \ln \mathbb{P}[\bar{X}_T^{\alpha^*,n} \geq x] \geq -\Gamma^*(x).$$

This last inequality combined with (86) ends the proof. □

Remark 5.9 Notice that in Theorem 5.5, the duality relation (82) holds for $x < \Gamma'(\bar{\theta})$. When $\Gamma'(\bar{\theta}) = \infty$, we say that fonction Γ is steep, so that (82) holds for all $x \in \mathbb{R}$. We illustrate in the next section different cases where Γ is steep or not.

Remark 5.10 In financial applications, X_t is the logarithm of an investor's wealth V_t^α at time t, α_t is the proportion of wealth invested in q risky assets S and Y is some economic factor influencing the dynamics of S and the savings account S^0. Hence, in a diffusion model, we have

$$dX_t = \left[r(Y_t) + \alpha_t'(\mu(Y_t) - r(Y_t)e_q) - \frac{1}{2} |\alpha_t' \vartheta(Y_t)|^2 \right] dt + \alpha_t' \vartheta(Y_t) dW_t,$$

where $\mu(y)$ (resp. $\vartheta(y)$) is the rate of return (resp. volatility) of the risky assets, $r(y)$ is the interest rate, and e_q is the unit vector in $\mathbb{R}^q$.

Notice that the value function of the dual problem can be written as :

$$\Gamma(\theta) = \lim_{T\to\infty} \frac{1}{T} \ln E \left[U_\theta \left(V_T^{\hat{\alpha}(\theta)} \right) \right],$$

where $U_\theta(c) = c^\theta$ is a power utility function with Constant Relative Risk Aversion (CRRA) $1 - \theta > 0$ provided that $\theta < 1$. Then, Theorem 5.5 means that for any target level x, the optimal overperformance probability of growth rate is (approximately) directly related, for large T, to the expected CRRA utility of wealth, by :

$$P[\bar{X}_T^{\alpha^*} \geq x] \approx E\left[U_{\theta(x)}\left(V_T^{\alpha^*}\right)\right] e^{-\theta(x)xT}, \tag{93}$$

with the convention that $\theta(x) = 0$ for $x \leq \Gamma'(0)$. Hence, $1 - \theta(x)$ can be interpreted as a constant degree of relative risk aversion for an investor who has an overperformance target level x. Moreover, by strict convexity of function Γ^* in (83), it is clear that $\theta(x)$ is strictly increasing for $x > \Gamma'(0)$. So an investor with a higher target level x has a lower degree of relative risk aversion $1 - \theta(x)$. In summary, Theorem 5.5 (or relation (93)) inversely relates the target level of growth rate to the degree of relative risk aversion in expected utility theory.

Explicit Calculations for the Dual Risk-Sensitive Control Problem

We now show that the dual control problem (80) may be transformed via a change of probability measure into a risk-sensitive control problem. We consider the framework of a general diffusion model for $Z = (X, Y)$:

$$dX_t = b(X_t, Y_t, \alpha_t)dt + \sigma(X_t, Y_t, \alpha_t)dW_t \quad \text{in } \mathbb{R} \tag{94}$$
$$dY_t = \eta(X_t, Y_t, \alpha_t)dt + \sigma(X_t, Y_t, \alpha_t)dW_t \quad \text{in } \mathbb{R}^d, \tag{95}$$

where W is a m-dimensional brownian motion on a filtered probability space $(\Omega, \mathcal{F}, \mathbb{F} = (\mathcal{F}_t)_{t \geq 0}, \mathbb{P})$, and $\alpha = (\alpha_t)_{t \geq 0}$, the control process, is $\mathbb{F}$-adapted and valued in some subset A of $\mathbb{R}^q$. We denote $\mathcal{A}$ the set of control processes. The coefficients b, η, σ and γ are measurable functions of their arguments, and given $\alpha \in \mathcal{A}$ and an initial condition, we assume the existence and uniqueness of a strong solution to (94)-(95), which we also write by setting $Z = (X, Y)$:

$$dZ_t = B(Z_t, \alpha_t)dt + \Sigma(Z_t, \alpha_t)dW_t. \tag{96}$$

From the dynamics of X in (94), we may rewrite the Laplace transform of X_T as :

$$\mathbb{E}\left[\exp(\theta X_T)\right] = e^{\theta X_0} \mathbb{E}\left[\exp\left(\theta \int_0^T b(Z_t, \alpha_t)dt + \theta \int_0^T \sigma(Z_t, \alpha_t)dW_t\right)\right]$$
$$= e^{\theta X_0} \mathbb{E}\left[\xi_T^\alpha(\theta) \exp\left(\int_0^T \ell(\theta, Z_t, \alpha_t)dt\right)\right], \tag{97}$$

where

$$\ell(\theta, z, a) = \theta b(z, a) + \frac{\theta^2}{2}|\sigma(z, a)|^2,$$

and $\xi_t^\alpha(\theta)$ is the Doléans-Dade exponential local martingale

$$\xi_t^\alpha(\theta) = \mathcal{E}\left(\theta \int \sigma(Z_u, \alpha_u) dW_u\right)_t$$

$$:= \exp\left(\theta \int_0^t \sigma(Z_u, \alpha_u) dW_u - \frac{\theta^2}{2} \int_0^t |\sigma(Z_u, \alpha_u)|^2 du\right), \quad t \geq 0. \quad (98)$$

If $\xi^\alpha(\theta)$ is a "true" martingale, it defines a probability measure $\mathbb{Q}$ under which, by Girsanov's theorem, the dynamics of Z is given by :

$$dZ_t = G(\theta, Z_t, \alpha_t) dt + \Sigma(Z_t, \alpha_t) dW_t^\mathbb{Q},$$

where $W^\mathbb{Q}$ is a $\mathbb{Q}$-Brownian motion and

$$G(\theta, z, a) = \begin{pmatrix} b(z, a) + \theta |\sigma(z, a)|^2 \\ \eta(z, a) + \theta \gamma \sigma'(z, a) \end{pmatrix}.$$

Hence, the dual problem may be written as a stochastic control problem with exponential integral cost criterion :

$$\Gamma(\theta) = \sup_{\alpha \in \mathcal{A}} \limsup_{T \to \infty} \frac{1}{T} \ln \mathbb{E}^\mathbb{Q}\left[\exp\left(\int_0^T \ell(\theta, Z_t, \alpha_t) dt\right)\right], \quad \theta \geq 0. \quad (99)$$

For fixed θ, this is an ergodic risk-sensitive control problem which has been studied by several authors, see e.g. [18], [10] or [39] in a discrete-time setting. It admits a dynamic programming equation :

$$\Lambda(\theta) = \sup_{a \in \mathcal{A}} \left[\frac{1}{2} \mathrm{tr}\left(\Sigma \Sigma'(z, a) D^2 \phi_\theta\right) + G(\theta, z, a). \nabla \phi_\theta \right.$$

$$\left. + \frac{1}{2} |\Sigma'(z, a) \nabla \phi_\theta|^2 + \ell(\theta, z, a)\right], \quad z \in \mathbb{R}^{d+1}. \quad (100)$$

The unknown is the pair $(\Lambda(\theta), \phi_\theta) \in \mathbb{R} \times C^2(\mathbb{R}^{d+1})$, and $\Lambda(\theta)$ is a candidate for $\Gamma(\theta)$. The above P.D.E. is formally derived by considering the finite horizon problem

$$u_\theta(T, z) = \sup_{\alpha \in \mathcal{A}} \mathbb{E}^\mathbb{Q}\left[\exp\left(\int_0^T \ell(\theta, Z_t, \alpha_t) dt\right)\right],$$

by writing the Bellman equation for this classical control problem and by making the logarithm transformation

$$\ln u_\theta(T, z) \simeq \Lambda(\theta) T + \phi_\theta(z),$$

for large T.

One can prove rigorously that a pair solution $(\Lambda(\theta), \phi_\theta)$ to the PDE (100) provides a solution $\Lambda(\theta) = \Gamma(\theta)$ to the dual problem (80), with an optimal control given by the argument max in (100). This is called a verification theorem in stochastic control theory. Actually, there may have multiple solutions ϕ_θ to (100) (even up to a constant), and we need some ergodicity condition to select the good one that satisfies the verification theorem. We refer to [37] for the details, and we illustrate our purpose with an example with explicit calculations.

We consider a one-factor model where the bond price S^0 and the stock price S evolve according to :

$$\frac{dS_t^0}{S_t^0} = (a_0 + b_0 Y_t)dt, \quad \frac{dS_t}{S_t} = (a + bY_t)dt + \sigma dW_t,$$

with a factor Y as an Ornstein-Uhlenbeck ergodic process:

$$dY_t = -kY_t dt + dB_t,$$

where a_0, b_0, a, b are constants, k, σ are positive constants, and W, B are two brownian motions, supposed non correlated for simplicity. This includes Black-Scholes, Platen-Rebolledo or Vasicek models. The (self-financed) wealth process V_t with a proportion α_t invested in stock, follows the dynamics : $dV_t = \alpha_t V_t \frac{dS_t}{S_t} + (1 - \alpha_t)V_t \frac{dS_t^0}{S_t^0}$, and so the logarithm of the wealth process $X_t = \ln V_t$ is governed by a linear-quadratic model :

$$dX_t = (\beta_0 Y_t^2 + \beta_1 \alpha_t^2 + \beta_2 Y_t \alpha_t + \beta_3 Y_t + \beta_4 \alpha_t + \beta_5)dt \\ + (\delta_0 Y_t + \delta_1 \alpha_t + \delta_2)dW_t, \qquad (101)$$

where in our context, $\beta_0 = 0$, $\beta_1 = -\sigma^2/2$, $\beta_2 = b - b_0$, $\beta_3 = b_0$, $\beta_4 = a - a_0$, $\beta_5 = a_0$, $\delta_0 = 0$, $\delta_1 = \sigma$ and $\delta_2 = 0$. Without loss of generality, we may assume that $\sigma = 1$ and so $\beta_1 = -1/2$ (embedded into α) and $\beta_5 = 0$ (embedded into x). The P.D.E. (100) simplifies into the search of a pair $(\Lambda(\theta), \phi_\theta)$ with ϕ_θ depending only on y and solution to :

$$\Lambda(\theta) = \frac{1}{2}\phi_\theta'' - ky\phi_\theta' + \frac{1}{2}|\phi_\theta'|^2 + \theta \left(\beta_0 + \theta \frac{\delta_0^2}{2} \right) y^2 + \theta(\beta_3 + \theta \delta_0 \delta_2)y + \theta^2 \frac{\delta_2^2}{2} \\ + \frac{1}{2}\frac{\theta}{1 - \theta \delta_1^2}[(\beta_2 + \theta \delta_0 \delta_1)y + \beta_4 + \theta \delta_1 \delta_2]^2. \qquad (102)$$

Moreover, the maximum in $a \in \mathbb{R}$ of (100) is attained for

$$\hat{\alpha}(\theta, y) = \frac{(\beta_2 + \theta \delta_0 \delta_1)y + \beta_4 + \theta \delta_1 \delta_2}{1 - \theta \delta_1^2}. \qquad (103)$$

The above calculations are valid only for $0 \leq \theta < 1/\delta_1^2$. We are looking for a quadratic solution to the ordinary differential equation (102) :

$$\phi_\theta(y) = \frac{1}{2}A(\theta)y^2 + B(\theta)y.$$

By substituting into (102), and cancelling terms in y^2, y and constant terms, we obtain

- a polynomial second degree equation for $A(\theta)$
- a linear equation for $B(\theta)$, given $A(\theta)$
- $\Lambda(\theta)$ is then expressed explicitly in function of $A(\theta)$ and $B(\theta)$ from (102).

The existence of a solution to the second degree equation for $A(\theta)$, through the nonnegativity of the discriminant, allows to determine the bound $\bar\theta$ and so the interval $[0,\bar\theta)$ on which Λ is well-defined and finite. Moreover, we find two possible roots to the polynomial second degree equation for $A(\theta)$, but only one satisfies the ergodicity condition. From Theorem 5.5, we deduce that

$$v(x) = - \sup_{\theta \in [0,\bar\theta)} \left[\theta x - \Lambda(\theta)\right], \quad \forall x < \Lambda'(\bar\theta), \tag{104}$$

with a sequence of nearly optimal controls given by:

$$\alpha_t^{*,n} = \begin{cases} \hat\alpha\left(\theta\left(x+\frac{1}{n}\right), Y_t\right), & \Lambda'(0) < x < \Lambda'(\bar\theta) \\ \hat\alpha\left(\theta\left(\Lambda'(0)+\frac{1}{n}\right), Y_t\right), & x \leq \Lambda'(0), \end{cases}$$

with $\theta(x) \in (0,\bar\theta)$ s.t. $\Lambda'(\theta(x)) = x$. In the one-factor model described above, the function Λ is steep, i.e. $\Lambda'(\bar\theta) = \infty$, and so (104) holds for all $x \in \mathbb{R}$. For example, in the Black-Scholes model, i.e. $b_0 = b = 0$, we obtain

$$\Gamma(\theta) = \Lambda(\theta) = \frac{1}{2}\frac{\theta}{1-\theta}\left(\frac{a-a_0}{\sigma^2}\right)^2, \quad \text{for } \theta < \bar\theta = 1,$$

$$v(x) = - \sup_{\theta \in [0,1)} \left[\theta x - \Gamma(\theta)\right]$$

$$= \begin{cases} -(\sqrt{x} - \sqrt{\bar x})^2, & \text{if } x \geq \bar x := \Gamma'(0) = \frac{1}{2}\left(\frac{a-a_0}{\sigma^2}\right)^2 \\ 0, & \text{if } x < \bar x, \end{cases}$$

$\theta(x) = 1 - \sqrt{\bar x/x}$ if $x \geq \bar x$, and 0 otherwise, and

$$\alpha_t^* = \begin{cases} \sqrt{2x}, & \text{if } x \geq \bar x \\ \frac{a-a_0}{\sigma^2}, & \text{if } x < \bar x. \end{cases}$$

We observe that for an index value x small enough, actually $x < \bar x$, the optimal investment for our large deviations criterion is equal to the optimal investment of the Merton's problem for an investor with relative risk aversion one. When the value index is larger than $\bar x$, the optimal investment is increasing with x, with a degree of relative risk aversion $1 - \theta(x)$ decreasing in x.

In the more general linear-quadratic model (101), Λ may be steep or not depending on the parameters β_i and δ_i. We refer to [37] for the details. Some variants and extensions of this large deviations control problem are studied in [30] and [1].

6 Conclusion

In these notes, we developed some applications and emphasized methods of large deviations in finance and insurance. These applications are multiple, and our presentation is by no means exhaustive. There are numerous works dealing with large deviations techniques in the context of insurance, see e.g. [14], [4], or more recently [33] and [35]. We also cite the paper [6], which develops asymptotic formula for calculating implied volatility of index options. Large deviation principle for backward stochastic differential equations is used by [5] in a setting motivated by credit risk management. Other papers using large deviations in portfolio management are [38] and [8]. Some aspects of large deviations applied to problems in macroeconomics are studied in [41].

From a general viewpoint, questions related to extremal events are embedded into the extreme value theory, and we refer to the classical book [16] for a development of this subject, especially regarding applications in finance and insurance.

References

1. Akian M., Gaubert S. and V. Kolokoltsov (2005) : "Solutions of max-plus linear equations and large deviations", Proceedings of the joint 44th IEEE Conference on Decision and Control and European Control Conference ECC 2005.
2. Arouna B. (2004) : "Adaptative Monte-Carlo methods, a variance reduction technique", *Monte-carlo Methods and Applications*, **10**, 1-24.
3. Asmussen S. (2000) : Ruin Probabilities, World Scientific.
4. Asmussen S. and C. Kluppelberg (1996) : "Large deviation results for subexponential tails, with applications to insurance risk", *Stoch. Proc. Appl.*, **64**, 103-125.
5. Astic F. and S. Rainero (2006) : "Conditional large deviations and applications to credit risk management", Preprint University Paris Dauphine.
6. Avellaneda M., Boyer-Olson D., Busca J. and P. Friz (2003) : "Méthodes de grandes déviations et pricing d'options sur indice", *C.R. Acad. Sci. Paris*, **336**, 263-266.
7. Baldi P., Caramellino L. and M. Iovino (1999) : "Pricing general barrier options: a numerical approach using sharp large deviations", *Mathematical Finance*, **9**, 293-322.
8. Bares P., Cont R., Gardiol L., Gibson R. and S. Gyger (2000) : "A large deviation approach to portfolio management", *International Journal of Theoretical and Applied Finance*, **3**, 617-639.
9. Barles G. (1994) : Solutions de viscosité des équations d'Hamilton-Jacobi, Springer Verlag.
10. Bielecki T. and S. Piska (2004) : "Risk-sensitive ICAPM with application to fixed-income management", *IEEE Transactions on automatic control*, **49**, 420-432.
11. Browne S. (1999) : "Beating a moving target : optimal portfolio strategies for outperforming a stochastic benchmark", *Finance and Stochastics*, **3**, 275-294.
12. Dembo A., Deuschel J.D. and D. Duffie (2004), "Large portfolio losses", *Finance and Stochastics*, **8**, 3-16.
13. Dembo A. and O. Zeitouni (1998) : Large deviations techniques and applications, 2nd edition, Springer Verlag.
14. Djehiche B. (1993) : "A large deviation estimate for ruin probabilities", *Scandinavian Actuarial Journal*, **1**, 42-59.
15. Dupuis P. and R. Ellis (1997) : A weak convergence approach to the theory of large deviations, Wiley Series in Probability and Statistics.

16. Embrechts P., Kluppelberg C. and T. Mikosch (2003), Modelling extremal events for insurance and finance, 4th edition, Springer Verlag.
17. Fleming W. and M. James (1992) : "Asymptotic series and exit time probability", *Annals of Probability*, **20**, 1369-1384.
18. Fleming W. and W. McEneaney (1995) : "Risk sensitive control on an infinite time horizon", *SIAM Journal on Control and Optimization*, **33**, 1881-1915.
19. Fleming W. and M. Soner (1994) : Controlled Markov processes and viscosity solutions, Springer Verlag.
20. Föllmer H. and P. Leukert (1999) : "Quantile hedging", *Finance and Stochastics*, **3**, 251-273.
21. Fournié E., Lasry J.M. and P.L. Lions (1997) : "Some nonlinear methods to study far-from-the-money contingent claims", *Numerical Methods in Finance*, L.C.G. Rogers et D. Talay, eds, Cambridge University Press.
22. Fournié E., Lasry J.M. and N. Touzi (1997) : "Monte Carlo methods for stochastic volatility models", *Numerical Methods in Finance*, L.C.G. Rogers et D. Talay, eds, Cambridge University Press.
23. Gaier J., Grandits P. and W. Schachermayer (2003) : "Asymptotic ruin probabilities and optimal investment", *Annals of Applied Probability*, **13**, 1054-1076.
24. Gerber H. (1973) : "Martingales in risk theory", *Mitt. Ver. Schweiz. math.*, **73**, 205-216.
25. Glasserman P., Heidelberger P. and P. Shahabuddin (1999), "Asymptotically optimal importance sampling and stratification for pricing path-dependent options", *Mathematical finance*, **9**, 117-152.
26. Glasserman P., Kang W. and P. Shahabuddin (2006), "Large deviations in multifactor portfolio credit risk", to appear in *Mathematical Finance*.
27. Glasserman P. and J. Li (2005), "Importance sampling for portfolio credit risk", *Management science*, **51**, 1643-1656.
28. Gobet E. (2000) : "Weak approximations of killed diffusion using Euler schemes", *Stochastic Processes and their Applications*, **87**, 167-197.
29. Guasoni P. and S. Robertson (2006) : "Optimal importance sampling with explicit formulas in continuous-time", Preprint, Boston University.
30. Hata H. and J. Sekine (2005) : "Solving long term invesmtment problems with Cox-Ingersoll-Ross interest rates", *Advances in Mathematical Economics*, **8**, 231-255.
31. Hipp C. and H. Schmidli (2004) : "Asymptotics of ruin probabilities for controlled risk processes in the small claims case", *Scandinavian Actuarial Journal*, 321-335.
32. Huh J. and A. Kolkiewicz (2006) : "Efficient computation of multivariate barrier crossing probability and its applications in credit risk models", Preprint University of Waterloo.
33. Kaas R. and Q. Tang (2005) : "A large deviation result for aggregate claims with dependent claims occurences", *Insurance Mathematics and Economics*, **36**, 251-259.
34. Lasry J.M. and P.L. Lions (1995) : "Grandes déviations pour des processus de diffusion couplés par un processus de sauts", *CRAS*, t. 321, 849-854.
35. Macci C. and G. Stabile (2006) : "Large deviations for risk processes with reinsurance", *Journal of Applied Probability*, **43**, 713-728.
36. Pham H. (2003a) : "A large deviations approach to optimal long term investment", *Finance and Stochastics*, **7**, 169-195.

37. Pham H. (2003b) : "A risk-sensitive control dual approach to a large deviations control problem", *Systems and Control Letters*, **49**, 295-309.
38. Sornette D. (1998) : "Large deviations and portfolio optimization", *Physica A : Statistical and Theoretical Physics*, **256**, 251-283.
39. Stettner L. (2004) : "Duality and risk sensitive portfolio optimization", in *Mathematics of Finance*, Proceedings AMS-IMS-SIAM, eds G. Yin and Q. Zhang, 333-347.
40. Stutzer M. (2003) : "Portfolio choice with endogenous utility : a large deviations approach", *Journal of Econometrics*, **116**, 365-386.
41. Williams N. (2004) : "Small Noise Asymptotics for a Stochastic Growth Model", *Journal of Economic Theory*, **119**, 271-298.

Lecture Notes in Mathematics

For information about earlier volumes
please contact your bookseller or Springer
LNM Online archive: springerlink.com

Vol. 1724: V. N. Kolokoltsov, Semiclassical Analysis for Diffusions and Stochastic Processes (2000)
Vol. 1725: D. A. Wolf-Gladrow, Lattice-Gas Cellular Automata and Lattice Boltzmann Models (2000)
Vol. 1726: V. Marić, Regular Variation and Differential Equations (2000)
Vol. 1727: P. Kravanja M. Van Barel, Computing the Zeros of Analytic Functions (2000)
Vol. 1728: K. Gatermann Computer Algebra Methods for Equivariant Dynamical Systems (2000)
Vol. 1729: J. Azéma, M. Émery, M. Ledoux, M. Yor (Eds.) Séminaire de Probabilités XXXIV (2000)
Vol. 1730: S. Graf, H. Luschgy, Foundations of Quantization for Probability Distributions (2000)
Vol. 1731: T. Hsu, Quilts: Central Extensions, Braid Actions, and Finite Groups (2000)
Vol. 1732: K. Keller, Invariant Factors, Julia Equivalences and the (Abstract) Mandelbrot Set (2000)
Vol. 1733: K. Ritter, Average-Case Analysis of Numerical Problems (2000)
Vol. 1734: M. Espedal, A. Fasano, A. Mikelić, Filtration in Porous Media and Industrial Applications. Cetraro 1998. Editor: A. Fasano. 2000.
Vol. 1735: D. Yafaev, Scattering Theory: Some Old and New Problems (2000)
Vol. 1736: B. O. Turesson, Nonlinear Potential Theory and Weighted Sobolev Spaces (2000)
Vol. 1737: S. Wakabayashi, Classical Microlocal Analysis in the Space of Hyperfunctions (2000)
Vol. 1738: M. Émery, A. Nemirovski, D. Voiculescu, Lectures on Probability Theory and Statistics (2000)
Vol. 1739: R. Burkard, P. Deuflhard, A. Jameson, J.-L. Lions, G. Strang, Computational Mathematics Driven by Industrial Problems. Martina Franca, 1999. Editors: V. Capasso, H. Engl, J. Periaux (2000)
Vol. 1740: B. Kawohl, O. Pironneau, L. Tartar, J.-P. Zolesio, Optimal Shape Design. Tróia, Portugal 1999. Editors: A. Cellina, A. Ornelas (2000)
Vol. 1741: E. Lombardi, Oscillatory Integrals and Phenomena Beyond all Algebraic Orders (2000)
Vol. 1742: A. Unterberger, Quantization and Non-holomorphic Modular Forms (2000)
Vol. 1743: L. Habermann, Riemannian Metrics of Constant Mass and Moduli Spaces of Conformal Structures (2000)
Vol. 1744: M. Kunze, Non-Smooth Dynamical Systems (2000)
Vol. 1745: V. D. Milman, G. Schechtman (Eds.), Geometric Aspects of Functional Analysis. Israel Seminar 1999-2000 (2000)
Vol. 1746: A. Degtyarev, I. Itenberg, V. Kharlamov, Real Enriques Surfaces (2000)
Vol. 1747: L. W. Christensen, Gorenstein Dimensions (2000)

Vol. 1748: M. Ruzicka, Electrorheological Fluids: Modeling and Mathematical Theory (2001)
Vol. 1749: M. Fuchs, G. Seregin, Variational Methods for Problems from Plasticity Theory and for Generalized Newtonian Fluids (2001)
Vol. 1750: B. Conrad, Grothendieck Duality and Base Change (2001)
Vol. 1751: N. J. Cutland, Loeb Measures in Practice: Recent Advances (2001)
Vol. 1752: Y. V. Nesterenko, P. Philippon, Introduction to Algebraic Independence Theory (2001)
Vol. 1753: A. I. Bobenko, U. Eitner, Painlevé Equations in the Differential Geometry of Surfaces (2001)
Vol. 1754: W. Bertram, The Geometry of Jordan and Lie Structures (2001)
Vol. 1755: J. Azéma, M. Émery, M. Ledoux, M. Yor (Eds.), Séminaire de Probabilités XXXV (2001)
Vol. 1756: P. E. Zhidkov, Korteweg de Vries and Nonlinear Schrödinger Equations: Qualitative Theory (2001)
Vol. 1757: R. R. Phelps, Lectures on Choquet's Theorem (2001)
Vol. 1758: N. Monod, Continuous Bounded Cohomology of Locally Compact Groups (2001)
Vol. 1759: Y. Abe, K. Kopfermann, Toroidal Groups (2001)
Vol. 1760: D. Filipović, Consistency Problems for Heath-Jarrow-Morton Interest Rate Models (2001)
Vol. 1761: C. Adelmann, The Decomposition of Primes in Torsion Point Fields (2001)
Vol. 1762: S. Cerrai, Second Order PDE's in Finite and Infinite Dimension (2001)
Vol. 1763: J.-L. Loday, A. Frabetti, F. Chapoton, F. Goichot, Dialgebras and Related Operads (2001)
Vol. 1764: A. Cannas da Silva, Lectures on Symplectic Geometry (2001)
Vol. 1765: T. Kerler, V. V. Lyubashenko, Non-Semisimple Topological Quantum Field Theories for 3-Manifolds with Corners (2001)
Vol. 1766: H. Hennion, L. Hervé, Limit Theorems for Markov Chains and Stochastic Properties of Dynamical Systems by Quasi-Compactness (2001)
Vol. 1767: J. Xiao, Holomorphic Q Classes (2001)
Vol. 1768: M. J. Pflaum, Analytic and Geometric Study of Stratified Spaces (2001)
Vol. 1769: M. Alberich-Carramiñana, Geometry of the Plane Cremona Maps (2002)
Vol. 1770: H. Gluesing-Luerssen, Linear Delay-Differential Systems with Commensurate Delays: An Algebraic Approach (2002)
Vol. 1771: M. Émery, M. Yor (Eds.), Séminaire de Probabilités 1967-1980. A Selection in Martingale Theory (2002)
Vol. 1772: F. Burstall, D. Ferus, K. Leschke, F. Pedit, U. Pinkall, Conformal Geometry of Surfaces in S^4 (2002)

Vol. 1773: Z. Arad, M. Muzychuk, Standard Integral Table Algebras Generated by a Non-real Element of Small Degree (2002)
Vol. 1774: V. Runde, Lectures on Amenability (2002)
Vol. 1775: W. H. Meeks, A. Ros, H. Rosenberg, The Global Theory of Minimal Surfaces in Flat Spaces. Martina Franca 1999. Editor: G. P. Pirola (2002)
Vol. 1776: K. Behrend, C. Gomez, V. Tarasov, G. Tian, Quantum Comohology. Cetraro 1997. Editors: P. de Bartolomeis, B. Dubrovin, C. Reina (2002)
Vol. 1777: E. García-Río, D. N. Kupeli, R. Vázquez-Lorenzo, Osserman Manifolds in Semi-Riemannian Geometry (2002)
Vol. 1778: H. Kiechle, Theory of K-Loops (2002)
Vol. 1779: I. Chueshov, Monotone Random Systems (2002)
Vol. 1780: J. H. Bruinier, Borcherds Products on $O(2,1)$ and Chern Classes of Heegner Divisors (2002)
Vol. 1781: E. Bolthausen, E. Perkins, A. van der Vaart, Lectures on Probability Theory and Statistics. Ecole d' Eté de Probabilités de Saint-Flour XXIX-1999. Editor: P. Bernard (2002)
Vol. 1782: C.-H. Chu, A. T.-M. Lau, Harmonic Functions on Groups and Fourier Algebras (2002)
Vol. 1783: L. Grüne, Asymptotic Behavior of Dynamical and Control Systems under Perturbation and Discretization (2002)
Vol. 1784: L. H. Eliasson, S. B. Kuksin, S. Marmi, J.-C. Yoccoz, Dynamical Systems and Small Divisors. Cetraro, Italy 1998. Editors: S. Marmi, J.-C. Yoccoz (2002)
Vol. 1785: J. Arias de Reyna, Pointwise Convergence of Fourier Series (2002)
Vol. 1786: S. D. Cutkosky, Monomialization of Morphisms from 3-Folds to Surfaces (2002)
Vol. 1787: S. Caenepeel, G. Militaru, S. Zhu, Frobenius and Separable Functors for Generalized Module Categories and Nonlinear Equations (2002)
Vol. 1788: A. Vasil'ev, Moduli of Families of Curves for Conformal and Quasiconformal Mappings (2002)
Vol. 1789: Y. Sommerhäuser, Yetter-Drinfel'd Hopf algebras over groups of prime order (2002)
Vol. 1790: X. Zhan, Matrix Inequalities (2002)
Vol. 1791: M. Knebusch, D. Zhang, Manis Valuations and Prüfer Extensions I: A new Chapter in Commutative Algebra (2002)
Vol. 1792: D. D. Ang, R. Gorenflo, V. K. Le, D. D. Trong, Moment Theory and Some Inverse Problems in Potential Theory and Heat Conduction (2002)
Vol. 1793: J. Cortés Monforte, Geometric, Control and Numerical Aspects of Nonholonomic Systems (2002)
Vol. 1794: N. Pytheas Fogg, Substitution in Dynamics, Arithmetics and Combinatorics. Editors: V. Berthé, S. Ferenczi, C. Mauduit, A. Siegel (2002)
Vol. 1795: H. Li, Filtered-Graded Transfer in Using Noncommutative Gröbner Bases (2002)
Vol. 1796: J.M. Melenk, hp-Finite Element Methods for Singular Perturbations (2002)
Vol. 1797: B. Schmidt, Characters and Cyclotomic Fields in Finite Geometry (2002)
Vol. 1798: W.M. Oliva, Geometric Mechanics (2002)
Vol. 1799: H. Pajot, Analytic Capacity, Rectifiability, Menger Curvature and the Cauchy Integral (2002)
Vol. 1800: O. Gabber, L. Ramero, Almost Ring Theory (2003)
Vol. 1801: J. Azéma, M. Émery, M. Ledoux, M. Yor (Eds.), Séminaire de Probabilités XXXVI (2003)

Vol. 1802: V. Capasso, E. Merzbach, B. G. Ivanoff, M. Dozzi, R. Dalang, T. Mountford, Topics in Spatial Stochastic Processes. Martina Franca, Italy 2001. Editor: E. Merzbach (2003)
Vol. 1803: G. Dolzmann, Variational Methods for Crystalline Microstructure – Analysis and Computation (2003)
Vol. 1804: I. Cherednik, Ya. Markov, R. Howe, G. Lusztig, Iwahori-Hecke Algebras and their Representation Theory. Martina Franca, Italy 1999. Editors: V. Baldoni, D. Barbasch (2003)
Vol. 1805: F. Cao, Geometric Curve Evolution and Image Processing (2003)
Vol. 1806: H. Broer, I. Hoveijn. G. Lunther, G. Vegter, Bifurcations in Hamiltonian Systems. Computing Singularities by Gröbner Bases (2003)
Vol. 1807: V. D. Milman, G. Schechtman (Eds.), Geometric Aspects of Functional Analysis. Israel Seminar 2000-2002 (2003)
Vol. 1808: W. Schindler, Measures with Symmetry Properties (2003)
Vol. 1809: O. Steinbach, Stability Estimates for Hybrid Coupled Domain Decomposition Methods (2003)
Vol. 1810: J. Wengenroth, Derived Functors in Functional Analysis (2003)
Vol. 1811: J. Stevens, Deformations of Singularities (2003)
Vol. 1812: L. Ambrosio, K. Deckelnick, G. Dziuk, M. Mimura, V. A. Solonnikov, H. M. Soner, Mathematical Aspects of Evolving Interfaces. Madeira, Funchal, Portugal 2000. Editors: P. Colli, J. F. Rodrigues (2003)
Vol. 1813: L. Ambrosio, L. A. Caffarelli, Y. Brenier, G. Buttazzo, C. Villani, Optimal Transportation and its Applications. Martina Franca, Italy 2001. Editors: L. A. Caffarelli, S. Salsa (2003)
Vol. 1814: P. Bank, F. Baudoin, H. Föllmer, L.C.G. Rogers, M. Soner, N. Touzi, Paris-Princeton Lectures on Mathematical Finance 2002 (2003)
Vol. 1815: A. M. Vershik (Ed.), Asymptotic Combinatorics with Applications to Mathematical Physics. St. Petersburg, Russia 2001 (2003)
Vol. 1816: S. Albeverio, W. Schachermayer, M. Talagrand, Lectures on Probability Theory and Statistics. Ecole d'Eté de Probabilités de Saint-Flour XXX-2000. Editor: P. Bernard (2003)
Vol. 1817: E. Koelink, W. Van Assche (Eds.), Orthogonal Polynomials and Special Functions. Leuven 2002 (2003)
Vol. 1818: M. Bildhauer, Convex Variational Problems with Linear, nearly Linear and/or Anisotropic Growth Conditions (2003)
Vol. 1819: D. Masser, Yu. V. Nesterenko, H. P. Schlickewei, W. M. Schmidt, M. Waldschmidt, Diophantine Approximation. Cetraro, Italy 2000. Editors: F. Amoroso, U. Zannier (2003)
Vol. 1820: F. Hiai, H. Kosaki, Means of Hilbert Space Operators (2003)
Vol. 1821: S. Teufel, Adiabatic Perturbation Theory in Quantum Dynamics (2003)
Vol. 1822: S.-N. Chow, R. Conti, R. Johnson, J. Mallet-Paret, R. Nussbaum, Dynamical Systems. Cetraro, Italy 2000. Editors: J. W. Macki, P. Zecca (2003)
Vol. 1823: A. M. Anile, W. Allegretto, C. Ringhofer, Mathematical Problems in Semiconductor Physics. Cetraro, Italy 1998. Editor: A. M. Anile (2003)
Vol. 1824: J. A. Navarro González, J. B. Sancho de Salas, $\mathscr{C}^\infty$ – Differentiable Spaces (2003)

Vol. 1825: J. H. Bramble, A. Cohen, W. Dahmen, Multiscale Problems and Methods in Numerical Simulations, Martina Franca, Italy 2001. Editor: C. Canuto (2003)
Vol. 1826: K. Dohmen, Improved Bonferroni Inequalities via Abstract Tubes. Inequalities and Identities of Inclusion-Exclusion Type. VIII, 113 p, 2003.
Vol. 1827: K. M. Pilgrim, Combinations of Complex Dynamical Systems. IX, 118 p, 2003.
Vol. 1828: D. J. Green, Gröbner Bases and the Computation of Group Cohomology. XII, 138 p, 2003.
Vol. 1829: E. Altman, B. Gaujal, A. Hordijk, Discrete-Event Control of Stochastic Networks: Multimodularity and Regularity. XIV, 313 p, 2003.
Vol. 1830: M. I. Gil', Operator Functions and Localization of Spectra. XIV, 256 p, 2003.
Vol. 1831: A. Connes, J. Cuntz, E. Guentner, N. Higson, J. E. Kaminker, Noncommutative Geometry, Martina Franca, Italy 2002. Editors: S. Doplicher, L. Longo (2004)
Vol. 1832: J. Azéma, M. Émery, M. Ledoux, M. Yor (Eds.), Séminaire de Probabilités XXXVII (2003)
Vol. 1833: D.-Q. Jiang, M. Qian, M.-P. Qian, Mathematical Theory of Nonequilibrium Steady States. On the Frontier of Probability and Dynamical Systems. IX, 280 p, 2004.
Vol. 1834: Yo. Yomdin, G. Comte, Tame Geometry with Application in Smooth Analysis. VIII, 186 p, 2004.
Vol. 1835: O.T. Izhboldin, B. Kahn, N.A. Karpenko, A. Vishik, Geometric Methods in the Algebraic Theory of Quadratic Forms. Summer School, Lens, 2000. Editor: J.-P. Tignol (2004)
Vol. 1836: C. Năstăsescu, F. Van Oystaeyen, Methods of Graded Rings. XIII, 304 p, 2004.
Vol. 1837: S. Tavaré, O. Zeitouni, Lectures on Probability Theory and Statistics. Ecole d'Eté de Probabilités de Saint-Flour XXXI-2001. Editor: J. Picard (2004)
Vol. 1838: A.J. Ganesh, N.W. O'Connell, D.J. Wischik, Big Queues. XII, 254 p, 2004.
Vol. 1839: R. Gohm, Noncommutative Stationary Processes. VIII, 170 p, 2004.
Vol. 1840: B. Tsirelson, B. Werner, Lectures on Probability Theory and Statistics. Ecole d'Eté de Probabilités de Saint-Flour XXXII-2002. Editor: J. Picard (2004)
Vol. 1841: W. Reichel, Uniqueness Theorems for Variational Problems by the Method of Transformation Groups (2004)
Vol. 1842: T. Johnsen, A. L. Knutsen, K_3 Projective Models in Scrolls (2004)
Vol. 1843: B. Jefferies, Spectral Properties of Noncommuting Operators (2004)
Vol. 1844: K.F. Siburg, The Principle of Least Action in Geometry and Dynamics (2004)
Vol. 1845: Min Ho Lee, Mixed Automorphic Forms, Torus Bundles, and Jacobi Forms (2004)
Vol. 1846: H. Ammari, H. Kang, Reconstruction of Small Inhomogeneities from Boundary Measurements (2004)
Vol. 1847: T.R. Bielecki, T. Björk, M. Jeanblanc, M. Rutkowski, J.A. Scheinkman, W. Xiong, Paris-Princeton Lectures on Mathematical Finance 2003 (2004)
Vol. 1848: M. Abate, J. E. Fornaess, X. Huang, J. P. Rosay, A. Tumanov, Real Methods in Complex and CR Geometry, Martina Franca, Italy 2002. Editors: D. Zaitsev, G. Zampieri (2004)
Vol. 1849: Martin L. Brown, Heegner Modules and Elliptic Curves (2004)
Vol. 1850: V. D. Milman, G. Schechtman (Eds.), Geometric Aspects of Functional Analysis. Israel Seminar 2002-2003 (2004)

Vol. 1851: O. Catoni, Statistical Learning Theory and Stochastic Optimization (2004)
Vol. 1852: A.S. Kechris, B.D. Miller, Topics in Orbit Equivalence (2004)
Vol. 1853: Ch. Favre, M. Jonsson, The Valuative Tree (2004)
Vol. 1854: O. Saeki, Topology of Singular Fibers of Differential Maps (2004)
Vol. 1855: G. Da Prato, P.C. Kunstmann, I. Lasiecka, A. Lunardi, R. Schnaubelt, L. Weis, Functional Analytic Methods for Evolution Equations. Editors: M. Iannelli, R. Nagel, S. Piazzera (2004)
Vol. 1856: K. Back, T.R. Bielecki, C. Hipp, S. Peng, W. Schachermayer, Stochastic Methods in Finance, Bressanone/Brixen, Italy, 2003. Editors: M. Fritelli, W. Runggaldier (2004)
Vol. 1857: M. Émery, M. Ledoux, M. Yor (Eds.), Séminaire de Probabilités XXXVIII (2005)
Vol. 1858: A.S. Cherny, H.-J. Engelbert, Singular Stochastic Differential Equations (2005)
Vol. 1859: E. Letellier, Fourier Transforms of Invariant Functions on Finite Reductive Lie Algebras (2005)
Vol. 1860: A. Borisyuk, G.B. Ermentrout, A. Friedman, D. Terman, Tutorials in Mathematical Biosciences I. Mathematical Neurosciences (2005)
Vol. 1861: G. Benettin, J. Henrard, S. Kuksin, Hamiltonian Dynamics – Theory and Applications, Cetraro, Italy, 1999. Editor: A. Giorgilli (2005)
Vol. 1862: B. Helffer, F. Nier, Hypoelliptic Estimates and Spectral Theory for Fokker-Planck Operators and Witten Laplacians (2005)
Vol. 1863: H. Führ, Abstract Harmonic Analysis of Continuous Wavelet Transforms (2005)
Vol. 1864: K. Efstathiou, Metamorphoses of Hamiltonian Systems with Symmetries (2005)
Vol. 1865: D. Applebaum, B.V. R. Bhat, J. Kustermans, J. M. Lindsay, Quantum Independent Increment Processes I. From Classical Probability to Quantum Stochastic Calculus. Editors: M. Schürmann, U. Franz (2005)
Vol. 1866: O.E. Barndorff-Nielsen, U. Franz, R. Gohm, B. Kümmerer, S. Thorbjønsen, Quantum Independent Increment Processes II. Structure of Quantum Lévy Processes, Classical Probability, and Physics. Editors: M. Schürmann, U. Franz, (2005)
Vol. 1867: J. Sneyd (Ed.), Tutorials in Mathematical Biosciences II. Mathematical Modeling of Calcium Dynamics and Signal Transduction. (2005)
Vol. 1868: J. Jorgenson, S. Lang, $Pos_n(R)$ and Eisenstein Series. (2005)
Vol. 1869: A. Dembo, T. Funaki, Lectures on Probability Theory and Statistics. Ecole d'Eté de Probabilités de Saint-Flour XXXIII-2003. Editor: J. Picard (2005)
Vol. 1870: V.I. Gurariy, W. Lusky, Geometry of Müntz Spaces and Related Questions. (2005)
Vol. 1871: P. Constantin, G. Gallavotti, A.V. Kazhikhov, Y. Meyer, S. Ukai, Mathematical Foundation of Turbulent Viscous Flows, Martina Franca, Italy, 2003. Editors: M. Cannone, T. Miyakawa (2006)
Vol. 1872: A. Friedman (Ed.), Tutorials in Mathematical Biosciences III. Cell Cycle, Proliferation, and Cancer (2006)
Vol. 1873: R. Mansuy, M. Yor, Random Times and Enlargements of Filtrations in a Brownian Setting (2006)
Vol. 1874: M. Yor, M. Émery (Eds.), In Memoriam Paul-André Meyer - Séminaire de Probabilités XXXIX (2006)

Vol. 1875: J. Pitman, Combinatorial Stochastic Processes. Ecole d'Eté de Probabilités de Saint-Flour XXXII-2002. Editor: J. Picard (2006)

Vol. 1876: H. Herrlich, Axiom of Choice (2006)

Vol. 1877: J. Steuding, Value Distributions of L-Functions (2007)

Vol. 1878: R. Cerf, The Wulff Crystal in Ising and Percolation Models, Ecole d'Eté de Probabilités de Saint-Flour XXXIV-2004. Editor: Jean Picard (2006)

Vol. 1879: G. Slade, The Lace Expansion and its Applications, Ecole d'Eté de Probabilités de Saint-Flour XXXIV-2004. Editor: Jean Picard (2006)

Vol. 1880: S. Attal, A. Joye, C.-A. Pillet, Open Quantum Systems I, The Hamiltonian Approach (2006)

Vol. 1881: S. Attal, A. Joye, C.-A. Pillet, Open Quantum Systems II, The Markovian Approach (2006)

Vol. 1882: S. Attal, A. Joye, C.-A. Pillet, Open Quantum Systems III, Recent Developments (2006)

Vol. 1883: W. Van Assche, F. Marcellàn (Eds.), Orthogonal Polynomials and Special Functions, Computation and Application (2006)

Vol. 1884: N. Hayashi, E.I. Kaikina, P.I. Naumkin, I.A. Shishmarev, Asymptotics for Dissipative Nonlinear Equations (2006)

Vol. 1885: A. Telcs, The Art of Random Walks (2006)

Vol. 1886: S. Takamura, Splitting Deformations of Degenerations of Complex Curves (2006)

Vol. 1887: K. Habermann, L. Habermann, Introduction to Symplectic Dirac Operators (2006)

Vol. 1888: J. van der Hoeven, Transseries and Real Differential Algebra (2006)

Vol. 1889: G. Osipenko, Dynamical Systems, Graphs, and Algorithms (2006)

Vol. 1890: M. Bunge, J. Funk, Singular Coverings of Toposes (2006)

Vol. 1891: J.B. Friedlander, D.R. Heath-Brown, H. Iwaniec, J. Kaczorowski, Analytic Number Theory, Cetraro, Italy, 2002. Editors: A. Perelli, C. Viola (2006)

Vol. 1892: A. Baddeley, I. Bárány, R. Schneider, W. Weil, Stochastic Geometry, Martina Franca, Italy, 2004. Editor: W. Weil (2007)

Vol. 1893: H. Hanßmann, Local and Semi-Local Bifurcations in Hamiltonian Dynamical Systems, Results and Examples (2007)

Vol. 1894: C.W. Groetsch, Stable Approximate Evaluation of Unbounded Operators (2007)

Vol. 1895: L. Molnár, Selected Preserver Problems on Algebraic Structures of Linear Operators and on Function Spaces (2007)

Vol. 1896: P. Massart, Concentration Inequalities and Model Selection, Ecole d'Eté de Probabilités de Saint-Flour XXXIII-2003. Editor: J. Picard (2007)

Vol. 1897: R. Doney, Fluctuation Theory for Lévy Processes, Ecole d'Eté de Probabilités de Saint-Flour XXXV-2005. Editor: J. Picard (2007)

Vol. 1898: H.R. Beyer, Beyond Partial Differential Equations, On linear and Quasi-Linear Abstract Hyperbolic Evolution Equations (2007)

Vol. 1899: Séminaire de Probabilités XL. Editors: C. Donati-Martin, M. Émery, A. Rouault, C. Stricker (2007)

Vol. 1900: E. Bolthausen, A. Bovier (Eds.), Spin Glasses (2007)

Vol. 1901: O. Wittenberg, Intersections de deux quadriques et pinceaux de courbes de genre 1, Intersections of Two Quadrics and Pencils of Curves of Genus 1 (2007)

Vol. 1902: A. Isaev, Lectures on the Automorphism Groups of Kobayashi-Hyperbolic Manifolds (2007)

Vol. 1903: G. Kresin, V. Maz'ya, Sharp Real-Part Theorems (2007)

Vol. 1904: P. Giesl, Construction of Global Lyapunov Functions Using Radial Basis Functions (2007)

Vol. 1905: C. Prévôt, M. Röckner, A Concise Course on Stochastic Partial Differential Equations (2007)

Vol. 1906: T. Schuster, The Method of Approximate Inverse: Theory and Applications (2007)

Vol. 1907: M. Rasmussen, Attractivity and Bifurcation for Nonautonomous Dynamical Systems (2007)

Vol. 1908: T.J. Lyons, M. Caruana, T. Lévy, Differential Equations Driven by Rough Paths, Ecole d'Eté de Probabilités de Saint-Flour XXXIV-2004. (2007)

Vol. 1909: H. Akiyoshi, M. Sakuma, M. Wada, Y. Yamashita, Punctured Torus Groups and 2-Bridge Knot Groups (I) (2007)

Vol. 1910: V.D. Milman, G. Schechtman (Eds.), Geometric Aspects of Functional Analysis. Israel Seminar 2004-2005 (2007)

Vol. 1911: A. Bressan, D. Serre, M. Williams, K. Zumbrun, Hyperbolic Systems of Balance Laws. Lectures given at the C.I.M.E. Summer School held in Cetraro, Italy, July 14–21, 2003. Editor: P. Marcati (2007)

Vol. 1912: V. Berinde, Iterative Approximation of Fixed Points (2007)

Vol. 1913: J.E. Marsden, G. Misiołek, J.-P. Ortega, M. Perlmutter, T.S. Ratiu, Hamiltonian Reduction by Stages (2007)

Vol. 1914: G. Kutyniok, Affine Density in Wavelet Analysis (2007)

Vol. 1915: T. Bıyıkoğlu, J. Leydold, P.F. Stadler, Laplacian Eigenvectors of Graphs. Perron-Frobenius and Faber-Krahn Type Theorems (2007)

Vol. 1916: C. Villani, F. Rezakhanlou, Entropy Methods for the Boltzmann Equation. Editors: F. Golse, S. Olla (forthcoming)

Vol. 1917: I. Veselić, Existence and Regularity Properties of the Integrated Density of States of Random Schrödinger (2007)

Vol. 1918: B. Roberts, R. Schmidt, Local Newforms for GSp(4) (2007)

Vol. 1919: R.A. Carmona, I. Ekeland, A. Kohatsu-Higa, J.-M. Lasry, P.-L. Lions, H. Pham, E. Taflin, Paris-Princeton Lectures on Mathematical Finance 2004. Editors: R.A. Carmona, E. Çinlar, I. Ekeland, E. Jouini, J.A. Scheinkman, N. Touzi (2007)

Recent Reprints and New Editions

Vol. 1618: G. Pisier, Similarity Problems and Completely Bounded Maps. 1995 – 2nd exp. edition (2001)

Vol. 1629: J.D. Moore, Lectures on Seiberg-Witten Invariants. 1997 – 2nd edition (2001)

Vol. 1638: P. Vanhaecke, Integrable Systems in the realm of Algebraic Geometry. 1996 – 2nd edition (2001)

Vol. 1702: J. Ma, J. Yong, Forward-Backward Stochastic Differential Equations and their Applications. 1999 – Corr. 3rd printing (2007)

Vol. 830: J.A. Green, Polynomial Representations of GL_n, with an Appendix on Schensted Correspondence and Littelmann Paths by K. Erdmann, J.A. Green and M. Schocker 1980 – 2nd corr. and augmented edition (2007)

Printing: Krips bv, Meppel
Binding: Stürtz, Würzburg